SUSTAINABLE MICRO IRRIGATION MANAGEMENT FOR TREES AND VINES

Research Advances in Sustainable Micro Irrigation

VOLUME 3

SUSTAINABLE MICRO IRRIGATION MANAGEMENT FOR TREES AND VINES

Edited by

Megh R. Goyal, PhD, PE

Apple Academic Press

TORONTO NEW JERSEY

Apple Academic Press Inc.
3333 Mistwell Crescent
Oakville, ON L6L 0A2
Canada

Apple Academic Press Inc.
9 Spinnaker Way
Waretown, NJ 08758
USA

Printed in the United States of America on acid-free paper

International Standard Book Number-13: 978-1-77188-025-1 (Hardcover)

Library of Congress Control Number: 2014943872

Library and Archives Canada Cataloguing in Publication

Sustainable micro irrigation management for trees and vines/edited by Megh R. Goyal, PhD, PE.

(Research advances in sustainable micro irrigation; volume 3)

Includes bibliographical references and index.
ISBN 978-1-77188-025-1 (bound)
1. Microirrigation. 2. Trees--Irrigation. 3. Climbing plants--Irrigation. 4. Sustainable agriculture. I. Goyal, Megh Raj., author, editor II. Series: Research advances in sustainable micro irrigation; v. 3

S619.T74S95 2014 631.5'87 C2014-904548-4

Apple Academic Press also publishes its books in a variety of electronic formats. Some content that appears in print may not be available in electronic format. For information about Apple Academic Press products, visit our website at **www.appleacademicpress.com** and the CRC Press website at **www.crcpress.com**

CONTENTS

PART II: RESEARCH ADVANCES AND APPLICATIONS

LIST OF CONTRIBUTORS

B. Keith Bellingham, PhD, PH

Certified Professional Hydrologist, and Soil Scientist, Stevens Water Monitoring Systems Inc., 12067 NE Glenn Widing Drive Suite 106, Portland-Oregon-97220-USA. Phone: (800) 452-5272, Email: kbellingham@stevenswater.com http://www.stevenswater.com/contact.aspx

Vincent F. Bralts, PhD, PE

Professor Purdue University, Agricultural and Biological Engineering Department, 225 South University Drive, West Lafayette, IN 47907–2093. Phone: 7654941177, Fax: 765-496-1115, Email: bralts@perdue.edu; website: https://www.purdue.edu/abe

Amy Fulcher, PhD

Assistant Professor, Sustainable Ornamental Plant Production and Landscape Management, University of Tennessee, Knoxville, TN 37996, Tel. office: 865-974-7152, Email: afulcher@utk.edu,

Robert Geneve, PhD

Professor, Dept. of Horticulture, University of Kentucky, Lexington, Kentucky, 40546, Tel. office: 859-257-8610, Email: rgeneve@uky.edu.

Megh R. Goyal, PhD, PE

Retired Professor in Agricultural and Biomedical Engineering, University of Puerto Rico at Mayaguez Campus; and Senior Technical Editor-in-Chief in Agriculture Sciences and Biomedical Engineering, Apple Academic Press Inc., PO Box 86, Rincon, PR – 00677, USA. Email: goyalmegh@gmail.com

Marvin E. Jensen, PhD, PE

Retired Research Program Leader at USDA-ARS and Irrigation Consultant. 1207 Spring Wood Drive, Fort Collins, Colorado 80525, USA. Email: mjensen419@aol.com

Pradeep Kumar, PhD

Scientist, National Institute of Hydrology, Roorkee, India

Lata, PhD

Principal Scientist, Water Technology Centre (WTC), Indian Agricultural Research Institute (IARI), New Delhi, Delhi 110012, India

Miguel A. Muñoz-Muñoz

Ex-President of University of Puerto Rico, University of Puerto Rico, Mayaguez Campus, College of Agriculture Sciences, Call Box 9000, Mayagüez, PR. 00681-9000. Tel. 787-265-3871, Email: miguel.munoz3@upr.edu

Susmitha Nambuthiri, PhD

Post doctoral Scholar, Dept. of Horticulture, University of Kentucky, Lexington, KY 40546. Email: ssnamb2@uky.edu

Neelam Patel, PhD

Senior Scientist, Water Technology Centre (WTC), Indian Agricultural Research Institute (IARI), New Delhi, Delhi 110012, India, Email: neelam@iari.res.in

T. B. S. Rajput, PhD

Principal Scientist, Water Technology Centre (WTC), Indian Agricultural Research Institute (IARI), New Delhi, Delhi 110012, India, Email: tbsraj@iari.res.in

Parameshwar S. Shirgure, PhD
Senior Scientist for Soil and Water Engineering, National Research Center for Citrus, Indian Council of Agricultural Research, Amravati Road, Nagpur, India, 440010. Email: shirgure@gmail.com

Gajendra Singh
Former Vice-President, Asian Institute of Technology, Thailand. C-86, Millennium Apartments, Plot E-10A, Sector 61, NOIDA – U.P. – 201301, India, Mobile: (91)9971087591, Email: prof.gsingh@gmail.com.

Shyam Singh, PhD
Consultant Horticulture at Government of Mizoram India, Former Director and Mission Leader (Technology Mission on Citrus) at National Research Center for Citrus, Indian Council of Agricultural Research, Amravati Road, Nagpur, India, 440010.

R. K. Sivanappan
Former Dean and Professor, Tamil Nadu Agricultural University at Coimbatore – Tamil Nadu, India; and Consultant, 14, Bharathi Park, 4th Cross Road, Coimbatore 641043, India. Email: sivanappan@hotmail.com

Anoop K. Srivastava, PhD
Principal Scientist, Jawaharlal Nehru Agricultural University, Jabalpur, Madhya Pradesh, India – 482 004

Vinod Kumar Tripathi, PhD
Assistant Professor, Centre for Water Engineering and Management, Central University of Jharkhand, Ratu-Loharghat road, Brambe, Ranchi- Jharkhand – 835205, India, Mobile: +918987661439; Email: tripathiwtcer@gmail.com

LIST OF ABBREVIATIONS

°C	degree Celsius
AC-FT	acre foot
ASABE	American Society of Agricultural and Biological Engineers
B-C	Blaney-Criddle
BCR	benefit-cost ratio
cfs	cubic feet per second
cfsm	water depth, cubic feet per second per square mile
CGDD	cumulative growing degree days
CIAE	Central Institute of Agricultural Engineering Bhopal – India
cm	centimeter(s)
CU	coefficient of uniformity
CWSI	crop water stress index
DAP	days after planting
DAT	days after transplanting
DOY	day of the year
DU	distribution uniformity
EPAN	pan evaporation
ET	evapotranspiration
ET_c	crop evapotranspiration
FAO	Food and Agricultural Organization, Rome
FC	field capacity
GOI	Government of India
gph	gallons per hour
gpm	gallons per minute
ha	hectare(s)
HRG	Hargreaves
IARI	Indian Agricultural Research Institute
ICAR	Indian Council of Agricultural Research
IDE	International Development Enterprises
ISAE	Indian Society of Agricultural Engineers
k_c	crop coefficient
K_p	pan coefficient
KSA	Kingdom of Saudi Arabia
LAI	leaf area index
LEPA	low energy pressure irrigation system
lps	liters per second
MAD	maximum allowable depletion

Mha	million hectare(s)
MI	micro irrigation
MSL	mean sea level
NGO	non government organization
PE	polyethylene
PET	potential evapotranspiration
pH	acidity/alkalinity measurement scale
PM	Penman-Monteith
ppb	one part per billion
ppm	one part per million
psi	pounds per square inch
PVC	poly vinyl chloride
PWP	permanent wilting point
RA	Extraterrestrial radiation
RH	relative humidity
RMAX	maximum relative humidity
RMIN	minimum relative humidity
RMSE	root mean squared error
RS	solar radiation
SAR	sodium absorption rate
SCS-BC	SCS Blaney-Criddle
SDI	subsurface drip irrigation
SWB	soil water balance
TE	transpiration efficiency
TEW	total evaporable water
TMAX	maximum temperature
TMIN	minimum temperature
TR	temperature range
TUE	transpiration use efficiency
USDA	US Department of Agriculture
USDA-SCS	US Department of Agriculture-Soil Conservation Service
VPD	vapor pressure deficit
VWC	volumetric water content
WATBAL	water balance
WISP	wind speed
WS	Weather Stations
WSEE	weighed standard error of estimate
WUE	water use efficiency
μg/g	micrograms per gram
μg/L	micrograms per liter

LIST OF SYMBOLS

CV_q	coefficient of variation
D	trunk diameter
Ef	sprinkler efficiency
ET	evapotranspiration rate
ET”	water consumption
ET_0	evapotranspiration
H	pressure head
H	tree height
K	potassium
K_c	crop coefficient
lph	emitter discharge
N	nitrogen
n	total number of emitters
P	fertigation
P	phosphorus
q_i	measured discharge of emitter
q_{ini}	corresponding mean discharge
q_{max}	maximum flow rate
q_{min}	minimum flow rate
q_{var}	variation in flow rate
SD	standard deviation
T_{max}	maximum time
T_{min}	minimum time
U	uniformity
V	coefficient of variation
x	reticulata
θ_{AC}	available water capacity
θ_{FC}	field capacity
θ_{PW}	permanent wilting point

PREFACE

Due to increased agricultural production, irrigated land has increased in the arid and subhumid zones around the world. Agriculture has started to compete for water use with industries, municipalities and other sectors. This increasing demand along with increments in water and energy costs have made it necessary to develop new technologies for the adequate management of water. The intelligent use of water for crops requires understanding of evapotranspiration processes and use of efficient irrigation methods.

The http://newindianexpress.com/cities/bangalore/Micro-irrigation-to-be-promoted/2013/08/17/ weblink published an article on the importance of micro irrigation in India. Every day, similar news appear all around the world indicating that government agencies at central/state/local level, research and educational institutions, industry, sellers and others are aware of the urgent need to adopt micro irrigation technology that can have an irrigation efficiency up to 90% compared to 30–40% for the conventional irrigation systems. I share here with readers the news on 17 August of 2013 by Indian Express Newspaper: "*In its efforts to increase the irrigated area by efficiently distributing the available water in the Cauvery basin, The Cauvery Neeravari Nigama Limited (CNNL) is planning to undertake pilot projects on micro irrigation at four places. The CNNL Managing Director Kapil Mohan said, 'the Cauvery water disputes tribunal has permitted the state to irrigate up to 18.85 lakh acres of land in the Cauvery basin. Therefore, we have to judiciously use the available water to increase the irrigated area. In the conventional irrigation method, a lot of water is required to irrigate even a small piece of land. Therefore, we are planning to undertake pilot projects to introduce micro irrigation in four or five places in the Cauvery basin.' Kapil further said that unless the farmers are willing to embrace micro irrigation, it would be difficult for the project to succeed. Therefore, the CNNL is holding discussions with the farmers in different villages of the basin to select the villages in which the project would be undertaken. The CNNL is also in the process of finalizing the technology that should be adopted while undertaking the pilot project. 'If everything goes as planned we should implement the pilot project within this financial year. If the project yields the desired result, we will think of extending it to the other areas in the basin,' Kapil added. According to the official sources, water would be supplied through micro sprinklers instead of canals in the micro irrigation system. Therefore, one can irrigate more than two acres of land through the system with the water that is used to irrigate one acre of land in the conventional canal irrigation system.*"

Evapotranspiration (ET) is a combination of two processes: evaporation and transpiration. Evaporation is a physical process that involves conversion of liquid water into water vapor and then into the atmosphere. Evaporation of water into the atmosphere occurs on the surface of rivers, lakes, soils and vegetation. Transpiration is a physical process that involves the flow of liquid water from the soil (root zone) through the trunk, branches and surface of leaves through the stomates. An energy gradient is created during the evaporation of water, which causes the water movement into and out of the plant stomates. In the majority of green plants, stomates remain open during the day and stay closed during the night. If the soil is too dry, the stomates will remain closed during the day in order to slow down the transpiration.

Evaporation, transpiration and ET processes are important for estimating crop water requirements and for irrigation scheduling. To determine crop water requirements, it is necessary to estimate ET by on site measurements or by using meteorological data. On site measurements are very costly and are mostly employed to calibrate ET methods using climatological data. There are a number of proposed mathematical equations that require meteorological data and are used to estimate the ET for periods of one day or more. Potential ET is the ET from a well-watered crop, which completely covers the surface. Meteorological processes determine the ET of a crop. Closing of stomates and reduction in transpiration are usually important only under drought or under stress conditions of a plant. The ET depends on four factors: (1) climate, (2) vegetation, (3) water availability in the soil and (4) behavior of stomates. Vegetation affects the ET in various ways. It affects the ability of the soil surface to reflect light. The vegetation affects the amount of energy absorbed by the soil surface. Soil properties, including soil moisture, also affect the amount of energy that flows through the soil. The height and density of vegetation influence efficiency of the turbulent heat interchange and the water vapor of the foliage.

Micro irrigation, also known as trickle irrigation or drip irrigation or localized irrigation or high frequency or pressurized irrigation, is an irrigation method that saves water and fertilizer by allowing water to drip slowly to the roots of plants, either onto the soil surface or directly onto the root zone, through a network of valves, pipes, tubing, and emitters. It is done through narrow tubes that deliver water directly to the base of the plant. It is a system of crop irrigation involving the controlled delivery of water directly to individual plants and can be installed on the soil surface or subsurface. Micro irrigation systems are often used in farms and large gardens, but are equally effective in the home garden or even for houseplants or lawns. They are easily customizable and can be set up even by inexperienced gardeners. Putting a drip system into the garden is a great do-it-yourself project that will ultimately save the time and help the plants grow. It is equally used in landscaping and in green cities.

The mission of this compendium is to serve as a textbook or a reference manual for graduate and undergraduate students of agricultural, biological and civil engineering, horticulture, soil science, crop science and agronomy. I hope that it will be a valuable reference for professionals who work with micro irrigation and water management; for professional training institutes, technical agricultural centers, irrigation centers, Agricultural Extension Services, and other agencies that work with micro irrigation programs.

After my first textbook, *Drip/Trickle or Micro Irrigation Management* by Apple Academic Press Inc., and response from international readers, I was motivated to bring out for the world community this series on *Research Advances in Sustainable Micro Irrigation.* This book series will complement other books on micro irrigation that are currently available on the market, and my intention is not to replace any one of these. This book series is unique because it is complete and simple, a one-stop manual, with worldwide applicability to irrigation management in agriculture. Its coverage of the field of micro irrigation includes historical review; current status and potential; basic principles and applications; research results for vegetable/row/tree crops; research studies from Chile, Colombia, Egypt, India, Mexico, Puerto Rico, Saudi Arabia, Spain, and USA; research results on simulation of micro irrigation and wetting patterns; development of software for micro irrigation design; micro irrigation for small farms and marginal farmers; studies related to agronomical crops in arid, humid, semiarid, and tropical climates; and methods and techniques that can be easily applied to other locations (not included in this book).

This book offers basic principles, knowledge and techniques of micro irrigation management that are necessary to understand before designing/developing and evaluating an agricultural irrigation management system. This book is a must for those interested in irrigation planning and management, namely, researchers, scientists, educators and students.

Volume 1 in this book series is titled *Sustainable Micro Irrigation: Principles and Practices* and includes 16 chapters.

And likewise, volume 2 in this book series is titled *Research Advances and Applications in Subsurface Micro Irrigation and Surface Micro Irrigation* and includes 16 chapters.

Volume three in this book series is titled *Sustainable Micro Irrigation Management for Trees and Vines* and includes 14 chapters.

Volume 4 in this book series is titled *Management, Performance, and Applications of Micro Irrigation.*

The contribution by all cooperating authors to this book series has been most valuable in the compilation of this multi-volume compendium. Their names are mentioned in each chapter. This book would not have been written without the valuable cooperation of these investigators, many of whom are renowned scientists who have worked in the field of evapotranspiration throughout their professional careers.

I would like to thank the AAP staff, Sandy Jones Sickels, Vice President, and Ashish Kumar, Publisher and President at Apple Academic Press, Inc., (http://appleacademicpress.com) for making every effort to publish the book when diminishing water resources are a major issue worldwide.

We request that the reader offer us your constructive suggestions that may help to improve the next edition.

I express my deep admiration to my family for understanding and collaboration during the preparation of this book series. With my whole heart and best affection, I dedicate this book series to my wife, Subhadra Devi Goyal, who has supported me during the last 44 years. We both have been trickling on to add our drop to the ocean of service to the world of humanity. Without her patience and dedication, I would not have been a teacher with vocation and zeal for service to others. As an educator, there is a piece of advice to one and all in the world: "*Permit that our almighty God, our Creator and excellent Teacher, irrigate the life with His Grace of rain trickle by trickle, because our life must continue trickling on . . .*"

—Megh R. Goyal, PhD, PE, Senior Editor-in-Chief
February 14, 2014

FOREWORD

Since 1978, I have been a research assistant at the Agricultural Experiment Substation – Juana Diaz; soil scientist; Chairman of Department of Agronomy and Soils in the College of Agricultural Sciences at the University of Puerto Rico – Mayaguez Campus; and President of University of Puerto Rico (February 2011 to June 2013). I was also an Under-Secretary (1993–1997) and Secretary of the Puerto Rico Agriculture Department (1997–2000). I am privileged to write a foreword for Goyal's book series that is titled *Research Advances in Sustainable Micro Irrigation.*

I have known Dr. Megh R. Goyal since October of 1979 when he came from Columbus, Ohio (later I went to study at the OSU to complete my MSc and PhD in Soil Fertility during 1981–1988) to Puerto Rico with his wife and three children. According to his oral story, he had job offers from Texas A&M Kenya, Nigeria; University of Guelph, and my university. He accepted the lowest paid job in Puerto Rico. I asked why he did so. His straight-forward reply was the challenges in drip irrigation offered by this job. With no knowledge of Spanish, Megh survived. He also started learning the Spanish language and tasting Puerto Rican food (of course no meat, as he with his family is vegetarian till today).

Within four months of his arrival in Puerto Rico, first drip irrigation system in our university for research on water requirements of vegetable crops was in action. Soon, he formed the State Drip Irrigation Committee consisting of experts from university, suppliers, and farmers. He published his first 22-page Spanish publication titled "Tensiometers: Use, service and maintenance for drip irrigation." Soon, he would have graduate students for their MSc research from our College of Agricultural Sciences. I saw him working in the field and laboratory hand in hand with his students. These students would later collaborate with Megh to produce a Spanish book on drip irrigation management in 1990. I have personally read this book and have found that it can be easily adopted by different groups of readers with a high school diploma or a PhD degree: farmers, technicians, agronomists, drip irrigation suppliers and designers, extension workers, scientists. It is a great contribution for Spanish speaking users!

Megh is a fluent writer. His research studies and results started giving fruit with at least one peer-reviewed publication on drip irrigation per month. Soon, our university researchers would have available basic information on drip irrigation in vegetable and tree crops so that they could design their field experiments. Megh produced research publications not only on different aspects of drip irrigation, but also on crop evapotranspiration estimations, crop coefficients, agroclimatic data, crop water requirements, etc. I had a chance to review his two latest books by Apple Academic Press Inc.: *Management of Drip/Trickle or*

Micro Irrigation (published 2013) and *Evapotranspiration: Research Advances and Applications for Water Management* (2014) and wrote a foreword for both books. I am impressed with professional organization of the contents in each book, which indicates his relationship with the world educational community. Now he is publishing a multi-volume series, "Research Advances in Sustainable Micro Irrigation. My appreciation to Megh for his good work and contribution on micro irrigation; and for this he will always be remembered among the educational fraternity today, tomorrow and forever.

The disadvantages of drip irrigation are: Initial cost can be more than overhead systems; the sun can affect the tubes used for drip irrigation, shortening their usable life; if the water is not properly filtered and the equipment not properly maintained, it can result in clogging; drip irrigation might be unsatisfactory if herbicides or top dressed fertilizers need sprinkler irrigation for activation; drip tape causes extra cleanup costs after harvest; and users need to plan for drip tape winding, disposal, recycling or reuse; waste of water, time and harvest, if not installed properly; highly technical; in lighter soils subsurface drip may be unable to wet the soil surface for germination; requires careful consideration of the installation depth; and the PVC pipes often suffer from rodent damage, requiring replacement of the entire tube and increasing expenses.

Modern drip irrigation has arguably become the world's most valued innovation in agriculture. Drip irrigation may also use devices called microspray heads, which spray water in a small area, instead of emitters. These are generally used on tree and vine crops with wider root zones. Subsurface drip irrigation (SDI) uses permanently or temporarily buried dripper-line or drip tape located at or below the plant roots. It is becoming popular for row crop irrigation, especially in areas where water supplies are limited or recycled water is used for irrigation. Careful study of all the relevant factors like land topography, soil, water, crop and agro-climatic conditions are needed to determine the most suitable drip irrigation system and components to be used in a specific installation.

The main purpose of drip irrigation is to reduce the water consumption by reducing the leaching factor. However, when the available water is of high salinity or alkalinity, the field soil becomes gradually unsuitable for cultivation due to high salinity or poor infiltration of the soil. Thus drip irrigation converts fields in to fallow lands when natural leaching by rain water is not adequate in semiarid and arid regions. Most drip systems are designed for high efficiency, meaning little or no leaching fraction. Without sufficient leaching, salts applied with the irrigation water may build up in the root zone, usually at the edge of the wetting pattern. On the other hand, drip irrigation avoids the high capillary potential of traditional surface-applied irrigation, which can draw salt deposits up from deposits below.

This multi-volume series brings academia, researchers, suppliers and industry partners together to present micro irrigation technology to partially solve water scarcity problems in agriculture sector. The series includes key aspects of micro irrigation principles and applications. I find it user-friendly and easy-to-read and recommend its being to be shelf of each library. My hat is held high to Apple Academic Press, Inc. and Dr. Megh R. Goyal, my longtime colleague.

Miguel A. Muñoz-Muñoz, PhD
Ex-President of University of Puerto Rico, USA
Professor and Soil Scientist
University of Puerto Rico – Mayaguez Campus
Call Box 9000
Mayaguez, P.R., 00681-9000, USA
Email: miguel.munoz3@upr.edu

February 14, 2014

FOREWORD

With only a small portion of cultivated area under irrigation and with the scope of the additional area that can be brought under irrigation, it is clear that the most critical input for agriculture today is water. It is important that all available supplies of water should be used intelligently to the best possible advantage. Recent research around the world has shown that the yields per unit quantity of water can be increased if the fields are properly leveled, the water requirements of the crops as well as the characteristics of the soil are known, and the correct methods of irrigation are followed. Significant gains can also be made if the cropping patterns are changed so as to minimize storage during the hot summer months when evaporation losses are high, if seepage losses during conveyance are reduced, and if water is applied at critical times when it is most useful for plant growth.

Irrigation is mentioned in the Holy Bible and in the old documents of Syria, Persia, India, China, Java, and Italy. The importance of irrigation in our times has been defined appropriately by N.D Gulati: "In many countries irrigation is an old art, as much as the civilization, but for humanity it is a science, the one to survive." The need for additional food for the world's population has spurred rapid development of irrigated land throughout the world. Vitally important in arid regions, irrigation is also an important improvement in many circumstances in humid regions. Unfortunately, often less than half the water applied is used by the crop – irrigation water may be lost through runoff, which may also cause damaging soil erosion, deep percolation beyond that required for leaching to maintain a favorable salt balance. New irrigation systems, design and selection techniques are continually being developed and examined in an effort to obtain high practically attainable efficiency of water application.

The main objective of irrigation is to provide plants with sufficient water to prevent stress that may reduce the yield. The frequency and quantity of water depends upon local climatic conditions, crop and stage of growth, and soil-moisture-plant characteristics. Need for irrigation can be determined in several ways that do not require knowledge of evapotranspiration (ET) rates. One way is to observe crop indicators such as change of color or leaf angle, but this information may appear too late to avoid reduction in the crop yield or quality. Other similar methods of scheduling include determination of the plant water stress, soil moisture status, or soil water potential. Methods of estimating crop water requirements using ET and combined with soil characteristics have the advantage of not only being useful in determining when to irrigate, but also enables us to know the quantity of water needed. ET estimates have not been made for the developing countries though basic information on weather data is available. This has contributed to one of the existing problems that the vegetable crops are over irrigated and tree crops are under irrigated.

Water supply in the world is dwindling because of luxury use of sources; competition for domestic, municipal, and industrial demands; declining water quality; and losses through seepage, runoff, and evaporation. Water rather than land is one of the limiting factors in our goal for self-sufficiency in agriculture. Intelligent use of water will avoid problem of sea water seeping into aquifers. Introduction of new irrigation methods has encouraged marginal farmers to adopt these methods without taking into consideration economic benefits of conventional, overhead, and drip irrigation systems. What is important is "net in the pocket" under limited available resources. Irrigation of crops in tropics requires appropriately tailored working principles for the effective use of all resources peculiar to the local conditions. Irrigation methods include border-, furrow-, subsurface-, sprinkler-, sprinkler, micro, and drip/trickle, and xylem irrigation.

Drip irrigation is an application of water in combination with fertilizers within the vicinity of plant root in predetermined quantities at a specified time interval. The application of water is by means of drippers, which are located at desired spacing on a lateral line. The emitted water moves due to an unsaturated soil. Thus, favorable conditions of soil moisture in the root zone are maintained. This causes an optimum development of the crop. Drip/micro or trickle irrigation is convenient for vineyards, tree orchards, and row crops. The principal limitation is the high initial cost of the system that can be very high for crops with very narrow planting distances. Forage crops may not be irrigated economically with drip irrigation. Drip irrigation is adaptable for almost all soils. In very fine textured soils, the intensity of water application can cause problems of aeration. In heavy soils, the lateral movement of the water is limited, thus more emitters per plant are needed to wet the desired area. With adequate design, use of pressure compensating drippers and pressure regulating valves, drip irrigation can be adapted to almost any topography. In some areas, drip irrigation is used successfully on steep slopes. In subsurface drip irrigation, laterals with drippers are buried at about 45 cm depth, with an objective to avoid the costs of transportation, installation, and dismantling of the system at the end of a crop. When it is located permanently, it does not harm the crop and solve the problem of installation and annual or periodic movement of the laterals. A carefully installed system can last for about 10 years.

The publication of this book series and this volume is an indication that things are beginning to change, that we are beginning to realize the importance of water conservation to minimize the hunger. It is hoped that the publisher will produce similar materials in other languages.

In providing this resource in micro irrigation, Megh Raj Goyal, as well as the Apple Academic Press, are rendering an important service to the farmers, and above all to the poor marginal farmers. Dr. Goyal, Father of Irrigation Engineering in Puerto Rico, has done an unselfish job in the presentation of this compendium that is simple and thorough. I have known Megh Raj since 1973 when we were working together at Haryana Agricultural University on an ICAR research project in "Cotton Mechanization in India."

Gajendra Singh, PhD
prof.gsingh@gmail.com, Tel. +91 99 7108 7591
Adjunct Professor, Indian Agricultural Research Institute, New Delhi
Ex-President (2010-2012), Indian Society of Agricultural Engineers
Former Vice Chancellor, Doon University, Dehradun, India.
Former Deputy Director General (Engineering), Indian Council of Agricultural Research (ICAR), New Delhi.
Former Vice-President/Dean/Professor and Chairman, Asian Institute of Technology, Thailand

New Delhi
February 14, 2014

FOREWORD

In the world, water resources are abundant. The available fresh water is sufficient even if the world population is increased by four times the present population, that is, about 25 billion. The total water present in the earth is about 1.41 billion Km^3 of which 97.5% is brackish and only about 2.5% is fresh water. Out of 2.5% of fresh water, 87% is in ice caps or glaciers, in the ground or deep inside the earth. According to Dr. Serageldin, 22 of the world's countries have renewable water supply of less than 1000 cubic meter per person per year. The World Bank estimates that by the year 2025, one person in three in other words 3.25 billion people in 52 countries will live in conditions of water shortage. In the last two centuries (1800–2000) the irrigated area in the world has increased from 8 million-ha to 260 million-ha for producing the required food for the growing population. At the same time, the demand of water for drinking and industries has increased tremendously. The amount of water used for agriculture, drinking, and industries in developed countries are 50% in each and in developing countries it is 90% and 10%, respectively. The average quantity of water is about 69% for agriculture and 31% for other purposes. Water scarcity is now the single threat to global food production. To overcome the problem, there is a compulsion to use the water efficiently and at the same time increase the productivity from unit area. It will involve spreading the whole spectrum of water thrifty technologies that enable farmers to get more crops per drop of water. This can be achieved only by introducing drip/trickle/micro irrigation in large scale throughout the world.

Micro irrigation is a method of irrigation with high frequency application of water in and around the root zone of plant (crop) and consists of a network of pipes with suitable emitting devices. It is suitable for all crops except rice especially for widely spaced horticultural crops. It can be extended to wastelands, hilly areas, coastal sandy belts, water scarcity areas, semi arid zones, and well-irrigated lands. By using micro irrigation, the water saving compared to conventional surface irrigation is about 40–60% and the yield can be increased up to 100%. The overall irrigation efficiency is 30–40% for surface irrigation, 60-70% for sprinkler irrigation, and 85–95% for micro irrigation. Apart from this, one has the advantage of saving of costs related to labor and fertilizer, and weed control. The studies conducted and information gathered from various farmers in India has revealed that micro irrigation is technically feasible, economically viable, and socially acceptable. Since the allotment of water is going to be reduced for agriculture, there is a compulsion to change the irrigation method to provide more area under irrigation and to increase the required food for the growing population.

The farmers in the developing countries are poor and hence it is not possible for them to adopt/install the micro irrigation with fertigation though it is economically viable and profitable. In Tamil Nadu – India, the number of marginal farmers (holding less than 1.0 hectare) and small farmers (holding 1 to 2 ha) has increased from 50,76,915 in 1967–1968 to 71,84,940 in 1995–1996 and area owned by them has also decreased in the same period from 0.63 ha to 0.55 ha. In addition, the small farmers category is about 89.68% in 1995–1996 of the total farmers in the state. At the same time if micro irrigation is used in all crops, yield can be increased and water saving will be 50%. In the case of sugarcane crop, the yield can be increased to 250 tons/ha from the present average yield of 100 tons/ha, which is highest at present in India. Therefore, to popularize the micro irrigation system among this group of farmers, more books like this, not only in English but also in the respective national languages, should be published.

Volumes 1 and 2 in this book series cover micro irrigation status and potentials, reviews of the system, principles of micro irrigation, the experience of micro irrigation in desert region—mainly in the Middle East, and application in the field for various crops, especially in water requirements, like banana, papaya, plantations, tanier, etc. The chapters are written by experienced scientists from various parts of the world bringing their findings, which will be useful for all the micro irrigation farmers in the world in the coming years. I must congratulate Dr. Goyal for taking trouble in contacting and collecting papers from experts on their subjects and publishing nicely in a short time.

Professor Megh R. Goyal is a reputed agricultural engineer in the world and has wide knowledge and experience in soil and water conservation engineering, particularly micro irrigation. After a big success for his first book titled, *Management of Drip/Trickle or Micro Irrigation* by Apple Academic Press Inc., this compendium is unique. Dr Goyal, Senior Editor-in-Chief of this book series, has taken into account the fate of marginal farmers and is thus serving the poor. He has contacted/consulted many experts who are involved in the subject matter to bring the experience and knowledge about micro irrigation to this book. He has also given many figures, illustrations and tables to understand the subject. I congratulate the author for writing this valuable book series. The information provided in this book series will go a long way in bringing micro irrigation the world especially in water scarcity countries. On behalf of Indian scientists and agricultural engineers on micro irrigation, I am indebted to Dr. Megh R. Goyal and Apple Academic Press for undertaking this project.

Professor (Dr.) R. K. Sivanappan
Email: sivanappan@hotmail.com
Former Dean-cum-Professor of College of Agricultural Engineering and Founding Director of Water Technology Centre at Tamil Nadu Agricultural University [TAMU], Coimbatore – India. Ex-member of Tamil Nadu State Planning Commission (2005–2006).
Father of Micro irrigation in India as mentioned by Mrs. Sandra Postel in her book *Pillar of Sand — Can the Irrigation Miracle Last* by W. W. Norton and Company – New York.
Recipient of Honorary PhD degree by Linkoping University – Sweden; and conferment of the honorary DSc degree by the TAMU-India.

February 14, 2014
Coimbatore—India

FOREWORD

The micro irrigation system, more commonly known as the drip irrigation system, was one of the greatest advancements in irrigation system technology developed over the past half century. The system delivers water directly to individual vines or to plant rows as needed for transpiration. The system tubing may be attached to vines, placed on or buried below the soil surface.

This book, written by experienced system designers/scientists, describes various systems that are being used around the world, the principles of micro irrigation, chemigation, filtration systems, water movement in soils, soil-wetting patterns, crop water requirements and crop coefficients for a number of crops. It also includes chapters on hydraulic design, emitter discharge and variability, and pumping station. Irrigation engineers will find this book to be a valuable reference.

Marvin E. Jensen, PhD, PE
Retired Research Program Leader at USDA-ARS; and
Irrigation Consultant
1207 Spring Wood Drive, Fort Collins, Colorado 80525, USA.
Email: mjensen419@aol.com

February 14, 2014

BOOK SERIES: RESEARCH ADVANCES IN SUSTAINABLE MICRO IRRIGATION

Volume 1: Sustainable Micro Irrigation: Principles and Practices
Senior Editor-in-Chief: Megh R. Goyal, PhD, PE

Volume 2: Sustainable Practices in Surface and Subsurface Micro Irrigation
Senior Editor-in-Chief: Megh R. Goyal, PhD, PE

Volume 3: Sustainable Micro Irrigation Management for Trees and Vines
Senior Editor-in-Chief: Megh R. Goyal, PhD, PE

Volume 4: Management, Performance, and Applications of Micro Irrigation
Senior Editor-in-Chief: Megh R. Goyal, PhD, PE

ABOUT THE SENIOR EDITOR-IN-CHIEF

Megh R. Goyal received his BSc degree in Engineering in 1971 from Punjab Agricultural University, Ludhiana, India; his MSc degree in 1977 and PhD degree in 1979 from the Ohio State University, Columbus; his Master of Divinity degree in 2001 from Puerto Rico Evangelical Seminary, Hato Rey, Puerto Rico, USA. He spent a one-year sabbatical leave in 2002–2003 at Biomedical Engineering Department, Florida International University, Miami, USA.

Since 1971, he has worked as Soil Conservation Inspector; Research Assistant at Haryana Agricultural University and the Ohio State University; and Research Agricultural Engineer at Agricultural Experiment Station of UPRM. At present, he is a Retired Professor in Agricultural and Biomedical Engineering in the College of Engineering at University of Puerto Rico – Mayaguez Campus; and Senior Acquisitions Editor and Senior Technical Editor-in-Chief in Agriculture and Biomedical Engineering for Apple Academic Press, Inc.

He was the first agricultural engineer to receive the professional license in Agricultural Engineering in 1986 from the College of Engineers and Surveyors of Puerto Rico. On September 16, 2005, he was proclaimed as "Father of Irrigation Engineering in Puerto Rico for the twentieth century" by the ASABE, Puerto Rico Section, for his pioneer work on micro irrigation, evapotranspiration, agroclimatology, and soil and water engineering. During his professional career of 45 years, he has received awards such as Scientist of the Year, Blue Ribbon Extension Award, Research Paper Award, Nolan Mitchell Young Extension Worker Award, Agricultural Engineer of the Year, Citations by Mayors of Juana Diaz and Ponce, Membership Grand Prize for ASAE Campaign, Felix Castro Rodriguez Academic Excellence, Rashtrya Ratan Award and Bharat Excellence Award and Gold Medal, Domingo Marrero Navarro Prize, Adopted son of Moca, Irrigation Protagonist of UPRM, Man of Drip Irrigation by Mayor of Municipalities of Mayaguez/Caguas/Ponce and Senate/Secretary of Agriculture of ELA, Puerto Rico.

He has authored more than 200 journal articles and textbooks including *Elements of Agroclimatology* (Spanish) by UNISARC, Colombia; two *Bibliographies on Drip Irrigation*. Apple Academic Press Inc. (AAP) has published his books, namely, *Biofluid Dynamics of Human Body*, *Management of Drip/Trickle or Micro Irrigation*, *Evapotranspiration: Principles and Applications for Water*

Management, and *Biomechanics of Artificial Organs and Prostheses*. With this volume, AAP will publish 10-volume set on *Research Advances in Sustainable Micro Irrigation*. Readers may contact him at: goyalmegh@gmail.com.

WARNING/DISCLAIMER

The goal of this compendium is to guide the world community on Sustainable micro irrigation management for trees and vines for economical crop production. The reader must be aware that the dedication, commitment, honesty, and sincerity are the most important factors in a dynamic manner for a complete success. It is not a one-time reading of this compendium. Read and follow every time, that it is needed. To err is human. However, we must do our best. Always, there is a space for learning new experiences.

The editor, the contributing authors, the publisher and the printer have made every effort to make this book as complete and as accurate as possible. However, there still may be grammatical errors or mistakes in the content or typography. Therefore, the contents in this book should be considered as a general guide and not a complete solution to address any specific situation in irrigation. For example, one size of irrigation pump does not fit all sizes of agricultural land and to all crops.

The editor, the contributing authors, the publisher and the printer shall have neither liability nor responsibility to any person, any organization or entity with respect to any loss or damage caused, or alleged to have caused, directly or indirectly, by information or advice contained in this book. Therefore, the purchaser/reader must assume full responsibility for the use of the book or the information therein.

The mention of commercial brands and trade names are only for technical purposes. It does not mean that a particular product is endorsed over another product or equipment not mentioned.

All weblinks that are mentioned in this book were active on October 31, 2013. The editors, the contributing authors, the publisher and the printing company shall have neither liability nor responsibility if any of the weblinks is inactive at the time of reading of this book.

PART I

PRINCIPLES OF SUSTAINABLE MICRO IRRIGATION FOR TREES AND VINES

CHAPTER 1

PRINCIPLES OF AUTOMATION

MEGH R. GOYAL

CONTENTS

Printed with permission from Goyal, Megh R., 2013. Chapter 7: Automation 143–165. In: Management of Drip/Trickle or Micro Irrigation edited by Megh R. Goyal. New Jersey: Apple Academic Press Inc.

1.1 INTRODUCTION

Drip irrigation is an artificial method to apply the essential water for the plant growth that the nature has failed to provide [1]. Typically the irrigation water is applied to supply moisture to root zone when most of the "water available" to the plant has been used. There are several methods of pressure irrigation, such as: Sprinkler irrigation, center pivot and LEPA; micro jets, drip/micro or trickle irrigation, surface or subsurface irrigation. These help to maintain the soil moisture that is adequate for the plant growth. Among these systems, drip irrigation is the most efficient in terms of water use efficiency. Drip irrigation system is used extensively in humid, arid and semiarid regions of the world. Any interruption or disturbance in an irrigation scheduling will cause a water stress to the crop. Therefore, the scheduling of drip (high frequency) irrigation should be automated so that it is able to respond to slower and faster changes in the soil moisture, the plant water or evapotranspiration. Automation of drip irrigation system has several advantages: Economy, saving of manual labor, increase in crop yield, conservation of energy and effective control of irrigation. This chapter presents basic concepts for automation of drip irrigation system, different methods of automation and irrigation programming [11].

1.2 PRINCIPLE OF AUTOMATION

Current technologies of irrigation programming consider several factors such as [2, 3]: Duration and stage of crop growth, allowable plant water stress, soil aeration, soil water potential, soil salinity, soil moisture available to the plant, class A pan evaporation and evapotranspiration. In most cases, programming of drip irrigation has been limited to a control system that uses duration or depth of irrigation. The irrigation controller is programmed to operate solenoid valves in sequence and to verify operating pressure and flow rates, wind, temperature and other indirect variables. To obtain the minimum cost-benefit and high efficiency of water use, it is necessary to achieve high crop yield. The water loss due to several processes (control of salinity, requirement of infiltration, evaporation, irrigation losses and runoff), must be reduced to a minimum so that the accurate application of the irrigation is limited only to the crop requirements. Four methods for automation of irrigation systems are based on: (1) Soil moisture, (2) Plant water stress, (3) Estimation of evapotranspiration, and (4) Combination of one or more of these methods.

1.2.1 SOIL MOISTURE METHOD [1, 6–8]

Irrigation based on soil water potential is perhaps the oldest method to program irrigation. Microprocessors along with sensors, tensiometers, heat transfer psychrometric methods, gypsum blocks and thermocouples have been used successfully for irrigation scheduling. The sensors can provide quick information to make decisions for application of irrigation depth. The microprocessor circuits combined with a computer programming can help to estimate the irrigation duration on the basis of field data, matrix potential of soil; and to calculate the number of days between two successive irrigation events.

A thermal method measures the matrix potential of soil, independent of soil texture, temperature or salinity. It is based on frequent measurements of ability of a porous ceramic sensor to dissipate a small amount of heat. With a good calibration, the sensor can be used in any soil to automatically watch the matrix potential of the soil and for irrigation scheduling. For closed circuit automated irrigation, the soil sensor is placed in the root zone. For an automatic control of an irrigation system based on matrix potential of a soil, we need equipments for the:

1. Automatic sampling from several sensors in sequence,
2. Comparison of the reading of each sensor at which the irrigation begins at a predetermined matrix potential of the soil, and
3 The operation of irrigation controller to control the irrigation depth. Desktop computers in combination with microprocessors have been successfully used. There is also a commercial equipment to measure the matrix potential of the soil and for an automatic control of a drip irrigation system.

1.2.2 WATER CONTENT IN THE PLANT [9, 10]

The water is frequently one of the limiting factors in agriculture. Transpiration loss occurs from the plant surface due to an evaporative demand of the atmosphere. Less than one percent of the absorbed water is retained by the plant. This small fraction of water is often used to replace the deficit between water use and transpiration. Thus any water deficiency can cause a plant water stress. The total water potential (the sum of turgor, matrix and osmotic potential) is used to indicate the condition of the plant water. The plant development and growth (cellular enlargement and photosynthesis), pollination, fruit formation, crop yield and fruits quality are affected by the water deficit. Probably, the cellular growth is most sensitive to the water deficit. There are several methods to estimate the condition of plant water. These include determination of relative water content, diffusive conductivity of the plant, water potential of the plant and surface temperature. The indirect or direct measurement of water potential is probably a good indicator of the plant water stress. There are several methods to measure the plant water stress such as: The total leaf water potential with a leaf psychrometer; temperature of leaf surface with an infrared thermometer, and the leaf water potential indirectly on the basis of the diameter of the stem.

1.2.3 LEAF WATER POTENTIAL

The leaf water potential can be measured by psychrometer or by adhering thermocouples to the leaves. Although the psychrometric measurements are taken routinely for research purpose, yet the instruments are expensive and not feasible for commercial purposes.

1.2.4 TEMPERATURE OF THE LEAF

Measurements of leaf temperature can indirectly indicate status of a water stress [10]. Plant water stress index can be used to automate the irrigation system, and to indicate when to irrigate. The operating system can be easily automated to take the data, to calculate the index of plant water stress, to make comparisons with predetermined values of irrigation depth and to make decisions for irrigation scheduling. Leaf temperature is measured with a noncontact infrared thermometer. The accuracy of temperature of the surface of leaf depends on the precision of calibration. The measurements are sensitive to changes in the ambient temperature, interactions with surrounding surfaces (such as soil), and leaf area index. Measurements of leaf area index of a crop vary from plant to plant. There is no standard value.

1.2.5 STEM DIAMETER

The diameter of stem and the leaf water potential are closely related to one another. The measurements of stem diameter can be used for continuous recording of the stem growth and the condition of plant water. The periodic calibration of the changes of diameter of stem versus leaf water potential can be conducted for each phenological stage of a plant. This technique can be used for the purpose of automation.

1.2.6 EVAPOTRANSPIRATION ESTIMATIONS

To program the irrigation, the evapotranspiration models have been successfully used throughout the world. The following information is needed for the evapotranspiration estimations and the criteria to decide when to irrigate.

1. Evapotranspiration of a reference crop, potential ET, etc.
2. Crop growth curve, crop coefficient and consumptive use of a crop.
3. Index to estimate the additional evaporation from the soil surface when the soil is wet or dry.
4. Index to estimate the effect of soil water loss in relation to ET.
5. Estimation of available soil moisture used by a crop: Consumptive water use.
6. Relation between expected crop yield and crop water use.

To estimate the ET, many of the variables are not well defined and must be estimated. Although the ET models can be useful to accurately estimate the irrigation needs, yet these are not viable for irrigation scheduling as available weather data are limited for a particular location.

1.2.7 DIRECT MEASUREMENT OF ESSENTIAL EVAPOTRANSPIRATION

The weighing lysimeter in a given crop can serve as a guide to provide an adequate irrigation depth for the crop need. A water tank is connected to a lysimeter so that the weight of the irrigation depth is included in the daily weight of lysimeter. Whenever one millimeter of ETc is registered, lysimeter is automatically watered by drip irrigation system to maintain the soil water potential. The tank is automatically filled daily to a constant depth. Therefore, the daily changes in the weight of lysimeter represent the crop growth. The water potential of the soil is almost maintained constant by the drip irrigation system [8, 11] (Fig. 1).

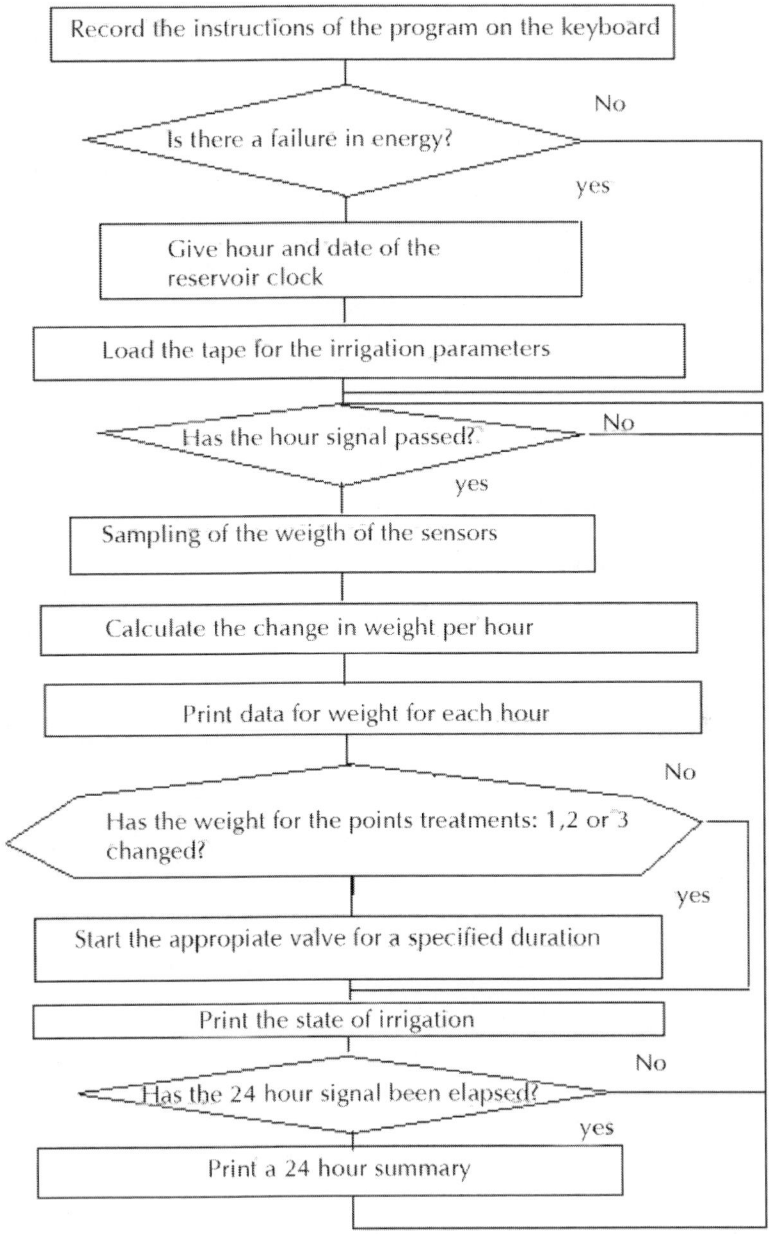

FIGURE 1 Logic diagram to measure weight of lysimeter sensors and to control the irrigation sequence with three depths of irrigation.

1.3 INSTRUMENTATION AND EQUIPMENTS [1–6]

The automation of a drip irrigation system at an operating pressure can potentially provide an optimum crop yield and optimum water use. A system of controls in an automated irrigation system must use sensors to measure variables, such as: Depth and frequency of irrigation, flow rate, operating pressure; and environmental conditions such as wind speed, ambient temperature, solar radiation, rain fall, soil moisture, leaf temperature, leaf area index, etc. Maximum irrigation efficiency is possible with the continuous monitoring and control of the operation of the system with measurements of flow (solenoid valves) and operating pressure (pressure regulators) at strategically important locations in the field. The data or control of functions can be transmitted by electrical cables, laser or hydraulic lines, rays, radio frequency signals, remote control or by satellites. A wide variety of instrumentation and equipments with characteristics are available commercially. These can be subdivided in six categories: (1) Controls, (2) Valves, (3) Flow meters, (4) Filter, (5) Chemical injectors, and (6) Environmental Sensors.

1.3.1 CONTROLS [2, 3, 5]

The controls receive feedback about the volume of water for the field, pressure in the line, flow rates, climatic data, soil water, plant water stress and from the field sensors. This information is then compared with the predetermined values and the irrigation is reprogrammed to adjust for the new values, if necessary. The controls, volumetric valves, hydraulic valves, fertilizer or chemical injectors, flushing of filters, etc., can be operated automatically or manually.

1.3.2 VALVES [8, 11]

Automatic valves (Figs. 2–9) can be activated electrically, hydraulically or pneumatically and these are used to release or to stop the water in the lines; to flush the mains and laterals; to continue the water from one field to another field and to regulate flow or pressure in main, submain or lateral lines. The type of valve will depend on the desired purpose. Valves receive feedback to verify the precision of operation.

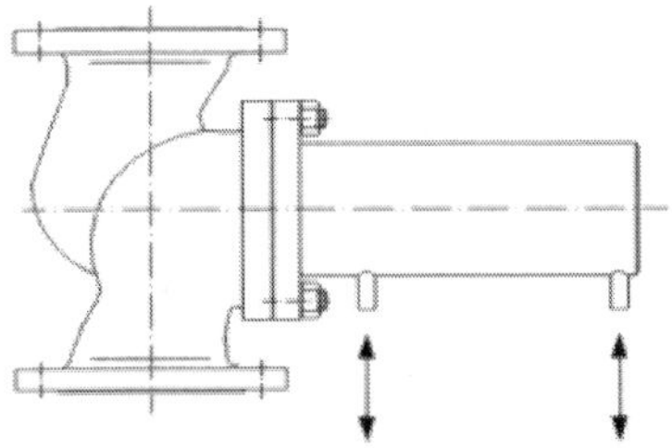

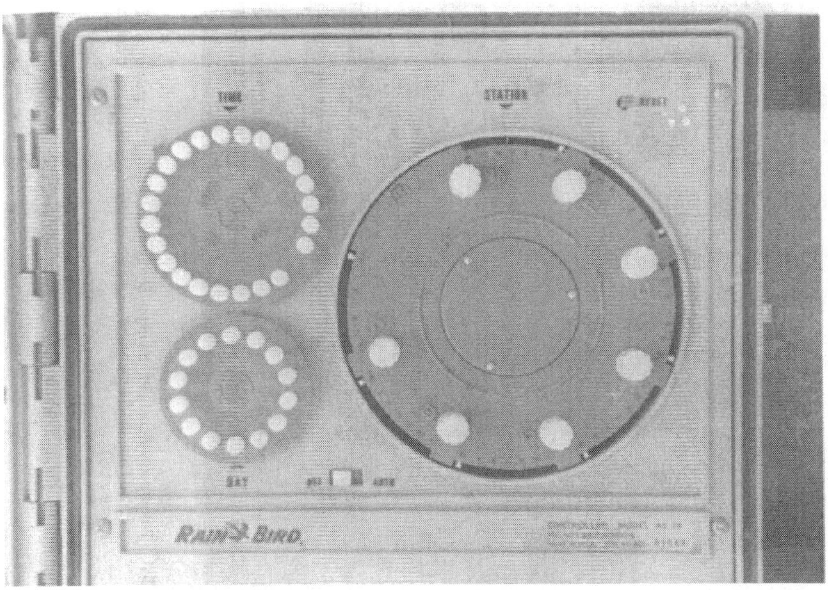

FIGURE 2 Automatic irrigation controller (Rain Bird).

FIGURE 3 Logic hydraulic valve.

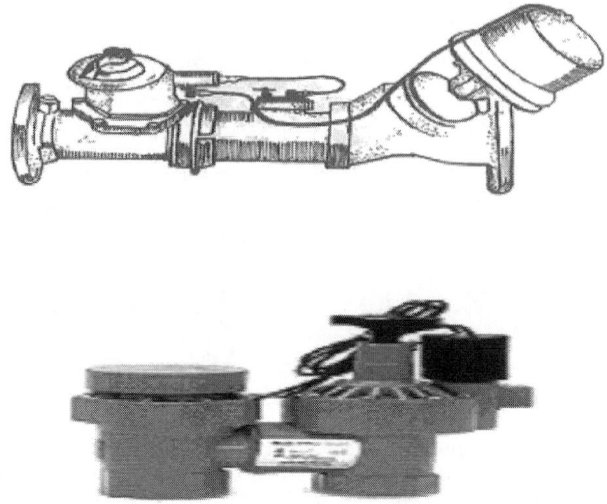

FIGURE 4 Automatic metering valve along with a hydraulic valve.

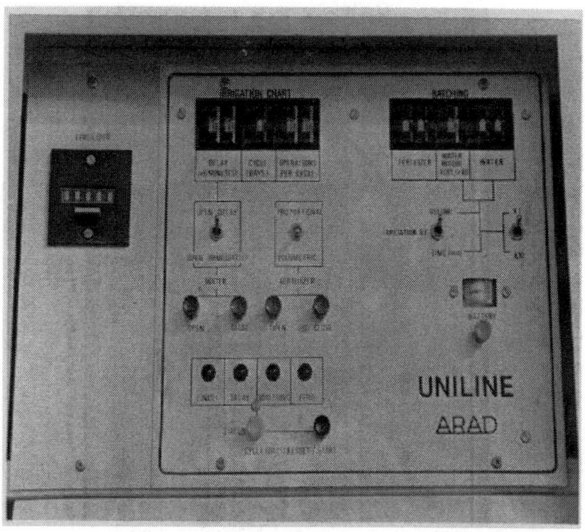

FIGURE 5 Fertilization and irrigation programmer for six different valves (for green house or field).

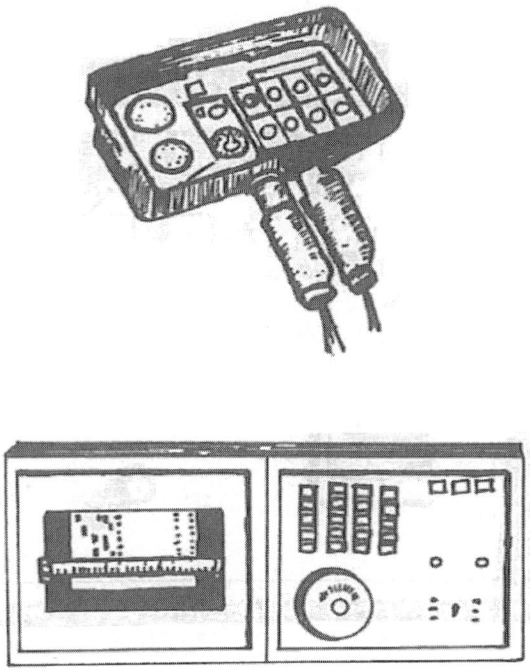

FIGURE 6 Automatic controller (Nirim electronics), using a programmer with a perforated tape or card.

FIGURE 7 Fertigation and chemigation equipments.

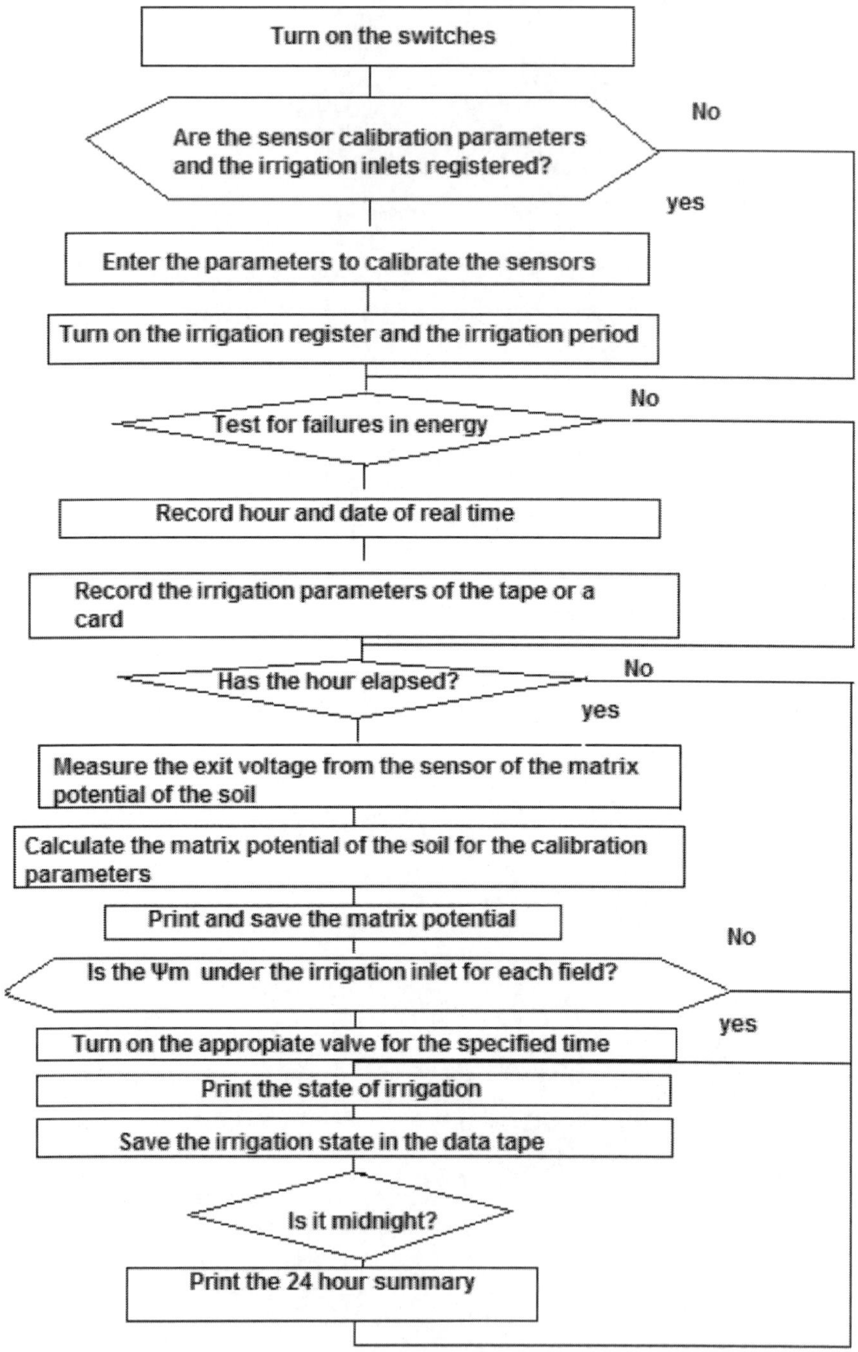

FIGURE 8 Logic diagram for an automatic controller in a drip irrigated field.

http://www.bermad.com/

FIGURE 9 Bermad automatic volumetric valve.

1.3.3 AUTOMATIC VOLUMETRIC VALVE: FLOW METERS

The flow-metering valve (Fig. 10) allows programming the predetermined values. Usually these meters are calibrated to measure applied volume of water or to measure the flow rate.

FIGURE 10 Bermad automatic volumetric valve: Field installation.

1.3.4 AMBIENT SENSORS [8–12]

Various types of instruments are available to determine the soil moisture (ceramic densitometry, ceramic cup, heat dissipater sensor, soil psychrometer); to measure climatic parameters (weather station, automated evaporation tank, etc.), plant water stress or leaf temperature of the crop (leaf psychrometer, porometer for stomate diffusion, infrared and sensorial thermometer to measure stem diameter). These can be used as feedback for the management of irrigation. If the soil at a particular field station is wet, the sensor opens the circuit of the hydraulic or solenoid valve and this station is bypassed. If the soil at this field station is dry, the closer the circuit and the field at this station are irrigated for a specified duration.

1.3.5 FILTERS

The obstruction in the drippers caused by clogging agents (physical, chemical or biological) is a common problem and is considered a serious problem in the maintenance of the drip irrigation systems. The suspended solids may finally clog or reduce the filtration efficiency. The automatic flushing valve is available for different types of filters. The flushing is done by means of back flow of water [4, 8, 11], thus allowing the water to move through the filter in an opposite direction (Fig. 11).

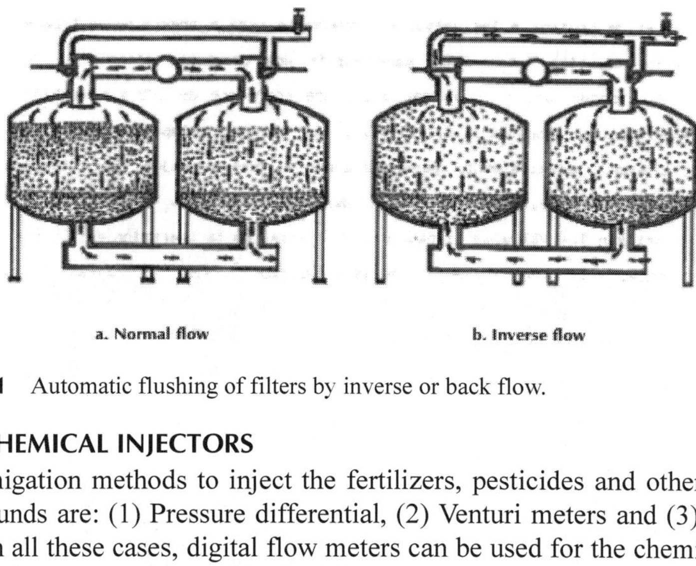

FIGURE 11 Automatic flushing of filters by inverse or back flow.

1.3.6 CHEMICAL INJECTORS

The chemigation methods to inject the fertilizers, pesticides and other inorganic compounds are: (1) Pressure differential, (2) Venturi meters and (3) Injection Pumps. In all these cases, digital flow meters can be used for the chemigation by allowing a known amount of chemicals in a known amount of water to maintain a constant concentration of chemicals-in-the-irrigation-water [8, 11].

1.4 AUTOMATIC SYSTEMS [1–3, 8, 11, 12]

With the exception of a volumetric metering valve that operates according to the time or the discharge rate, the automatic irrigation systems can be divided in three groups on the basis of operation: (1) Sequential hydraulically operated system, (2) Sequential electrically or hydraulically electrically operated systems and (3) Non-sequential electrically operated system with or without programming: With the possibility of using information of the field (feedback) by remote control.

1.4.1 SEQUENTIAL HYDRAULICALLY OPERATED SYSTEM

This system controls the valves in sequence (Fig. 12). The valves open and close based on the water pressure in the line. The pressure arrives at the valve by means of a flexible hydraulic tube (micro tube: polyethylene tube of small diameter) to provide a required pressure. The diameter of the micro tube is generally between 6 and 12 mm and is connected to the hydraulic valve at one end and the other end is connected to the automatic control or the line of water.

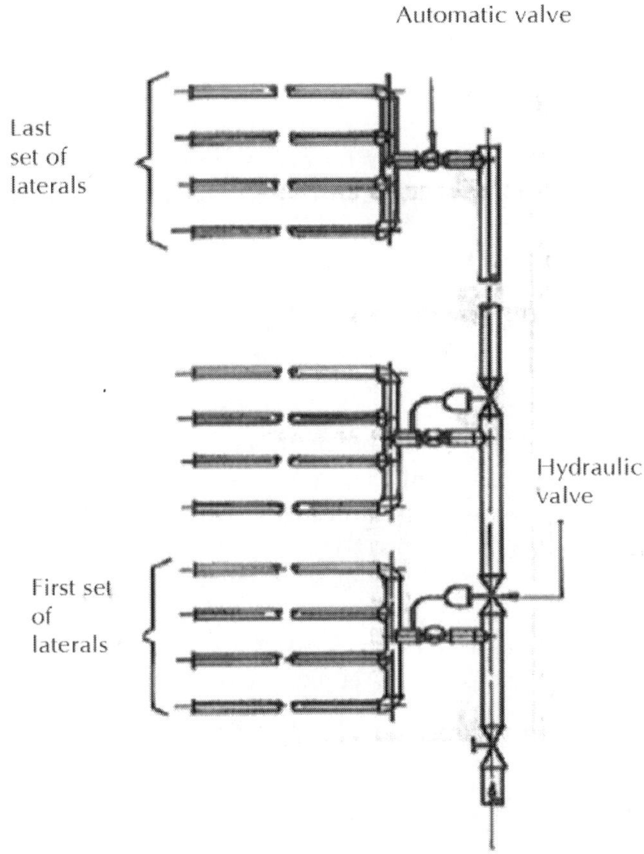

FIGURE 12 Sequential hydraulically operated system for green houses, gardens, nurseries and fruit orchards.

Some hydraulic systems can be connected to the main valve of the line or to the system that replaces the water. In this case, the main valve is connected automatically to open when the system in series is in operation and to close at the end of the irrigation cycle. Electrically operated automatic system activates the pump and deactivates the pump, when the irrigation cycle is over.

Sequential hydraulically operated system is controlled by a predetermined amount of water. The amount of water can be different for each valve and can be adjusted by a regulator mounted in same valve. The hydraulically sequential system can be used to water fruit orchards, gardens, green houses and nurseries, establishing low flow rates through tubes of small diameter and for flow rates in any diameter of tube. The system includes automatic metering valve, hydraulic valve and hydraulic tube.

1.4.2 SEQUENTIAL SYSTEM: OPERATED ELECTRICALLY OR OPERATED HYDRAULICALLY ELECTRICALLY

These systems supply an electrical current through cables for the remote control of the valves (Fig. 12). The current from the "control panel" to the valves, usually passes through a step down transformer to supply a voltage of 24 V. For safety reasons, a current of 220 V should not be used when the subsurface cables extend to the field valves. The regular solenoid valves are mainly used for low flow rates. For pipes of larger diameters, the solenoid valves are used only as controls to activate the hydraulic valves and all the automation process is hydraulically electrical. The control of the second valve is always hydraulic. In the hydraulic sequential system, the opening is controlled electrically by a timer mounted next to the main valve. In such cases, the current source is direct and not alternating.

1.4.2.1 PROGRAMMING IRRIGATION WITH SOLENOID VALVES

The solenoid valves can be used to program the irrigation (Fig. 13). In order to calculate the crop water use, a computer program can be used with the information such as: The soil moisture, evapotranspiration, the date of the next irrigation and the amount of water to be applied. The irrigation programs are based on evapotranspiration estimations, complex water budgets in several dimensions or crop growth models. The ET models use crop and climatic data such as crop coefficient, root zone depth, allowable depletion, drainage rates, air temperature, sun radiation, precipitation and constants in the evapotranspiration equations, and so forth.

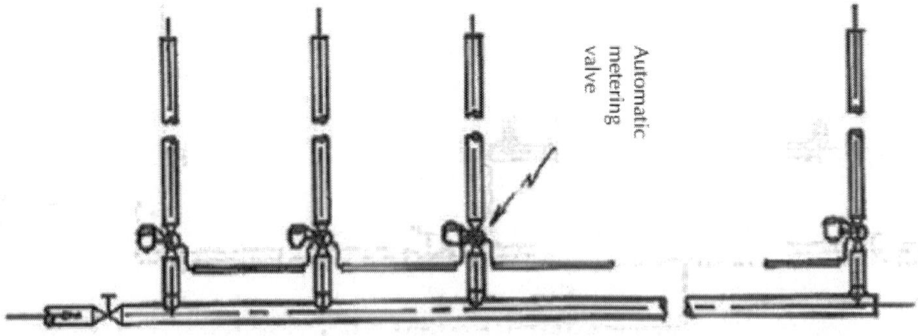

FIGURE 13 Electrically operated tensiometer and solenoid valve.

Then the model incorporates the climatic information to calculate the evapotranspiration rates and to adjust the water balance in the soil as the water is being used. The evapotranspiration model requires an irrigation criteria based on the allowable depletion or the irrigation interval. Actual field data after the irrigation can be helpful to compute the infiltration and immediate drainage for correction of estimated soil moisture. The rate of computed ET can be used to indicate the required amount of irrigation or to specify the time for irrigation interval. This method is more practical for drip irrigation than for other irrigation methods. The records of field data are kept in the office files for the irrigation programming, so that the data can be used to-update-the-inputs-in-the-program.

1.4.2.2. AUTOMATIC VALVES

The automatic valves are commonly used for the pump house and filters; for regulating the pressure in the main line; to control the flushing cycles in the filters, or to control the volume of water through the secondary or lateral lines. The solenoid valves can be used in the secondary or lateral lines to control the volume of water to the individual blocks. The primary function of a solenoid valve is to switch on or switch off the system. However, these valves can be equipped with pressure regulators and check valves. The solenoid valves are operated electrically from the "Central Control Panel." Automatic control valves can also be equipped with manual valves for better efficiency. Automatic valves require periodic maintenance to assure a satisfactory operation. The maintenance program depends on the use of the valve and the flushing operations.

At least, it is recommended that all the diaphragm valves are disarmed and cleaned at least once a year. It is important to clean the deposits on the stem of the valve. Almost all the manufacturers provide a service or fast replacement of most of the components. This can usually be done without removing the valve from the irrigation line. A number of auxiliary controls can be adapted to the diaphragm valves to provide flexibility and convenience.

PRESSURE REDUCING VALVE

This valve responds to changes in the pressure at the exit of the main valve and adjusts to the pressure in the cap or valve cover to compensate for any change. A trouble in the operation can be caused by contamination, obstructions, incorrect assembly, and damages or worn out parts.

PRESSURE-REGULATOR-VALVE

This valve is used to separate the system from the pressure in the main line. It must be open during the normal operation. Whenever the pressure exceeds a preset value, the valve releases the excess pressure.

CONTROLS-TO-ADJUST-THE-VELOCITY-OF-THE-MAIN-VALVE

These are small adjustable controls in the pilot control system. These regulate the speed of opening and closing of the main valve by blocking or strangling the flow that enters or leaves the casing. These can be subjected to obstructions by fine sediments if tightly fit.

1.4.2.3 CONTROLS

Several electromechanical and electronic controls in the drip irrigation system are automated. The controls with mechanical time clocks open and close only a single valve at one time. These are programmed based on series of climatic and soil sensors: to decide when to begin and to end the irrigation cycle; start and to put off pump; to open and close the valves to supply an irrigation depth and to remember how much water and fertilizer was applied to each block within the field. The controls are also available to diagnose operation and identify the troubles and to take remedial steps. Others put off the system during rainfall and restart the system when necessary. A timer uses a clock to program the beginning and sequence of irrigation. The control is a source of electric or hydraulic signal to activate by remotely located valves to allow or to stop the flow.

The communication between the irrigation controller and the valves is by means of electrical wires, hydraulic lines or radio signals. The microprocessors and microcomputers also can be programmed using data of tensiometers, pan evaporation, thermocouples, soil moisture tension gages, anemometer, flow meter, pressure transducer, etc. These controls are based on the climatic and soil sensors or according to the program specified by the irrigator. Using these data, the controller uses a program to compute irrigation requirements for each crop and block within a field.

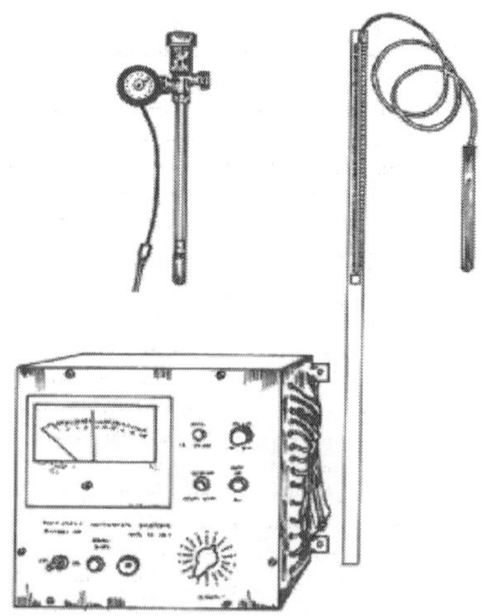

FIGURE 14 Automatic unit for control of irrigation based on the gypsum blocks (sensors).

The data from the flow meters and pressure gages is used to determine the flushing time and to detect any troubles in the system. In most of the cases, the controller has a calendar programmer, so that the cycle of irrigation begins automatically on a particular day of the week and at a particular time of the day. Most of the controllers can be programmed for 14 days, while others are only limited to seven days. Practically all-automatic controllers have a station selector on the outer surface of the panel (Figs. 4–7). This station selector shows a green light to show the station in operation. In addition, it can also be set manually so that the irrigation operator can start and put if off whenever desired.

1.4.3 SEQUENTIAL SYSTEM: ELECTRICALLY OPERATED

In these systems, the amount of water distributed to the different blocks is determined by a flow meter. A timer determines the duration of operation: 14 days and 24 h per day. Sensors based on tensiometers or pan evaporation can activate these. Although this type of system was developed mainly to water green houses, yet it can be used for the drip irrigation system.

1.4.4 NON-SEQUENTIAL SYSTEM

These systems are completely automatic and are controlled electrically. These nonsequential systems are controlled by hydraulic or electrical valves that can operate the valve in the desired block at random, and can supply known amount of water for a known duration to a desired block. Each unit can supply a known flow at different hours during the day, in response to soil moisture status in each block. The "Control Panel" consists of electrical circuits that operate the pump, main valve, adds fertilizer according to a pre-established schedule and measures the soil moisture to estimate the crop irrigation requirements. This system usually operates by a remote control system and is designed to provide feedback of field data, so that the automatic adjustment can be made and adjustments for changes in pressure and flow rates can be made to the discharge flow in the distribution lines.

1.4.4.1 CENTRAL PANEL

The central panel controls all the operations of the field, sending instructions to the valves and receiving continuous data on the operation of the irrigation system. It consists of a programmed unit of irrigation, a unit for transmission of information, a unit for the control of flow in the laterals and a unit for warning signals.

1.4.4.2 FIELD PANEL

The field panel is placed centrally in the field and operated by remote control unit. The signals of the main panel are sent by an individual communication channel and these are transmitted to individual field panel. The field panel can collect the data on water meters, operating pressures and warning signals. Then the data can be transmitted to the main panel (control panel).

1.4.5 USE OF SENSORS TO PROGRAM IRRIGATION

In addition to the above-mentioned instruments, sensors are available to determine the soil moisture tension or the soil moisture. Tensiometers and gypsum blocks are simple and economical to use. Another method is a neutron scattering method, but it is quite expensive and is used for research purpose only.

1.4.6 USE OF GYPSUM BLOCKS AND TENSIOMETERS

The gypsum blocks can measure the soil moisture tension in the range of 1–15 atmospheres. There are two electrodes inserted in each block and the changes in the soil moisture are calibrated with variations in the resistance. The precision of this method is based on the temperature, salt concentration in the soil solution, physical characteristics of the gypsum block and the electrical resistivity of the soil. For tensions of 80 cbars, a tensiometer is recommended instead of a gypsum block. Tensiometer (Fig. 14) measures the tension and the reading is given in cbars. The main disadvantage of a tensiometer is a relatively low critical tension of 85 cbars after which the air enters the plastic stem of a tensiometer. The soil moisture by any method will show variations in the soil moisture within the same field. A sample of the soil in a given location represents only the soil condition of that location. Therefore, several observations of soil moisture at various locations in the field are desirable.

1.4.7 NEUTRON SCATTERING METHOD

The neutron scattering method consists of a neutron radiation source of high energy and a neutron detector. Neutrons travel through the soil medium, loose energy, and the speed is reduced when these hit the elements that are present in the soil. The hydrogen, a component of the water, is dominant in the reduction of the speed of fast neutrons. Due to other factors that can affect the reading, the calibration of this method is done in a location where the equipment will be installed and used. The use of the neutron scattering method requires the installation of access tubes at the beginning of planting and removal of these tubes after the last harvest.

It is recommended to install one sensor at each 30 cm depth. Periodically the operator will obtain the readings of the tube at the desired depth. A minimum of three readings are taken: at shallow root depth, at middle depth, and at a deeper depth. The water content of these readings is added and the water content at field capacity is deducted from the sum. The difference between these two estimates will be the amount of water that should be applied. The readings can be recorded automatically and are stored in the memory of the neutron scattering equipment. Then these can be downloaded on the computer of the Control Panel. With this information, the computer will give the necessary commands to the drip irrigation system so that the crop water requirements are met in the desired block.

1.4.8 CLASS A PAN EVAPORATION TO AUTOMATE THE SYSTEM

The relationship between pan evaporation and the water loss have been well established. Both are exposed to similar climatic conditions in the same field. This correlation can be used to schedule the irrigation. If electrodes in the tank can be installed at a depth (based on previous experience), the irrigation can be controlled automatically. The irrigation will begin when the surface of the water in the class A pan lowers to a predetermined level and will stop when the level raises to certain level in the tank [7].

1.5 PREVENTIVE MAINTENANCE

1.5.1 PREPARATION AFTER THE LAST HARVEST

1. Clean the controllers, valves and sensors.
2. Examine the condition of the control panel and store it well.
3. Remove and store batteries.
4. Flush and drain the hydraulic tubes.
5. Disconnect the electrical wires in the field.
6. Examine for possible breakage and defects in electrical conductors.

1.5.2 PREPARATION FOR THE START OF A CROP SEASON

1. Be sure that all the electrical connections are cleaned and adjusted well.
2. Make sure that the electrical contacts are free of corrosion and dirt.
3. Inspect all the hydraulic lines and pneumatic lines for leakage or breakage.
4. Verify that the equipments, accessories and sensors operate properly.

1.5.3 DURING THE CROP SEASON

1. Visually examine all external components weekly.
2. Disconnect the electrical wires in the field during electric storms.
3. Disconnect the batteries when the control is out of service for one week or more than one week.

1.6 TROUBLE SHOOTING

Trouble	Cause	Remedy
	Controls	
1. The cycle of irrigation does not work at the pre-established time.	The clock of the control panel is outside pre-established calibration for the schedule of the cycle.	Calibrate clock at the pre-established time.
2. Some stations do not operate. Cables of the valves are not connected properly. Hydraulic tube is broken or missing.	Control station for time is off. Check connections between valves. Replace the hydraulic tube.	Place ignition control in "on" position.

3. Danger signal is in "on-position" Program-of-emergency-is-in operation due to bad operation of the system.	Battery is dead. To locate the source of the problem in the system and to correct it.	Recharge battery.
Filters		
4. Poor filtration	High difference in pressure is due to obstruction of filters by the clogging agents.	Flushing of filters by inverse (back) flow.
5. Pressure difference at the entrance and exit of a filter -exceeds the recommended -values.	Depth of filter media is not adequate. Valves are obstructed.	Add more media until it is at recommended level. Verify valves for obstruction.

1.7 SUMMARY

Principle of Automation includes factors such as: duration and stage of crop growth, allowable plant water stress, soil aeration, soil water potential, soil salinity and evapotranspiration. Leaf water potential can be measured by a psychrometer or by adhering thermocouples to the leaves. Leaf temperature is measured with a noncontact infrared thermometer. The accuracy of temperature of the surface of leaf depends on the precision of calibration. Measurements of leaf area index of a crop vary from plant to plant. The diameter of the stem can be used for continuous recording of the stem growth and the condition of plant water stress for each phenological stage of a plant. This technique can be used for the purpose of automation. The automation of a drip irrigation system provides an optimum crop yield and optimum water use. The system uses sensors to measure depth and frequency of irrigation, flow rate, operating pressure, wind speed, ambient temperature, solar radiation, rain fall, soil moisture, leaf temperature, leaf area index, etc. The instrumentation and equipments for automation can be subdivided in six categories: (1) Controls, (2) Valves, (3) Flow meters, (4) Filter, (5) Chemical injectors, and (6) Environmental.

There are three types of automatic irrigation systems. In sequential hydraulically operated system, the valves open and close in response to the application or elimination of water pressure. In sequential electrically or operated hydraulically electrically, the system supplies an electrical current for remote control of the valve. The automatic valves are commonly used for the pump house and filters; for regulating the pressure in the main line; to control the flushing cycles in the filters, or to control the volume of water through the secondary or lateral lines. Solenoid valves are used in the secondary or lateral lines to control the volume of pressure-regulator-valves are used to separate the system from the pressure in the main line. Whenever the pressure exceeds a preset value, the valve releases the excess pressure. The controls with mechanical time clocks open and close only a single valve at one time.

The communication between the irrigation controller and the valves is by means of wires, hydraulic lines or radio signals. In Electrically Operated Sequential System the amount of water distributed to the different blocks is determined by a flow meter. The nonsequential systems are controlled by hydraulic or electrical valves that can operate the valve in the desired block at random, and can supply known amount of water for a known duration to a desired block. The central panel allows control all the operations of the field. The field panel is operated by a remote control unit. The signals of the main panel are sent by an individual communication channel and are transmitted to individual field panel. Sensors are available to determine the soil moisture tension or the soil moisture. Tensiometer and gypsum blocks are simple and economical to use. Another method is a neutron scattering method but it is quite expensive and is for research purpose only. Preventive maintenance and trouble shooting of the system are also presented.

KEYWORDS

- **Atmosphere**
- **Automatic metering valve**
- **Automatic system**
- **Automation**
- **Back flow**
- **Bar**
- **Cap or plug**
- **Check valve**
- **Chemigation**
- **Class A pan**
- **Clogging**
- **Consumptive water use**
- **Control system**
- **Crop coefficient**
- **Crop water requirement**
- **Diffusion**
- **Drainage**
- **Dripper or emitter**
- **Evaporation**
- **Evaporation tank**
- **Evapotranspiration**
- **Fertigation**
- **Fertilizer**

- **Field capacity**
- **Filter**
- **Flow meter**
- **Flushing valve**
- **Gypsum blocks**
- **Hydraulic system**
- **Hydraulic valve**
- **Infiltration**
- **Irrigation, drip/trickle**
- **Irrigation frequency**
- **Irrigation requirement**
- **Irrigation, depth**
- **Irrigation, duration**
- **Irrigation, sprinkler**
- **Irrigation, subsurface**
- **Irrigation, surface**
- **Lysimeter**
- **Main line**
- **Main valve**
- **Microtube**
- **Neutron scattering**
- **Pan evaporation**
- **Photosynthesis**
- **Polyethylene (PE)**
- **Precipitation**
- **Psychrometer**
- **Pump**
- **Pump house**
- **Root zone**
- **Sequential electrically operated system**
- **Soil moisture**
- **Soil texture**
- **Solar radiation**
- **Solenoid valve**
- **Station selector**
- **Tensiometer**
- **Transpiration**
- **Volumetric valve**
- **Water content**
- **Water potential**

REFERENCES

1. Anonymous, (1983). Irrigation news in Israel (Spanish). Agricultura de las Américas. Febrero, paginas 6–17.
2. Anonymous, (1982). The automatization of irrigation (Spanish). Irrinews. ISSN 304-3606. Bet Degan-Israel. 25: 12–13.
3. Anonymous. *Computerized Irrigation Control System Handout.* Motorola Inc. Agro-Control Department, Fresno, CA, 93710.
4. Anonymous. *Instruction manual for installation and operation of "Free Flow" media filters.* Water Management Products Division. P.O. Bo× 352 Corona, CA, 91720.
5. Anonymous. *Instruction manual – Irrigation Controller Model AG-7.* Rain Bird, C. A.
6. Anonymous. *Irrigation Water Management Using the Neutron Probe.* Buchanan Circle Corp., Pacheco – CA, 94553.
7. Bauder, J. W., King, L. D., Westesen, G. L. (1982). *Scheduling irrigation with evaporation pans.* Coop. Ext. Ser., Montana State Univ., Bozeman, Bulletin. 1262, Form Circular. 1211, 1–23.
8. Goldberg, D., Gornat, B., Rimon, D. (1976). *Drip Irrigation: Principles, Designing and Agriculture Practices.* Israel Drip Irrigation Scientific Publications, 236–250.
9. Howell, T. A., Hatfield, J. L., Yamada, H., Davis, K. R. (1984). *Evaluation of cotton canopy temperature to detect crop water stress.* Trans. ASAE, 27(1): 84–88.
10. Jackson, R. D. (1982). Canopy temperature and crop water stress. In: *Advances in Irrigation,* editor D.I. Hillel. Volume I: 43–85. Academic Press, New York.
11. Nakayama, F. S., Bucks, D. A. (1986). *Trickle Irrigation for Crop Production Designing, Operation and Management.* Elsevier Science Publishers B. V. Amsterdam – The Netherlands. Pages 188–210, 308 and 311.
12. Phene, C. J., Howell, T. A. (1984). Soil sensor control of high frequency irrigation. Trans. ASAE, 27 (2): 386–391, 396.

CHAPTER 2

PRINCIPLES OF SERVICE AND MAINTENANCE

MEGH R. GOYAL

CONTENTS

Printed with permission from Goyal, Megh R., 2013. Chapter 11: Service and maintenance 213–218. In: *Management of Drip/Trickle or Micro Irrigation* edited by Megh R. Goyal. New Jersey: Apple Academic Press Inc.

Success of drip irrigation depends on the support from specialist for installation and maintenance.

Flow meter to measure the volume of water applied.

Water source, pump, and check valve assembly.

2.1 INTRODUCTION

The orifices in the drip lines or the emitters emit water to the soil. The emitters allow only the discharge of few liters or gallons per hour. Because the emitters have small orifices, these can be easily obstructed with clogging agents (physical, chemical, and biological). The obstruction can reduce degree of emission, the uniformity of water distribution, and therefore, this can reduce plant growth. Once the system has been obstructed, it becomes more difficult to restore the normal water flow. Therefore, we must prevent the obstructions in the filters, laterals, and emitters. The clogging can be prevented with a good maintenance and periodic service of the system. To operate and to maintain a drip irrigation system in a good working condition, the following considerations are important for an adequate operation [1–4]:

1. Pay strict attention to the filtration and flushing operation.
2. Maintain an adequate operating pressure in the main, sub main, and lateral lines.
3. Flushing and periodic inspection of the drip irrigation system.

2.2 MAINTENANCE OF FILTERS AND FLUSHING OPERATION [1–3]

For effective filtration efficiency, we must make sure that the system is maintained in good condition and it is not obstructed by the clogging agents. For this purpose, pressure gages are installed at the entrance and the exit of a filter. The pressure difference between these two gages should vary from 2 to 5 psi when the filter is clean and the mesh is free from obstructions. The filtration system should be cleaned and flushed, when the pressure difference is from 10 to 15 psi. The filters must be flushed before each irrigation operation. If the water contains high percentage of suspended solids, then the filters should be flushed more frequently. Entrance of dust and foreign material should be avoided, when the filters are open. Filters may not be able to remove the clay particles and algae.

2.3 FLUSHING METHOD [1–4]

The frequency of flushing depends on the water quality. For flushing of irrigation lines, the following procedure can be adopted:

1. Open the ends of the distribution and lateral lines. Allow the flow of water through the lines until all the sediments are thrown out of the lines.
2. Close the ends of the distribution lines. Begin to close the lines one after another, from one block to second, and so on. There must be a sufficient pressure to flush out all the sediments.

2.3.1 CLEANING WITH PRESSURIZED AIR

The clogging can be caused due to presence of organic matter in water. It may be necessary to use pressurized air to clean the drippers. Before beginning this process, the water is passed through the lines for a period of 15 min. When adequate operating pressure has been established, then the air at 7 bars of pressure is allowed through the system. The compressed air will clean the lines, laterals, and drippers from the accumulated organic matter.

2.3.2 CLEANING WITH ACIDS AND CHLORINE

The clogging may also be caused due to precipitation of salts. The cleaning with acids will help to dissolve the chemical deposits. This process is not effective to remove the organic matter. Sodium hypochlorite (at the rate of one ppm) can be injected on the suction side of a pump for 45–90 min before shutting off the pump. The best time of injection is after flushing the sand filters, because the chlorine prevents the growth of bacteria in the sand. The surface water containing iron can be treated with chlorine or commercial bleaching agent for 45 min for lowering the pH to <6.5. At pH > 6.5, certain reactions in combination with the precipitates of iron may gradually obstruct the irrigation lines.

One may use commercial grade phosphoric acid or hydrochloric acid. Before using the acid, the water is allowed to pass through the system at a pressure greater than the operating pressure. Fill the fertilizer tank up to two third (2/3) parts of the capacity of a tank. Add the acid at the rate of one liter per cubic meter per hour of flow rate. Inject the diluted acid into the system, as one will inject the fertilizer, in a normal process..

Remember: When using the chemigation tank, first pour the water and then add the acid.

2.4 METHODS TO REPAIR TUBES OR DRIP LINES

1. The orifices of Bi-wall tubing may be obstructed due to salts, and so forth. A polyethylene tubing of small diameter is used as a bypass method to repair these drip lines.
2. If the line is broken or there is an excessive escape of water, the pipe, or the tube is cut down and is connected with a union or a coupling.
3. If the main line is made of flexible nylon flat and is leaking, then use a small piece of plastic pipe of same diameter to insert into the flexible nylon tubing. The both ends are sealed with the use of pipe clamps.

2.5 SERVICE BEFORE THE SOWING SEASON

1. Clean and flush all the distribution system and the drip lines, with water.
2. Wash with water and clean the pump house system. Lubricate all valves and accessories.
3. Turn on the pump and activate the system. Check the pipes and drip lines for leakage. Repair if necessary.
4. If the system has been used previously, then cleaning and flushing should be carried out for a longer period of time. It is particularly important in sandy soils, as the sand can penetrate into the pipe during the removal of lines.

2.6 SERVICE AT THE END OF CROP SEASON

At the end of a crop season, following steps should be taken:

1. Flush the pipes. Clean the filters and other components of the system.
2. Lubricate all the gate valves and accessories.
3. If the pipes are permanently installed in the field and cannot be removed at the end of a crop season, keep these free of soil and weeds that can grow nearby.
4. If the system can be moved from one place to another (according to the season), the following procedure is adequate:
   a. Flush and clean the system.
   b. Remove the drip lines and collect these carefully.
   c. It is best to leave the main lines in place. If it is not possible or if there is a need for transfer to an area, then these should be rolled. Close both ends and store in the shaded area.
   d. It is advisable to label the hoses with tags. Distance between orifices and frequency of use should be indicated on the tag.

2.7 TROUBLE SHOOTING

Causes	Remedies
Pressure difference > Recommended value	
1. Filters are obstructed.	Flush the filters.
2. Lines are broken.	Repair or replace lines.
3. Pump is defective.	Repair or replace the pump.
4. Gate valve is blocked.	Fix or replace the gate valve.
5. Pressure regulator is defective.	Remove and replace the regulator.
Laterals (or drip lines) and drippers are clogged	
6. Sand is being accumulated in the drippers and lines.	Open ends of laterals and leave open for more than two minutes so that water at pressure passes through.
7. Formation of algae and bacteria.	Wash with chlorine. Paint the PVC pipes or install the lines below soil surface.
8. Sediments are being accumulated.	Wash with acid.
9. Precipitation of chemical compounds due to chemigation.	Wash with acid and conduct the chloration process.
10. Obstruction due to nest of insects.	Wash with insecticide.
Pressure is increased	
11. Orifices in the drip lines or drippers are clogged.	Flush the drip lines or laterals.

2.8 SUMMARY

The orifices in the drip lines or the emitters emit water to the soil. The emitters allow only the discharge of few liters or gallons per hour. The emitters have small orifices and these can be easily obstructed. For a trouble free operation, one should follow these considerations: Pay strict attention to filtration and flushing operation. Maintain an adequate operating pressure in the main, sub main and lateral lines. Flushing and periodic inspection of the drip irrigation system is a must.

For effective filtration efficiency, we must maintain the system in good condition and it is not obstructed by the clogging agents. For this, pressure gages are installed at the entrance and the exit of a filter. The frequency of flushing depends on the water quality. Some recommendations for an adequate maintenance are cleaning with pressurized air, acids, and chlorine. This chapter includes methods to repair tubes or drip lines. Also there is a procedure for the service before the sowing season and the service at the end of the crop season.

KEYWORDS

- **Bar**
- **Check valve**
- **Chemigation**
- **Clay**
- **Clogging**
- **Dripper**
- **Emitter**
- **Fertilizer**
- **Fertilizer tank**
- **Filter**
- **Filter, sand**
- **Filtration system**
- **Flow metering valve**
- **Gate valve**
- **Leaching**
- **Line, distribution**
- **Line, main**
- **Maintenance**
- **Orifices**
- **Poly Vinyl Chloride (PVC) pipe**
- **Polyethylene (PE)**

- **Precipitation**
- **Pressure regulator valve**
- **Pump**
- **Pump house**
- **Sodium hypochlorite**
- **Union**
- **Water quality**
- **Weed**

REFRENCES

1. Bar-Ram. *Instructions for the Maintenance of BAR-RAM Drip Irrigation Systems*. Israel.
2. *Design, Installation, and Performance of Trickle Irrigation Systems*, 2004. ASAE Engineering Standard No. ASAE-EP-405
3. *Drip Irrigation Management*. (1981). Division of Agricultural Sciences at University of California. Berkley – CA.
4. *Trickle irrigation in the Eastern United States*, (1980). Agricultural Cooperative Extension Service, Northeast Regional Agricultural Engineering.

CHAPTER 3

EVALUATION OF THE UNIFORMITY COEFFICIENTS

VINCENT F. BRALTS

CONTENTS

Printed with permission from "Bralts, Vincent F., 2013. Evaluation of the uniformity coefficients. Chapter 14: 261–270. In: *Management of Drip/Trickle or Micro Irrigation* edited by Megh R. Goyal. New Jersey: Apple Academic Press Inc."

3.1 INTRODUCTION

The uniformity of irrigation application is one of the most difficult factors to evaluate. The initial cost, the operational cost, and the plant response are related with the uniformity of water application. A considerable effort has been directed to this problem in the design and management of the irrigation system. This chapter presents a simple method to evaluate uniformity of water application in drip irrigation. This method does not require mathematical equations and sophisticated equipments [1, 2].

In an irrigation system, there is a direct relationship between the uniformity of water application and the initial cost. The pressure decreases as the water flows through the pipelines due to loss by friction. This results in reduction of water application rates at the farthest sections of the irrigation system. The water is distributed more uniformly and the loss by friction is reduced if the pipelines have a large diameter. Since the pipelines of larger diameter are more expensive, therefore a system with high uniformity is more expensive compared to poorly designed system with low uniformity. Before purchasing the system, the buyer should evaluate the cost of the system, its capabilities, and uniformity [1–3].

The operational cost of an irrigation system are directly associated with uniformity of water application. In many cases the water is applied uniformly when the system is operated at high pressures. This practice goes against one of the advantages of drip irrigation in relation to savings in energy consumption. The operational cost of a system with smaller diameter pipes is higher than an appropriate design. This is due to the fact that the small size pipes need more time to apply the desired quantity of water at the farthest end of the lateral.

The crop production efficiency is also related with the uniformity. In general, it is difficult to determine the loss in efficiency by low uniformity, because the efficiency is affected by many factors. In general, the efficiency losses are due to the fact that some plants do not receive the adequate amount of water while others receive in excess. Excessive applications of water may wash away the nutrients that are accessible to the plants.

To fulfill the objectives of the drip irrigation, the system must be designed to apply the water uniformly within the economical limits. This way, each plant in the field will receive the same amount of water. This facilitates the operator to adjust the quantity of water applications according to the crop requirements.

3.2 FACTORS THAT REDUCE THE UNIFORMITY OF WATER APPLICATION

The following factors can interact to reduce the uniformity of water application:

1. Defective irrigation pump.
2. Broken or twisted distribution lines.
3. Obstruction of drippers and/or filters by the physical, biological, and chemical agents.

4. Corrosion of some parts in the irrigation system.
5. Obstructed or defective valves.
6. Inadequate design.

3.3 PROCEDURE FOR THE EVALUATION OF UNIFORMITY

A simple procedure was developed to evaluate the uniformity of water application in a drip irrigation system. This procedure can be used by farmers, designers, and sales persons:

1. This method can be used by a potential customer to evaluate a system before acquiring it. In addition system can be evaluated to determinate if it complies with minimum requirements, before the final payment is made to the seller.
2. This method can be used by a designer or seller to determine if the system was designed and installed properly. The system components can also be evaluated.
3. The irrigation operator can use it to detect variations in the uniformity of water application. The operator can also detect problems due to the obstructed drippers and filters. The lack of uniformity of the water application due to changes in hydraulic characteristics of the drippers and other components of the system can be detected. Therefore, the defective parts can be repaired and replaced.

3.4 DEFINITION OF UNIFORMITY

The uniformity (U) of water application is defined in statistics [1, 2] by the following equation:

$$U = 100 \times (1.0 - V) \qquad (1)$$

where, U = Uniformity or the emitter discharge rate, intervals between 0 and 100%, V= Coefficient of variation.

The coefficient of variation (V) is a variation in flow of each dripper compared to average flow rate of all drippers. The uniformity is expressed in relative terms so that it does not depend on the magnitude of flow of drippers. Instead, it depends on the variation between the flow of an individual dripper and average flow.

A uniformity of 100% in Eq. (1) corresponds to a coefficient of variation of zero. This indicates a perfect uniformity, therefore there is no variation in the flow among the drippers. Uniformities of 100%, 90%, 80%, 70%, and 60% corresponds to coefficient of variation of 0.0, 0.1, 0.2, 0.3, and 0.4. This uniformity can be classified as shown in Table 1:

TABLE 1 Uniformity classification.

Classification	Statistical Uniformity	Emission Uniformity
Excellent	For U = 100–95%	100–94%
Good	For U = 90–85%	87–81%
Fair	For U = 80–75%	75–68%
Poor	For U = 70–65%	62–56%
Not Acceptable	For U < 60%	<50%

3.5 STANDARD FOR UNIFORMITY OF WATER APPLICATION

The American Society of Agricultural and Biological Engineers (ASABE) have developed a standard for the uniformity of water application in drip irrigation [1]. This standard establishes minimum acceptable uniformity for the design of a drip irrigation system. The standards for uniformity are presented in Figures 1 and 2 that show the efficient economical values of uniformity [2]. Table 2 shows acceptable intervals. Design for a uniformity level less than the design value will result in a reduction in the irrigation efficiency; and cause loss of water and fertilizer due to poor uniformity of water application. Design based on high values of uniformity will increase the initial cost.

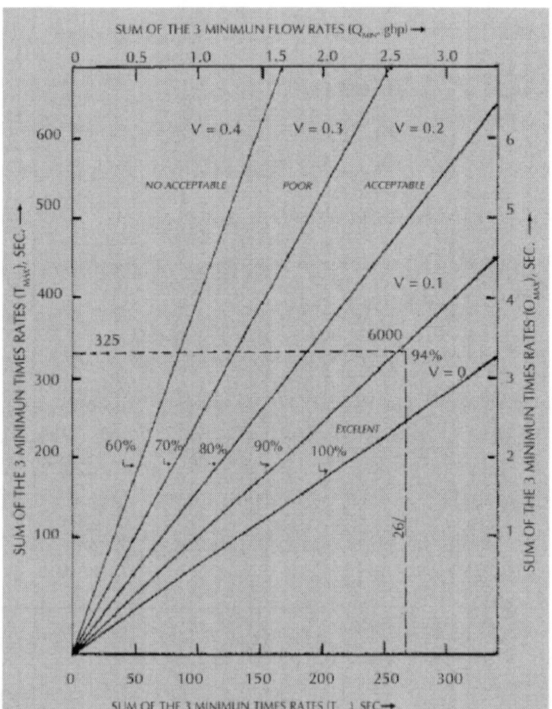

FIGURE 1 The field uniformity of an irrigation system based on the dripper times and the dripper flow rate, with an example in this chapter.

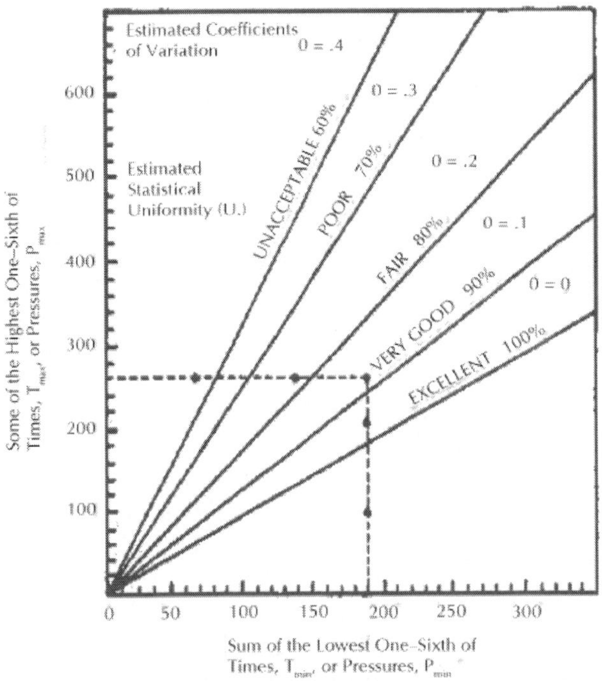

FIGURE 2 The field uniformity of a drip irrigation system based on the time to collect a known quantity of water or based on pressure for hydraulic uniformity.

TABLE 2 Acceptable intervals of uniformity in a drip irrigation system.

Type of dripper	Slope	Uniformity interval, %
Point Source: located in planting distance > 3.9 m.	Level*	90–95
	Inclined**	85–90
Point Source: located in planting distance < 3.9 m.	Level*	85–90
	Inclined**	80–90
Drippers inserted in the lines for annual row crops.	Level*	80–90
	Inclined**	75–85

* Level = Slope less that 2%.

** Inclined = Slope greater than 2%.

An irregular topography of the land affects the design and uniformity of water application. It will result in a high cost of the system. In soils with an irregular topography, allow a smaller uniformity to compensate for the initial and operational costs of the system.

The planting distance for a crop also affects the desired uniformity of the water application. The uniformity should be higher in crops with larger planting distance when one or two drippers per plant are used to apply water to a plant. In narrow planting and narrow dripper spacing, the variation in the flow per dripper reduces.

Each plant may have two or more drippers and this way the effects of random variation on the dripper emission rate are much less. However, this does not eliminate the effects of continuous reduction along the laterals due to loss by friction. Therefore, allowable reductions in uniformity for crops with narrow planting distance are smaller (10%) compared to the crops with wider planting distance.

3.6 EVALUATION PROCEDURE

The evaluation procedure uses a known size container to determine the uniformity of water application in drip irrigation. The time required to fill the container is used to calculate the flow rate and the uniformity. This required time can be measured by a stopwatch.

One must take at least 18 samples (one sample per dripper) by recording the time to fill the container. One can take more than 18 samples if it is necessary. The selected drippers must be a representative of the area (some at the beginning, others in the middle and others at the end of the lateral line). Some samples should also be taken at highest elevations of the field and at lowest elevations. The representative drippers should be recorded along with location for analysis. The sum of the three highest observations is denominated as maximum time (T_{max}). The sum of the three lowest observations is denominated as minimum time (T_{min}). T_{max} and T_{min} are used to determine the uniformity of water application using Fig. 1. Example: In Table 3, the three highest observations are 107, 110, and 108 seconds. The sum of these three observations = 325 seconds = T_{max}. The three lowest observations are 89, 87, and 91 seconds. The sum of these lowest observations = 267 seconds = T_{min}. The vertical line, T max and the horizontal line for T_{min} intersects at a point to the uniformity using Fig. 1. The interpolation between the uniformity lines for this particular point gives us a uniformity of 94%. This is interpreted as an excellent uniformity. If the data are representative of the field, then it can be concluded that the system is well designed and well constructed.

TABLE 3 Time required to fill the container in a given field, (example of the field data).

89 sec. (smaller)	**97 sec.**	**110 sec. (higher)**
104 sec.	107 sec.(higher)	93 sec.
92 sec.	100 sec.	103 sec.
96 sec.	94 sec.	108 sec. (higher)
100 sec.	98 sec.	91 sec. (smaller)
99 sec.	102 sec.	87 sec. (smaller)

TABLE 3 *(Continued)*

89 sec. (smaller)	97 sec.	110 sec. (higher)
$T_{max.}$ = 107 + 110 + 108 = 325 seconds.		
$T_{min.}$ = 89 + 91 + 87 = 267 seconds.		

3.7 CONFIDENCE INTERVALS

The example above is only for 18 drippers in a larger area. Therefore, the results may not be entirely accurate. The only way to determine the exact uniformity is to measure the flow of each dripper in the field. From statistical point of view, it can be proved that the values in Figs. 1 and 2 are precise. The Table 4 gives us confidence limit of 95% for the uniformity values presented in Fig. 1. Interpolations in Table 4 indicate that the confidence limit for a uniformity coefficient of 94% is U ± 1.7% for 18 measurements in the field. If we get a uniformity of 94% from Fig. 1, the true uniformity of the field is 94 ± 1.7%. Therefore, uniformity varies from 92.3 to 95.7%. This interval of 95% of confidence indicates that the value of the field uniformity should be within a confidence interval (92.3% a 95.7%), 95 times out of 100, if the sampling procedure is repeated. In Table 4, the confidence limit increases as the uniformity decreases. It implies that at low uniformity, the results are less accurate.

For example, if the field uniformity was only 60%, the Table 4 shows a confidence limit of 60% ± 13.3%.

A level of 95% shows that the field uniformity should be between the confidence interval of the exact uniformity of the field, 95 times out of 100, if the procedure is repeated.

Therefore, the true field uniformity has a confidence interval of 46.7–73.3%. There is wider range, because we took only 18 drippers in the whole field. The variability between the drippers is larger as shown by low uniformity. With a random selection of the drippers, there is a higher chance for a representative data.

In Table 4, the confidence limits are given for trials of 18, 36, and 72 drippers. The certainty for the results can be increased if more samples are taken. This way confidence interval can be reduced.

TABLE 4 Confidence limits for field uniformity (U).

Field uniformity	18 drippers		36 drippers		72 drippers	
	Confidence limit		Confidence limit		Confidence limit	
	N Sum*	%	N Sum	%	N Sum	%
100%	3	U ± 0.0	6	U ± 0.6%	12	U ± 0.0%
90%	3	U ± 2.9	6	U ± 2.0%	12	U ± 1.4%
80%	3	U ± 5.8	6	U ± 4.0%	12	U ± 2.8%
70%	3	U ± 9.4	6	U ± 6.5%	12	U ± 4.5%
60%	3	U ± 13.3	6	U ± 9.2%	12	U ± 6.5%

***N Sum** = 1/6 part of the total measured drippers. This is a number of samples that will be added to calculate T_{max} and T_{min}.

3.8 FLOW RATE MEASUREMENTS

The time to fill the container will be infinite if the dripper is obstructed completely. Therefore, one cannot use Fig. 1 directly. In this case, we shall add the three highest flow rates and the three lowest flow rates. Now we shall use flow rate units in Figure 1 to calculate the field uniformity.

Now the flow rate measurements will be more difficult to take and to calculate. It requires the use of a calibrated container to measure the volume of water in a given period of time. Then the flow rate is calculated.

3.9 TROUBLE SHOOTING

Causes	Remedies
Uniformity 60%	
1. Few samples were taken.	Take more than 18 drippers as a sample.
2. Clogging in the filters, lines or drippers.	Clean the filters flush the lines and drippers with acid. Clean or replace the clogged drippers.
Considerable difference in T max and T min.	
3. It is possible that the drippers lines are obstruct or broken.	Repair the broken lines and clean the obstructed lines.
Loss of pressure due to excess friction in the lines.	
4. Pipes are of small diameter than the design values.	Revise the design and use correct size of pipes.

3.10 SUMMARY: THE PROCEDURE FOR FIELD EVALUATION

In this chapter, the procedure to evaluate the uniformity coefficient for a trickle irrigation system is presented. The uniformity of water application is affected by the degree of clogging, accuracy of the design, and periodic maintenance of the system. Nomograph for the determination of uniformity is presented. The procedure involves taking water samples in a known time from the representative drippers. The three highest and lowest values are summed to give T_{max} and T_{min}. The evaluation procedure is summarized below:

1. Allow the system to operate at design operational pressure for enough time to remove all the air from the lines.
2. Measure the required time to fill up the containers in each of the 18 drippers. Be sure that the drippers represent all parts of the field.
3. Calculate Tmax adding the three highest times, or 1/6 of the total number of the required drippers to fill the container.
4. Calculate Tmin using the three lowest times, or 1/6 of the total number of the required drippers to fill the container.

5. Using Figure 1, determine the field uniformity for a point where lines of Tmax and Tmin intersect. Interpolate to calculate the uniformity if it is necessary.
6. If the uniformity of the field is lower or the confidence limit is higher, it is convenient take more data or repeats the procedure to make sure that the system is not poorly designed.

This method is particularly advantageous for use in the field due to a limited number of required data and the simplicity of the procedure. To facilitate the compilation of data, one may use the data sheet in Appendix I. Nomograph for statistical uniformity is shown Appendix II.

KEYWORDS

- **ASABE**
- **Clogging**
- **Coefficient of variation**
- **Confidence limits**
- **Dripper**
- **Emitter FLOW rate**
- **Fertilizer**
- **Filter**
- **Filter, mesh**
- **Interpolation**
- **Irregular topography**
- **Maintenance**
- **Plant nutrient**
- **Planting distance**
- **Pump**
- **Term**
- **Uniformity**

REFERENCES

1. ASABE, (1996). Field evaluation of micro irrigation systems. In: *ASABE Standards:* ASAE EP405.1 and EP458. St. Joseph, MI: American Society of Agricultural and Biological Engineers, 1–7.
2. Bralts, V. G., Keesme, C. D. (1982). Drip irrigation field Uniformity estimation. Paper No. 82-2062, Summer meeting of American Society of Agricultural and Biological Engineers.
3. Smajstrla, A. G., Boman, B. J., Haman, D. Z., Pitts, D. J., Zazueta, F. S. (2002). Field evaluation of micro irrigation water application uniformity. Cooperative Extension Service, Department of Agricultural Engineering, Gainesville, FL: University of Florida.

APPENDIX I: EVALUATION OF UNIFORMITY: DATA SHEET

Name of evaluator: ____

Date of evaluation: Month _______Day ___________Year___________

Name of farmer: __________________ Direction: __________________

Description of the trickle system:

High pressure/low pressure

Size of pump ____________KW or __________ HP

Size of the farm ___________ acres

Size of the filter mesh _____________

Area of block where sample are taken ___________ acres

Procedure:

1. Turn on the system to eliminate the air from the lines.
2. Measure the time (seconds) to fill the container in each of the 18 drippers.
3. Calculate the maximum time, sum three highest times.
4. Calculate the minimum time, sum three lowest times.
5. Determine the field uniformity (Fig. 1).
6. If the field uniformity is low or the confidence interval is high, take more samples.

Dripper #	**Time (Seconds)**	**Dripper #**	**Time (Seconds)**
1.		10.	
2.		11.	
3.		12.	
4.		13.	
5.		14.	
6.		15.	
7.		16.	
8.		17.	
9.		18.	

Sum of the three highest times = _______ + _______ + _______ = _______ T_{max}.

Sum of the three lowest times = _______ + _______ + _______ = _______ T_{min}.

Using Fig. 1, Uniformity = ______________________ % (where T_{max} and T_{min} intersect).

Observations and recommendations:

____

____

____

____

APPENDIX – II: NOMOGRAPH FOR STATISTICAL UNIFORMITY

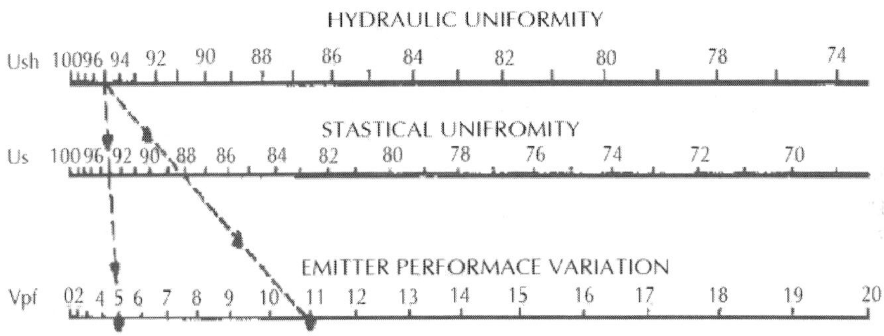

CHAPTER 4

WATER MANAGEMENT IN CITRUS: INDIA

P. S. SHIRGURE and A. K. SRIVASTAVA

CONTENTS

Published with permission and Modified from P. S. Shirgure, A. K. Srivastava and Shyam Singh, *Water Management in Citrus – A Review.* Agricultural Reviews, **2000,** *21,* 223–230.

4.1 INTRODUCTION

Water is an important natural resource, a basic human need and of vital requirement for all developmental activities. The demand of water is increasing with increase in population and economic activities. Irrigation has been practiced in India since long ago. It is considered as a very important input for agriculture and hence, continuous development has been taking in this field through the centuries. It is a means to mitigate the impact of irregular, uneven and inadequate or wide fluctuations in rainfall from year to year. India's annual rainfall is 117 cm and most of it occurs during monsoon. The irrigation potential in India has risen from 19.5 m-ha in 1950 to 67.89 m-ha in 1985.

The best estimates available indicate that the maximum amount of exploitable irrigation potential by all types of irrigation is 113.5 m-ha. This could be sufficient for 50% of the total cultivable area of the country and 50% of the area, would be left completely dependent on rainfed farming. Conjunctive use of rain and irrigation water offers scope for optimizing water use in areas having problems of surface drainage during, rainy season and water scarcity during the rest of the year.

Citrus is the third largest fruit crop grown in an area of 234,570 ha. Nagpur mandarin and acid lime occupies 40% and 25% of the total area under citrus cultivation in the country. The large scale drying of the citrus orchards is mainly due to scarce water resources, frequent drought and lowered water table in mandarin growing areas of Vidarbha (Maharashtra) and Central India [11].

The average yield of these orchards is 7 to 8 t/ha, which is 3 to 4 times less than other citrus producing countries of the world. Citrus plants are more extracting in their demand for irrigation. Direct contact of water with the trunk adversely affects the trees growth. Citrus being an evergreen fruit crop use moisture constantly throughout the year of course at a much slower rate during winter and faster in summer. There is a good amount of research available on irrigation water management of citrus from abroad but a little work has been done under Indian conditions.

There is a need for carrying out the research on estimating the water requirements of the Nagpur mandarin and acid lime under subtropical conditions of the Central India. The use of micro irrigation systems is gaining popularity among the citrus growers and it is necessary to standardize the best system for the citrus orchards. The moisture conservation techniques like mulching and fertigation are also equally important for water and fertilizer conservation point. So, the research in this regard is also required to be carried out for optimizing the productivity and efficient use of inputs including water.

4.1.1 IRRIGATION SCHEDULING AND WATER REQUIREMENT IN CITRUS

The literature on irrigation methods, irrigation systems, scheduling, water requirements and fertigation in citrus in International and under Indian conditions is reviewed. The literature cited related to water requirement and irrigation scheduling in Citrus is reviewed. The growth of 'Valencia' oranges slowed down at 32-cb and 55-cb soil suctions at 30 cm depth in light and medium textured soil, respectively

[22]. The preliminary studies on the effect of soil management system on soil moisture in Sweet orange orchard was initiated by Randhawa et al. [47]. Stolzy et al., [76] found that the treatments irrigated at 20 Kpa tensiometer readings were best as compared to calender schedule. Hashemi and Gerber [19] attempted correlation between actual evapotranspiration (AET) and potential evapotranspiration computed with Penman's model. Koo [25] advised Florida citrus growers to maintain soil moisture at 55 to 65% of field capacity from bloom the young fruit exceeds 1 inch in diameter. Retiz [51] estimated the water requirement of citrus at 40–45 inch/year. Richards and Warnke [53] studied the irrigation systems to lemon and irrigation at 60 cb and extrapolations to 150 cb resulted in no measured differential response in tree growth and fruit yield under coastal conditions. Leyden [28] found that 610 mm irrigation water applied via a drip system at 0, 200, 300 and 400 L/tree gave the significant difference in total yield and fruit size distribution.

Toledo et al., [79] found that irrigation at 65% field capacity caused drought injury symptoms, excessive defoliation and less water consumption. Best results were obtained with irrigation at 85% filed capacity. Evapotranspiration ranged form 3.78 to 4.42 and 1.46 to 1.3 mm/day for 85% and 65% field capacity irrigations respectively. Kelin [24] compared drip irrigation scheduling according to soil water potential to class A pan evaporation in different horticultural crops using a crop factor and concluded that 12 to 23% water could be conserved by using the irrigation scheduling based on soil water potentials. Moreshet et al., [34] compared the 100% and 40% of soil volume irrigation in 'Shamouti' orange and found that partially irrigated plot was 66% of that of the fully irrigated plot one.

Transpiration from the trees of partially irrigated plots was 72% of that of the fully irrigated plot and the evaporation from the soil surface was 58%. Fruit TSS and acid contents were higher in partially irrigated plots. Smajstrla et al., [71] found that greatest yields were obtained using spray-jet trickle irrigation. Yield increases were not linear with volume of rootzone irrigated but ranged from 39% for the drip irrigation treatments which irrigated 5–10% of the area beneath the tree canopies to 64% for 2-spray jet per tree, which irrigated 50.7% of the areas beneath the tree canopies.

Plessis [38] obtained the highest yields (190 kg/tree) and the largest average fruit size with irrigation at a crop factor of 0.9 on a 3-day cycle, with thin consumption micro irrigation gave better results than drip irrigation. Makhija et al., [31] obtained water need for 6 year old Kinnow mandarin varying from 539 to 1276 mm depending upon the level of irrigation with average consumptive use of water in 2 years as 61.5 cm. Smajstrla et al., [73] concluded that the tree growth of young 'Valencia' orange was greatest when irrigations were scheduled at 20 centibar for no-grass and 40 centibar for the grass treatments. Randhawa and Srivastava [48] emphasized on irrigation aspects in Citriculture in India. Autkar et al., [1] studied the distribution of active roots of Nagpur mandarin as it can be useful in planning irrigation nutrition, planting density and drainage management. The root depth and radial extent for trees aged 1–4 years was 7.5–8.0 cm deep and 5–12.5 cm respectively and for 10 years old age tree it was 2–3 m and 80–90 cm. Barbera and Carimin [4] studied the different levels of water stress on yield and quality of lemon tree and found that yield was lower in most stressed plot. The number of flower/m^3 of canopy was higher in most stressed treatment indicating a relationship between severity of stress and flowering response. Mageed et al., [30] carried the research on influence of irrigation and nitrogen on water use and growth of Kinnow mandarin receiving 4 levels of irrigation and three levels of Nitrogen (0, 115 or 230 Kg N./tree). The consumptive use varied from 66.7 cm to132.5 cm. Moreshet et al., [35] studied on water use and yield of a mature 'Shamouti' orange orchard submitted to the root volume restriction and intensive canopy pruning.

Du Plessis [40] with a mature 'Valencia' orange trees and field experiment shown that the water use pattern over the entire season reaching a maximum of 87 lit/day in January. Highest net income was obtained with tensiometer scheduling. He [1989] also demonstrated that 690 L irrigation when tensiometer-reaching fell to −50kpa gave the highest net income. Use of tensiometer rather than evaporation pan scheduling could save 2000 m^3 water/ha annually. The water requirement of citrus plants varies with species, season and age governed with different climatic conditions. Plant growth retards below certain critical level of available moisture depending upon soil type, climatic factor and plant genetic make up [46]. Autkar et al., [2] also studied the effect of Pan evaporation, canopy size and tree age on daily irrigation water requirement of 1–5, 5–8 and above 8 years old Nagpur mandarin trees over 9 months [October–June] and concluded that the requirement rose with age. Ghadekar et al., [17] estimated that the consumptive use of Nagpur mandarin by modified Penman equation using 40 years air temperature, relative humidity, wind velocity and Solar relation data. Under clean cultivation the water requirement of young, middle age and mature trees was 651.9, 849.0 and 997.3 mm/year, respectively. An equation for daily water use was proposed and it can be used for drip irrigation. Sanehez et al., [57] compared five flood irrigation treatments with daily drip irrigation at 0.475 Epan and concluded that the drip irrigation gave higher yields as compared to flood-irrigated plants. Castel and Buj [8] carried out trials on mature 'Satstuma' average trees grafted on Sour orange rootstocks. Plants were irrigated with 60% of the estimated ET losses from a class A pan and 80% of the control throughout the year. Irrigation treatments affected both yield and fruit quality.

Ray et al., [49] studied the response of young 'Kinnow' mandarin to irrigation. Irrigations were scheduled at −0.05, −0.1, −0.2, −0.4 and −0.8 MPa soil water potential 0.8 IW/CPE ratio and irrigation to replenish estimated crop ET. The water use increased as the frequency of irrigation increased as the frequency of irrigation increased. The highest bio-mass per plant was obtained when irrigation was scheduled at −0.05 MPa soil water potential (SWP) and 18–19 irrigations were required. The best tree growth in terms of trunk diameter, plant height, canopy volume, leaf number and shoot growth was also obtained at −0.05 MPa SWP using 182.4 cm water/tree/annum. He also studied [50] the effect of irrigation on plant water status and stomatal resistance in young Kinnow mandarin and found that the leaf water potential (LWP) and Relative water content (RWC) declined considerably with reduction in soil moisture in rootzone due to differential irrigation schedules. Reduction in RWC was more conspicuous where soil moisture dropped below 11% LWP measurements in early morning hours showed a significant curvilinear relationship with soil water status. Leaf stomatal values were lowest in September and highest in January. Shirgure et al., [60] initiated the irrigation scheduling based on depletion of available water content and fraction of open pan evaporation in acid lime in prebearing stage. He studied [68] the effect of different soil moisture regimes with irrigation scheduling based on available soil moisture depletion and open pan evaporation on soil moisture distribution and evapotranspiration in acid lime and it was concluded that the evapotranspiration varied from 213.6 mm to 875.6 mm in various irrigation schedules. It was also found that the change in soil-moisture distribution in the rootzone of acid lime plants varied from 195.9 mm to 321.3 mm with different irrigation schedules.

4.1.2 IRRIGATION METHODS AND DRIP IRRIGATION SYSTEMS IN CITRUS

The common methods of applying water to the orchards are basin, border strip, furrow, sprinkler and drip irrigation. Ring basin is generally followed in early establishment phase of fruit trees. Micro-irrigation to citrus is common in developed countries. Drip and microjet irrigation has the advantage over surface irrigation methods, for more uniform and complete wetting of the soil surface and adoption on sloppy terrain.

Faton [12] observed better tree growth and yield, less weed growth, evaporation and leaching with 16 gallon water applied through drip to each 4-year-old lime trees at two weeks interval compared to 320 gallon water in flood irrigation. Fritz [15] observed that all applied water is transferred directly to rootzone of plants and 20–50% water saving is reported depending on soil and climate. Raciti and Sckderi [44] compared drip irrigation with the basin and found that the fruits under drip system ware more acid and lower maturity ratio. Ronday et al., [55] observed better tree growth and less water consumption in Valencia orange under drip irrigation in sandy soil Sucderi and Raciti [58] compared basin irrigation with different combination of drip irrigation and measured number, weight, quality of fruits in Valencia orange. He also studied micronutrient levels in leaves, annual trunk increments. Drip irrigation gave the higher yields. Simpson [69] found that there is a shift from furrow irrigation and overhead sprinkler irrigation systems to under tree systems like microjets. Slack et al., [70] demonstrated that trickle irrigation on young orange trees used 5,400 L of water compared to 23,400 L of water per tree for dragline.

Raciti and Barbargallo [45] found that the yields of lemon were more with localized irrigation amounting to 227.23 q/h and 213.2 q/h for basin irrigation. Ozsan et al., [37] compared furrow, under tree, over tree and drip irrigation in lemons. Amounts of water applied were greatest (1,286 mm) with under tree method and least (207 mm) with drip irrigation system. Yield was more with over tree sprinkling and least with furrow. Water use efficiency was high in drip irrigation. Cevik and Yazar [9] demonstrated that a new irrigation system, that is, Bubbler irrigation for the orchards. He observed that under tree sprinkling and drip irrigation had the best pomological effects. Amounts of water applied per tree for over sprinkling, under sprinkling and drip irrigation were 22.01, 17.04 and 10.33 m^3/season. Pyle [43] appraised the use of micro irrigation in Citrus especially drip irrigation. Except the higher cost the advantages includes saving in labor, water and power, better orchard uniformity and immediate response to crop need, better soil-water relationships, rooting environment and better yield and quality.

Tash be kov et al., [78] studied different irrigation methods. Drip irrigation and under tree sprinkling produced the highest yield with the least water requirements. The application rate for drip irrigation of 4 years old lemon trees was 7400 m^3/ha annually. Capra and Nicosia [7] studied flooding, sprinkler, and subirrigation with sprays and concluded that the rates of water application affects the rate of growth of fruit diameter. Robinson and Alberts [54] compared under canopy sprinkler and drip irrigation systems in crop like Banana and found that the drip irrigation is superior to under canopy sprinklers. Increased tree growth and yield were recorded in young Valencia orange under drip irrigation method with emitter placed at distance of 1 meter from the trunk [3]. Greive [18] concluded that under tree microsprinklers increased yield by 12% and reduced water application by 9.3% compared to conventional full ground cover. Interligolo and Raciti [23] demonstrated that water saving with subsurface irrigation was 32% over the traditional basin irrigation. The yield was higher but fruit quality was not much different. Marler and Davies [32] studied the effect of microsprinkler irrigation scheduling on growth of young Hamlin orange trees and found that growth was not affected by pattern of irrigation, suggesting that 90% emitters are enough for root system. Zekri and Parsons [80] studied drip, microsprinkler and overhead sprinkler irrigation at two water application rate and found that fruit size and tree canopy area were 9 to 20% greater in the overhead sprinkler treatments. Marler and Davies [33] studied the growth response of micro irrigation on growth of young Hamlin orange and found that more than 90% of root dry weight was within 80 cm of the trunk at the end of first growing season.

Rumayor et al., [56] studied three irrigation systems (drip, microsprinkler and flooding) and found that yields were higher for sprinkler-irrigated trees and the fruits were smaller in flood irrigation. Smajstrala [74] researched on micro irrigation for citrus production in Florida. Gangwar et al., [16] studied the economics of investment on adoption of drip irrigation system in Nagpur mandarin orchards in Central India and concluded that the drip irrigation system is technically feasible and economically viable with Benefit to Cost ratio as 2.07. Shirgure et al., [64] initiated the research work on evaluation of micro irrigation systems in acid lime and a comparison was done with that of basin (ring) method of irrigation. Shirgure et al., [66] studied the effect of dripper 8 L per hour microjet 300°, microjet 180° and basin irrigation method on water use and growth of acid lime and found that microjet 300° recorded higher growth than rest of the systems. He also studied [67] the efficacy of these micro irrigation systems and basin irrigation on fruit quality and soil fertility changes in acid lime.

4.1.3 FERTIGATION IN CITRUS

Fertigation in application of liquid or water-soluble solid fertilizer along with irrigation trough the drip irrigation to the plants. It has many advantages like increasing fertilizer-use efficiency, ensured supply of water and nutrients, labor saving and improvement in yield and quality. It is a very new under Indian conditions but getting popular along with adoption of drip irrigation system.

The research related to injection of fertilizers through the drip irrigation systems was started during 1979 by Smith et al. [75]. Koo [26] appraised the potential advantage of micro irrigation systems and its usefulness to fertigation. Bielorai et al., [6] advocated use of fertigation technology in citrus as it resulted in higher production of good quality Shamouti oranges. He compared N. fertigation at 100, 170 and 310 Kg/ha with broadcast application at 170 kg/ha through irrigation system. Phosphatic and potash fertilizers were given at same rate by conventional method in all the treatments. Average yields for 4 years were 62, 73 and 82 mg/ha with 100, 170 and 310 kg N./ha, through fertigation.

Koo and Smjstrala [27] supplied 15% and 30% of crop N. and K requirements through fertigation and rest through conventional method to Valencia orange. Partial fertigation of N. and K resulted in lower N. contents of leaves. TSS and acid concentration in juice was also reduced but yield was not affected. Haynes [20] discussed the principles of fertilizer use for trickle-irrigated crops. Haynes [21] also studied the comparison of fertigation with broadcast applications of urea on levels of available soil nutrients and on growth and yield of trickle irrigated peppers. He found that growth and yields were greatest at the low rate of N. applied as fertigation or as a combination of broadcast plus fertigation.

Fouche and Bester [14] tried various fertilizer combinations through fertigation on 13-year-old Navel oranges. Fertigation was given with a soluble fertilizer 'Triosol' [3: 1: 5] + 350 gm Urea by broadcast, fertigation of N. and K with broadcast of single super phosphate and NPK through broadcast. Highest yields were obtained with fertigation of NPK through Triosol or by complete broadcasting of NPK fertilizers. No significant differences were observed as fruit size, acidity, percent juice content and TSS among treatments. Beridze [5] conducted trial on 5 year old lemon tree and fertilized 150 kg N. + 120 kg P_2O_5 + 90 kg K_2O per hectare as basal dressing. The highest yield of 6.6 ton per hectare was obtained from trees fertilized with basal dressing + 250 kg peat/tree as a mulch + FYM at 25 t/ha.

Ferguson [13] studied the fertigation as growth of 'Sunburst' tangerine trees. Two years old citrus *reticulata* x *C. paradisi cv.* Sunburst was fertilized with 0.66 or 1.32 lb N./tree during 1988–89 and it was 0.52 or 1.05 lb N./tree during 1990. Leaf analysis showed that low to deficient concentrations of N., K, Mn and Zn with both N. treatments. Zekri and Parsons [80] tried micronutrients through fertigation with different sources of various rates. Inorganic forms (NO_3 and SO_4) were ineffective in evaluating microelement levels in oranges. But chelated sources of Fe, Mn, Zn and Cu were very effective and their rates of application were comparable with rates through foliar applications. Neilsen et al., [36] studied that fertigation with calcium ammonium nitrate showed increased vigor and leaf Ca concentration but decreased leaf Mg and Mn compared to trees fertigated with Urea or ammonium nitrate (NH_4NO_3) in apple trees. Fertigation with *P* increased early tree vigor, leaf and fruit *P* concentration and decreased leaf Mn.

Syvevtsen and Smith [77] studied the nitrogen uptake efficiency and leaching losses from lysimeter grown trees fertilized at three nitrogen rates. He concluded that Average N. uptake efficiency decreased with increased N. application rates, overall canopy volume and leaf N. concentration increased with N. rate, but there was no effect of N. rate on fibrous root dry weight. In the first 5 years of the experimentation fertigation did not provide a significant enough yield advantage over banded application to warrant the added cost of the fertigation equipment and higher labor requirement. A very little work was done on fertigation in India. The fertigation research in citrus was initiated during 1995 at NRC for Citrus on acid lime. Shirgure et al., [61, 62] studied the effect of differential doses of Nitrogen fertigation in comparison with band placement of fertilizer application on leaf nutrients, plant growth and fruit quality of acid lime during prebearing stage. The percentage increase in plant height, stock girth and canopy volume was more with 100% N. fertigation followed by 80% N. of recommended dose in acid lime. He also [63, 64] studied that effect of N. fertigation on soil and leaf nutrient build-up and fruit quality of acid lime.

4.2 FUTURE WATER MANAGEMENT STRATEGIES IN CITRUS

a. Citrus is a very sensitive crop. Any excess or deficit of water even for a short duration adversely affects its growth and productivity. Irrigation scheduling based on scientific principles like available water content, soil water potential and potential crop evapotranspiration is in practice in developed countries. The efficiency of different methods of irrigation scheduling varies with climate, irrigation method and citrus species. Since micro irrigation systems are gaining popularity among the farmers due to scarcity of water resources and Govt. subsidy for these systems. A modern system of irrigation will effectively used if it is backed by scientific principles of irrigation application. The water management research pertaining to the citrus is still in a preliminary stage. There is urgent need to evolve efficient irrigation scheduling for citrus crops in different regions of India.

b. Irrigation scheduling definitely help in maximizing the utilization of water resources and boosting the productivity. Scheduling using tensiometers of various depths, neutron moisture probe, climatological approach like modified Penman equation and water balance approach should be studied. Irrigation scheduling based on canopy temperature and leaf water potential may also be studied for better yield and quality.
c. Method of irrigation scheduling varies with the irrigation system adopted. It needs to be standardized for both system adopted. It needs to be standardized for both conventional and modern methods like drip, sprinkler, microjet, etc. Infiltration rates, water distribution and retention parameters vary greatly with soil composition and structure. Thus study on these aspects will help in formulating the scientific water management.
d. Another aspect that requires immediate attention is that water requirement and root distribution of fruit crops increases with age. Therefore, suitable design needs to be evolved which should enable to irrigate the entire root zone with required quantity of water. A farmer should use the installed system for longer period without many modifications, which incur high cost otherwise.
e. Citrus growers in Central India give water stress to induce flowering. In absence of any scientific information, farmers apply water stress according to the past experience. Plants are subjected to stress to the extent where complete restoration of vigor may be possible in all the plants. Relationship needs to be established between water stress and flowering on one hand and water stress and plant growth on the other hand. These in turn should be related to soil characteristics. Farmers should have idea about the duration of stress required for different king of soils.
f. The modern micro irrigation systems have one potential advantage of giving soluble fertilizer through irrigation water known on 'fertigation.' It not only saves labors, fertilizer but also gives higher yield and better quality. The fertilizers through water are applied to the rootzone, which increases fertilizer use also. A comprehensive research related to different NPK soluble fertilizers, their rates and frequency of application at different growth stages of plants are required to be researched.
g. Since the water available for irrigation is becoming scarce day by day. The applied water to the tree root zone needs to be conserved for longer period and that is possible with mulching. The material available to the farmers from the farm itself like grass, leaf litter, straw and trashes can be used for mulching. It not only helps in moisture conservation but also in thermal regulation, disease control and weed control. The research is required on use of organic (grass, straws, leaf litter and trashes) and synthetic (polythene sheets) mulches. The basin area of the citrus trees will be covered with above mulch material and effect may be studied on water saving, growth and yield of the trees. The synthetic mulches are commercially

available. But in case of organic mulches around 5 cm thickness of the mulch is required to be maintained uniformly in all the basins. All these strategies mentioned above are definitely make efficient use of water and enhances productivity in citrus.

4.3 SUMMARY

Irrigation management is one of the prime concerns of modern citriculture irrespective of water resource availability. A variety of recommendations have emerged world over on irrigation scheduling based on analysis of meteorological pedigree, evapotranspiration, depletion of available water content, soil and leaf water potential. The review of literature has revealed best promising results on irrigation scheduling based on depletion pattern of soil available water content. Various micro irrigation systems have established their superiority over traditionally used flood irrigation with microjets having little edge over rest of the others. Similarly, fertigation has shown good responses on growth, yield, quality and uniform distribution pattern of applied nutrients with the rootzone compared to band placement on other methods involving localized fertilization. Automated fertigation in citrus orchards is a new concept, which would be the only solitary choice of among many irrigation-monitoring methods in near future.

KEYWORDS

- **acid lime**
- **acidity**
- **band fertilizer application**
- **basin irrigation**
- **black polythene mulch**
- **canopy volume**
- **Citrus**
- **drainage**
- **drip irrigation**
- **drippers**
- **fertigation**
- **fertilizer use efficiency**
- **field drains**
- **flowering**
- **fruit quality**
- **fruits**
- **grass mulch**
- **harvesting**

- **high density planting**
- **input use efficiency**
- **iron**
- **irrigation**
- **irrigation scheduling**
- **juice percent**
- **leaf nutrient composition**
- **lemons**
- **maturity period**
- **microjet irrigation**
- **microjets**
- **mulches**
- **mulching**
- **Nagpur mandarin (*C. reticulate* Blanco)**
- **net returns**
- **nitrogen**
- **nutrient management**
- **nutrient uptake**
- **orchard efficiency**
- **orchards**
- **organic farming**
- **organic mulches**
- **phosphorous**
- **plant growth**
- **potash**
- **production**
- ***Rangapur* lime**
- **rootstocks**
- **rough lemon**
- **scion girth**
- **stock girth**
- **surface irrigation method**
- **thermal balance**
- **total soluble solids**
- **trickle irrigation**

- **water requirement**
- **water stress period**
- **water use**
- **water use efficiency**
- **weed population**
- **weight**
- **yield**
- **Zinc**
- **rootstocks**
- **rough lemon**
- **scion girth**
- **stock girth**
- **surface irrigation method**
- **thermal balance**
- **total soluble solids**
- **trickle irrigation**
- **water requirement**
- **water stress period**
- **water use**
- **water use efficiency**
- **weed population**
- **weight**
- **yield**
- **Zinc**

REFERENCES

1. Autkar, V. N., Kolte, S. O., Bagade, T. R. (1988). Distribution of active rooting zones in Nagpur mandarin and estimates of WR for Vertisoles of Maharashtra. *Ann. Pl. Physiol. 2 (2),* 219–222.
2. Autkar, V. N., Patel, V. S., Deshpande, S. L., Bagade. T. R. (1989). Management of Drip irrigation in Nagpur mandarin. *Ann. Pl. Physiol. 3(10),* 74.
3. Azzena, M; Deidda, P., Dettori. (1988). Drip and microsprinkler irrigation for young Valencia orange trees. *Proc. Sixth Intern. Citrus Cong.,* Tel Aviv, Israel, *2,* 747–751.
4. Barbera, G., Carimin, F. (1988). Effect of different levels of water stress an yield and quality of lemon tree. *Sixth Intern. Citrus Cong,* Tel Aviv Israel, *2,* 717–722.
5. Beridze, T. R. (1990). The effect of organic fertilizers on lemon tree productivity. *Sub tropicheskie Kul'tury,* (3), 83–86.
6. Bielorai, H., Deshberg, Erner; Brum, M. (1984). The effect of fertigation and partial wetting of the rootzone on production of shamouti orange. *Proc. Int. Soc. of Citriculture, 1,* 118–120.
7. Capra, A., Nicosia, O. U. D. (1987). Irrigation management in citrus orchards. *Irrigazine. 34 (1),* 3–15.

8. Castel, J. R., Buj, A. (1990). Response of Satustinana oranges to high frequency deficit irrigation. *Irrig. Sci. 11 (20),* 121–127.
9. Cevik, B., Yazar, A. (1985). A new irrigation systems for orchards (bubbler irrig.). *Doga Bilim Dergisi, 9(3),* 419–424.
10. Cevik, B., Kaplankiran, M., Yurdakul, O. (1987). Studies for determining the most efficient irrigation method for growing lemons under Cukurova conditions. *Doga, Tarum ve, Ormaniciuk, 11(1),* 42–43.
11. Dass, H. C. (1989). National Research priorities in Citriculture. *Paper Presented at the Citrus Shows cum Seminar, Res, Station, R. F.,* Abhor, Jan 6–7.
12. Faton, J. (1970). Drip irrigation at Yuma. *Citrog. 55,* 173–175.
13. Ferguson, J. J., Davies, F. S., Bulger, J. M. (1990). Fertigation and growth of young 'Sunburst' tangerine trees. *Proc. Fla. Stat. Hort. Sci. 103,* 8–9.
14. Fouche, P. S., Bester, D. H. (1986). The influence of water soluble fertilizer on nutrition and productivity of Navel orange trees under microjet irrigation. *Citrus and Sub-Tropical Fruit J. 62,* 8–12.
15. Fritz, W. D. (1970). Citrus cultivation and fertigation. *RUHR, Stickstoff, A. G., Bochum.*
16. Gangwar, L. S., Shirgure P. S., Shyam Singh. (1997). Economic viability of investment on adoption of drip irrigation system in Nagpur mandarin orchards. *Proc. National Symp. Citriculture* Nov. 17–19, at NRCC, Nagpur, 246.
17. Ghadekar, S. R., Dixit, S. V., Patil, V. P. (1989). Climatological water requirement of the citrus under Nagpur agroclimatic conditions. *PKV Research J. 13 (2),* 143–148.
18. Grieve, A. M. (1988). Water use efficiency of microirrigated Citrus. *Proc. 4th Intern. micro irrigation congress, Oct 23–28, 1988. Albury-Nodonga,* Australia.
19. Hashemi, F., Gerber, J. F. (1968). Estimating evapo-transpiration from a citrus orchard with weather data. *Proc. Amer. Soci Hort. Sci. 916,* 173–179.
20. Haynes, R. J. (1985). Principles of fertilizer use for trickle irrigated crops. *Fert. Res. 6,* 235–255.
21. Haynes, R. J. (1988). Comparison of fertigation with Broad cast applications of Urea-N. On levels of available soil nutrients and on growth and yield of trickle irrigated peppers. *Scientia Hort. 35,* 189–198.
22. Hilgeman, R. H., Hewland, L. H. (1955). Fruit measurement and tensiometers can tell you when to irrigate Citrus trees. *Proc. Agri. Arizona. 7,* 10–11.
23. Intrigliolo, F., Raciti, G. (1989). Subsurface system experiments in citrus. *Irrigazione and Drenaggio 36(2),* 25–27.
24. Kelin, I. (1983). Drip irrigation based on soil matric potentials conserves water in peach and grape. *Hort Science. 18,* 942–944.
25. Koo, R. C. J. (1968). Evapotrarspiration and soil moisture distribution as guide to citrus irrigation. *Proc. First Intern. Citrus Symp. Riverside,* 269.
26. Koo, R. C. J. (1981). Results of Citrus fertigation studies. *Proc. Fla. State Hort. Sci. 93,* 33–36.
27. Koo, R. C. J., Smjstrala, A. G. (1984). Effect of trickle irrigation and fertigation on fruit production and fruit quality of Valencia orange. *Proc. Fla. State Hort. Sci., 97,* 8–10.
28. Leyden, R. F. (1977). Water requirement of grapefruit in Texas. *Proc. Int. Soc. of Citriculture. 3,*1037–1039.
29. Louse Ferguson. (1990). Nitrogen fertigation of citrus summery of citrus research. Citrus Research Centre and Agricultural station. *University of California Riverside.* 20–22.
30. Mageed, K. J. A., Sharma, B. B., Sinha, A. K. (1988). Influence of irrigation and nitrogen on water use and growth of Kinnow mandarin. *Indian. J., Agric. Sci. 58 (6),* 284–86.
31. Makhija, M., Sharma, B. B., Sinha, A. K. (1986). Estimating water requirements of Kinnow mandarin. *South Indian Hort. 34 (3),* 129.
32. Marler, T. E., Davies, F. S. (1989). Microsprinkler irrigation scheduling and pattern effects on growth of young Hanilin orange trees. *Proc. Fla. State. Hort. Sci. 102,* 57–60.
33. Marler, T. E., Davies, R. S. (1990). Microsprinkler irrigation and growth of young Hamlin' orange trees. *J. Amer. Soc. Hort. Sci. 115 (1),* 45–51.

34. Moreshet, S., Cohen, R., Fuchs, M. (1983). Response of mature 'Shomouti' orange trees to irrigation of different soil volumers of similar levels of available water. *Irrig. Sci. 3 (4),* 223–236.
35. Moreshet, S., Cohen, Y., Fuchs, M. (1988). Water use and yield of a mature Shamouti orange orchard submitted to root volume restriction and intensive canopy pruning. *Proc. Sixth Intern. Citrus Cong. Tel Aviv, Israel,* 2739–2746.
36. Nielsen, G. H; P., Parchomchuck; Wolk W. D., Lau, O. L. (1993). Growth and mineral composition of Newly planted apple trees following Fertigation with N., *Amer, P. J. Soc. Hort. Sci., 118,* 50–53.
37. Ozsan, M., Tekinel, O, Tuzcu, O., Cevik, B. (1983). Studies on determining the most efficient irrigation method for growing lemons under cukurova conditions. *Doga Bilim Dergisi Dz, Tarim Ve Ormancilik. 7 (1),* 63–69.
38. Plessis, S. F. Du. Irrigation of citrus. *Citrus and Sub tropical Fruit J.* (1985). *614,* 12.
39. Plessis, S. F. (1987). Du. Some factors affecting the fruit growth of citrus. *South African J. Pla. Soil. 4 (1),* 12.
40. Plessis, S. F. Du. (1988). Irrigation scheduling. *Proc. Sixth Intern. Citrus Congress,* Tel Aviv, Israel, *988,* 731.
41. Plessis, S. F. Du. (1989). Irrigation scheduling of citrus, Research results. *Water and Irrig. Rev. 9 (4),* 4–6.
42. Plessis, S. F. Du., Laere, H. C., Van M. E. (1991). Salt accumulation in citrus orchards as influenced by irrigation. *J., South African Soc. Hort. Sci. 1(1),* 29–32.
43. Pyle, K. R. (1985). An appraisal of micro irrigation for use in citrus with an emphasis on drip irrigation. *Citrus and Subtropical Fruit J., 612,* 4–7.
44. Raciti, G., Scuderi, A. (1977). Drip irrigation trial in citrus orchard. *Proc. Int. Soci. Citriculture* 31040–1045.
45. Raciti, G., Barbagallo, A. Localized irrigation in lemon forcing. *Informatore Agravio* (1982). *38 (41),* 22887–22891.
46. Rajput, R. K. (1989). Strategies of water management in tropical fruits. *Sixth Binenial workshop AICRP on fruits.* Tirupati, India.
47. Randhawa, Singh, G. S., Dudani, G. J. P. (1960). Preliminary studies on the effect of soil management systems on soil moisture in sweet orange orchard. *Ind. Jr. Hort., 7,* 246–249.
48. Randhawa, G. S., Srivastava, K. C. (1986). Citriculture in India. *Hindustan Publishing Cooperation (India).* 501.
49. Ray, P. K., Sharma, B. B. (1990). Studies on response of young Kinnow trees to irrigation. *Ind. J., Hort. 47 (3),* 291–296.
50. Ray, P. K., Sharma, B. B., Sinha, A. K. (1990). Effect of irrigation on plant status and stomatal resistance in young Kinnow mandarin trees. *South Indian Hort. 38 (3),* 123–128.
51. Retiz, H. J. (1968). How much water do Florida citrus trees use? *Citrus Ind. 49 (10),* 4–6.
52. Richards, L. A. (1954). Diagnosis and improvement of saline and alkali soils, USDA., Agri. Handbook No. 60. US Dept. of Agriculture, Washington, D.C., USA.
53. Richards, S. J., Warnke, J. E. (1968). Lemon irrigation management under coastal conditions. *Calif. Citrog. 53,* 378–384.
54. Robinson, J. C., Alberts, A. J. (1987). The influence of under canopy sprinkler and drip irrigation systems on growth and yield of bananas. *Scientia Hortic., 32,* 49–66.
55. Rondey, D. R., Roth, R. L., Gardner, B. R. (1977). Citrus response to irrigation methods. Proc. Int. Soc. of Citriculture, *1,* 106–110.
56. Rumayor-Rodriguez, A., Bravo-Lonzano, A. (1991). Effects of three systems and levels of irrigating apple trees. *Scientia Hort. 47,* 67–75.
57. Sanehez Blenco, M. J., Torrecillas, A., Leon, A., Del Amor, F. (1989). The effect of different irrigation treatments on yield and quality of verna lemon. *Pl. Soil 120 (2),* 299–302.
58. Scuderi, A., Raciti, G. (1978). Citrus trickle irrigation trials. *Proc. Int. Soc. Citriculture,* 244.
59. Shinde, B. N., Firake, N. N. (1998). Integrated water management for crop production. Dept. of Irrigation and Water management, M. P. K. V., Rahuri (Maharashtra).

60. Shirgure, P. S., Marathe, R. A., Lallan Ram; Shyam Singh. (1998a). Effect of irrigation scheduling on depth of irrigation, soil moisture distribution and Evapotranspiration in acid lime. *Proc. of National Seminar on Water Management held at, W. T. C., IARI,* New Delhi *from* April, 15–17. 30.
61. Shirgure, P. S., Marathe, R. A., Lallan Ram; Shyam Singh. (1998b). Leaf nutrient, growth and fruit quality of acid lime affected by nitrogen fertigation. *Proc. of National Seminar on Water management held at, W. T. C., IARI,* New Delhi *from* April, 15–17. 27.
62. Shirgure, P. S., Srivastava, A. K., Shyam Singh. (1998c). Response of fertigation verses band placement of Nitrogen in acid lime. *Presented in Seminar on new horizons in production and postharvest management of tropical and subtropical fruits*, Dec., 8–9, IARI, New Delhi. 26.
63. Shirgure, P. S., Lallan Ram, Marathe, R. A., Yadav, R. P. (1999). Effect of Nitrogen fertigation on vegetative growth and leaf nitrogen content of acid lime. *Indian Journal of Soil Conservation 27(1),* 45–49.
64. Shirgure, P. S., Srivastava A. K., Shyam Singh. (1999). Fruit quality and soil fertility changes in acid lime under drip, microjets and basin irrigation methods. Proc. of International Symposium on Citriculture held at NRCC, Nagpur (India) on 23–27th Nov., 358–366.
65. Shirgure, P. S., Srivastava A. K., Shyam Singh. (1999). Soil -leaf nutrient build-up and growth response of prebearing acid lime as affected by N. fertigation versus band placement method. Proc. of International Symposium on Citriculture held at NRCC, Nagpur (India) on 23–27th Nov., 551–557.
66. Shirgure, P. S., Lallan Ram, Shyam Singh, Marathe, R. A., Yadav, R. P. (2000a). Water use and growth of acid lime under different irrigation systems. *Indian J. of Agri. Sci. 70 (2),* 125–127.
67. Shirgure, P. S., Srivastava A. K., Shyam Singh. (2000b). Efficiency of micro irrigation systems in relation to growth and nutrient status of acid lime. Micro-irrigation: Eds. Singh, H. P., Kaushish, S. P., Murty, T. S., Jose C., Samuel. CBIP Publication No. 282, 262–269.
68. Shirgure, P. S., Marathe, R. A., Lallan Ram; Shyam Singh. (2000c). Irrigation scheduling in acid lime as affected by different soil moisture regimes. *Indian J., Agri. Sci. 70 (3),* 173–176.
69. Simpson, G. H. (1978). Developments in under tree irrigation systems in the Murray valley. *Proc. Int. Soc. Citriculture*. 234–235.
70. Slack, J., Turpin, J. W., Duncan, J. H., Mckay, O. L. (1978). Trickle irrigation of young citrus on coarse sands. *Proc. Int. Soc. Citriculture,* 236–237.
71. Smajstrla, A. G., Koo, R. C. J., Weldon, J. H. (1984). Effects of volume of the rootzone irrigated on water use and yield of citrus. *J. Am. Soc. of Agri. Eng., 84,* 2107.
72. Smajstrla, A. G., Koo, R. C. J. (1984). Effects of trickle irrigation methods and amounts of water applied on citrus yields. *Proc. Fla. State Hort. Soc. 97,* 3–7.
73. Smajstrla, A. G., Parsons, L. R., Aribi, K., Yelledis, G. (1986). Response of young citrus trees to irrigation. *Proc. Fla. State Hort. Soc. 98,* 25.
74. Smajstrla, A. G. (1993). Micro-irrigation for Citrus production in Florida. *HortScience. 28 (4),* 295–298.
75. Smith, M. W., Kenworthy, A. L., Bedford, C. L. (1979). The response of fruit trees to injections of nitrogen through a trickle irrigation system. *J. Am. Soc. Hort. Sci. 104,* 311–313.
76. Stolzy, L. H., Taylor, O. C., Gavber, M. J., Lambard, P. B. (1963). Previous irrigation treatments as factor in subsequent irrigation level studies in orange production. *Proc. Amer. Soc. Hort. Sci. 82,* 119–123.
77. Syvertsen, J. P., Smith, M. L. (1996). Nitrogen uptake efficiency and Leaching losses from Lysimeter grown citrus trees fertilized at three nitrogen rates. *J. Am. Soc. Hort. Sci. 121 (1),* 57–62.
78. Tash be kov, Kh. K., Saidov, I. I., Kreidik, B. M. (1986). Water requirement of young lemon plantations in central Tajikistan conditions. Tekhnologiya orosheniya; programmirovanie Urozhaya, Moscow, USSR, 147–150.
79. Toledo, P., Rey, E. M., Cardenas, O. (1982). Soil moisture level studies for up to 5-year-old 'Olinda Valencia' oranges. *Ciencia y Tecniica enla Agricultura Riego y Drenaje, 5 (1),* 17–30.
80. Zekri, M., Parsons, L. R. (1989). Grapefruit leaf and fruit growth in response to drip, micro-sprinkler and overhead sprinkler irrigation. *J. Am. Soc. Hort. Sci. 114,* 25–29.

CHAPTER 5

RESEARCH ADVANCES IN IRRIGATION AND FERTIGATION MANAGEMENT: CITRUS

P. S. SHIRGURE

CONTENTS

5.1 INTRODUCTION

The research advances of irrigation scheduling and water requirement of citrus cultivars are reviewed in this chapter. Crop annual nutrient needs are defined as the amount of nutrients consumed throughout the year by the tree, which are supposed to be enough for an optimum tree development and fruit production. This determination includes the needs of both new developing organs (reproductive and vegetative) and old permanent organs for growth consumption. This demand does not include annual old leaves requirements because these leaves, at the beginning of a new fertilization program, translocate mobile nutrients to new organs, before its abscission.

In citrus, many years ago, quantitative determinations of nutrients consumptions were determined through chemical analysis of young or aerial tree tissues Smith, [87]. However, these data did not properly reflect the annual nutritional needs of the tree since neither elements accumulated in perennial tissues (roots, trunk and older branches) nor the nutrients supplied by the storage tissues (internal remobilization) can be determined without extracting trees from soil. Legaz and Primo-Millo [42] and Martínez-Alcántara et al. [47] determined the total amount taken up by a citrus tree along one-year vegetative cycle by means of sequential destructive harvests of trees of different ages (2-, 6- and 12-years-old) along the cycle. In the case of N, these data were obtained by supplying nitrogen heavy isotope (^{15}N) in an inert soil-free medium (sand) or in soil. Annual nutrient requirements indicated that some nutrients are provided by the reserves of old leaves, except for Fe, which is scarcely mobile in the tree and its translocation from old leaves to new developing organs can be considered negligible. The difference between new and old organs nutrient demand and that covered by old leaves reserves represents net annual needs for the citrus tree.

Citrus is predominantly grown in tropical and subtropical areas of world at 40°latitude of either north or south of equator [15, 63, 64]. Basin irrigation is widely used in citrus orchards, especially in South-Asian countries [81]. However, it has several drawbacks in terms of losses through conveyance, percolation, evaporation, and distribution, yet without much adverse impact on growth, yield, and fruit quality [80, 82]. In light of growing scarcity of water and poor water use efficiency (WUE) of basin irrigation, micro irrigation has gained wide application in citrus orchards. However, the efficacy of micro irrigation is often questioned, especially where soil moisture deficit stress is used to regulate the stress for induction of flowering in the areas lacking low temperature deficit stress, e.g. central India [88]. The lack of uniformity in moisture distribution within the tree root zone due to variation in subsoil properties can adversely affect the development of desired fruit size [65, 75]. Any method of irrigation capable of replenishing the evapotranspiration demand of a tree, and simultaneously keeping the soil moisture within the desired limit during different ontogenic stages, will ensure a production sustainability of citrus orchards in addition to prolonged productive life of an orchard [56].

Many efforts have been made to enhance fruit yield with combined use of irrigation and fertigationand to compare it with the broadcast method of fertilization [65, 66, 101]. Bowman [13] evaluated the effects of conventional broadcast fertilization and of a combination of broadcast/fertigation in mature *grapefruit cv Ruby Red* trees established on Flatwood soils of Florida, USA. Conventional fertilization consisted of broadcast application: 3-times a year (January/February, May/June, and October/November), while combination of fertigation treatment received 33% of annual N and K in February/March followed by fertigation scheduling at 2-weeks interval to the remaining dosage beginning in April. The total soluble solids (cumulative of 4 years) in combination broadcast/fertigation were much higher (10.9 tons.ha^{-1}) compared to conventional fertilization (10.1 tons.ha^{-1}). In many citrus growing areas, low water use efficiency (WUE) and fertilizer use efficiency (FUE) are among the major production related constraints [30, 89]. Of the many components influencing the efficiency of applied fertilizers, application timing, method, and rate play an important role in affecting fruit yield and quality.

5.2 MICRO IRRIGATION SYSTEMS

The micro irrigation, under-tree sprinklers, microsprinklers, and microjets have been reported to be highly effective in commercial citrus cultivars like: Valencia orange [8], Navel orange [28], Hamlin orange [46], Satsuma mandarin [54], Clementine [19] and lemon [20]. Earlier studies in India comparing drip with flood irrigation in Nagpur mandarin [7, 68] sweet orange [40], and acid lime [70, 72] showed better performance using micro irrigation. Micro-irrigation systems are commonly used in citrus orchards throughout the world. The results have shown some distinct transformations. There is now a gradual shift in method of irrigation from furrow irrigation-overhead sprinkler irrigation systems to under-tree sprinkling systems like microjets [25, 60].

Basin irrigation for citrus trees is usually used in countries like India, Pakistan, Thailand, etc., in south Asia [31, 78], Argentina [18], Australia [85], Turkey [94], Italy [17], South Africa [55]. When basin irrigation is used in north-west and central India, temporary excess soil moisture condition occurs as well as the leaching of applied nutrients below the effective rootzone [19, 65, 76]. The problem is further compounded by the swelling and shrinking of montmorillonitic clay soils of central India where Nagpur mandarin (*Citrus reticulata* Blanco) is grown extensively. Therefore, a strategy which allows judicious use of water as well as nutrients in concurrence with tree demand is likely to impart an improvement in citrus production besides fruit quality. Fruit yield of Nagpur mandarin with different micro irrigation systems on Vertic Ustochrept was significantly higher (48.23–58.93 kg.tree^{-1}) over basin irrigation (32.3 kg.tree^{-1}) with corresponding WUE of 0.19–0.24 versus 0.109 t ha^{-1}cm^{-1} and leaf N content of 2.38–2.42% versus 2.01–2.12% [73]. The highest fruit yield of mandarin was 40.33 t/ha with irrigation system of microjet 180°(Fanjet, 2/tree) followed 39.89 t/ha with 270°

microjet (Rayjet, 2/tree); and the lowest fruit yield was 35.10 t/ha with 300° Rayjet (2/tree). The highest TSS (10.12° Brix) and juice content (43.05%) was found in microjet 180°(Fanjet) and microjet 300°(Rayjet), respectively [68, 73].

The Nagpur mandarin fruit yield was highest (30.91 tons/ha) with irrigation on alternate days with irrigation duration of 120 min three times, followed by irrigation scheduling with 90 min interval two times daily (30.11 tons/ha). Fruit weight (154.7 g), TSS (10.22–°Brix) and juice percent (40.77%) were highest with automatic irrigation at alternate day with 120 min three times. The automatic micro irrigation scheduling can be a better substitute for manual micro irrigation operation and enhancing the WUE and FUE [78].

5.3 FERTIGATION TECHNOLOGY

Fertigation is an application of nutrients through irrigation water. It is most effective and convenient means of maintaining optimum fertility level and water supply according to the specific nutrient requirement of each crop. In the area of scarce water resource and insufficient rainfall, fertigation offers the best and sometimes the only way of ensuring the nutrients enter the root zone of acid lime [62, 67]. Fertigation has improved the tree growth, fruit yield, quality, the reserve pool of soil nutrients, and consequently the tree nutritional status [84]. Besides the better mobility of nutrients, fertigation has been shown to have several advantages over broadcast method of application of granular fertilizers [99] with respect to growth response [35], nutrient uptake [36], effective placement of nutrients and flexibility in application frequency [27], development of uniform root distribution in wetted zone – an important prerequisite for better FUE – [6], fruit yield [39], and improvement in fruit quality [13]. Other research studies have shown superior results with fertigation in Spain [41], central India [65, 66] and in Arizona (USA) using microsprinklers over basal fertilizer application in flood irrigation [96]. However, studies from Zhang et al. [101] evaluating the effect of fertigation versus broadcast application of water soluble granular fertilizer on the root distribution of26-year-old 'White Marsh' grapefruit trees on sour orange rootstock, showed 94% of the root density in the top 0–30 cm soil depth with soluble granular fertilizers. These observations support the earlier observations that shallow depth of wetting and delivery of nutrient resulted in confining most of the roots within surface soil [6, 101].

Bester et al. [11] observed an increase in leaf nitrogen levels of young trees fertigated frequently with NPK solution compared to a broadcast fertilizer application using sprinkler irrigation system, but no significant difference was observed with respect to P and K levels. Similar observations were later made by Intriglio et al. [33] while comparing a single annual application of NPK to continuous fertigated application. Koo [37, 38] reported that the treatment having 37% coverage of ground and 82% of canopy area produced fruit yield higher than the broadcast fertilizer treatment covering 100% of ground surface and 53% canopy area. These observations suggest the importance of canopy coverage for high nutrient uptake efficiency and higher yield. Response of six year-old 'Hamlin' orange to fertigation frequency using 324 to 464 g of N.tree^{-1}, showed that nitrogen uptake efficiencies ranged from 24 to 41% of N applied, but no effect of fertigation frequency on the amount of N taken up by the trees, was observed when fertigation frequency was increased from 12 to 80 times.year^{-1} [92]. Alva and Paramasivam [2] found that 18 split fertigation applications through microsprinklers under the trees increased the fruit yield with fertigation than equivalent rates of granular fertilizer treatments due to greater FUE. The investigations on prebearing Acid lime (Citrus aurantifolia Swingle) during 1995–1997 having fertigation with 60%, 80% and 100% N of the recommended doses were compared to the research with band placement (100% N) method of fertilizer application. The percentage increase in tree height, tree girth and canopy volume was maximum with 100% N fertigation followed by with 80% N fertigation. The percentage increase in leaf Nitrogen content was more in case of 80% N fertigation (27.47%) followed by 100% N fertigation (24.32%), 60% N fertigation (20.23%) and band placement (7.5%). This study clearly indicates the advantage of N fertigation over the conventional method of fertilizer application [62].

Alva et al. [4] studied the comparative response of 32 months-old nonbearing 'Hamlin' orange trees on a Candler fine sand (Typic Quartzipsamments) using three methods of fertilization namely: fertigation (FRT), controlled release fertilizers (CRT), and water soluble granular fertilizers (WSG) at high and low fertilizers rates. Total N content in tress, which received the higher fertilizer rates were 82.3, 70.2, and 41.4 g.tree^{-1} for the FRT, CRF, and WSG sources, respectively. The corresponding values for the low-fertilizer rate treatments were 38.6, 50.4, and 28.4 g.tree^{-1}. However, the proportion of total N partitioned to leaves was greater for WSG than for the CRF and FRT sources at both the fertilizer rates. Similar observations were made through the response of 25 year-old 'Hamlin' orange in Highland county with varying N rates (112–180 kg ha^{-1}) and fertilizer management practices (WSG, CRF and FRT). Spring flush leaf N content increased with increasing N rates decreased in the order of FRT > WSG > CRF [53]. Other studies [14] involving CRF (one application per year), FRT (15 applications per year), and WSG (three applications per year) showed no response of fertilizer sources either on fruit yield of grapefruit or leaf nutrient composition on Arenic Glossaqualf soil.

These important breakthroughs indicate that fertigation is now increasingly gaining importance as a popular method of fertilizing citrus trees. According to Lekchiri [43], the phosphorus and potassium requirements of citrus trees are relatively high. However, soil conditions and restricted root colonization may limit the availability and uptake of soil nutrients. To overcome these difficulties, two alternatives can be adapted:

- **Using micro irrigation system**, fertilizer application using fertigation or by placement in furrow parallel to the dripping ramp where the soil is moist, thereby, improving the mobilities of P and K and enriching the soil where roots are concentrated to improve fertilizer uptake efficiency, and

- Application of fertilizers by placement in the zone receiving water, to improve the mobility of P and especially of K up to a depth of 60 cm.

Besides the mobility of nutrients, fertigation has several advantages over broadcasted granular fertilizers [98, 99] including effective placement of nutrients and flexibility in application frequency [27], in addition to development of uniform root distribution (an important prerequisite for better fertilizer use efficiency) under fertigation [101]. Fouche and Bester [28] evaluated various fertilizer combinations through fertigations on 13 year old Navel oranges. Fertigation was supplemented with: (1) Soluble fertilizer 'Trisol' (3:1:5) + 350 g urea by broadcast; (2) Fertigation of N and K with broadcast of single superphosphate; and (3) N P K through broadcast application. Highest yield was obtained with fertigation of N, P and K through Trisol or by complete broadcasting of N P K fertilizers. No significant differences were observed in fruit quality parameters: fruit size, acidity, juice content and TSS when compared within treatments.

Field experiments on response of prebearing acid lime trees to differential N-fertigation versus circular band placement (CBP) method of fertilizer application showed superiority compared to other treatments. The higher leaf N, P and K with 80% fertigation over 100% N through CBP further demonstrated that 20% saving of N is attainable [67]. Earlier studies carried out by Garcia-Petillo [29] demonstrated 50% higher leaf N content with 64% higher yield on cumulative basis in fertigation treated trees compared to conventional method of fertilization. All these studies suggest that fertigation is better than conventional basin or flood irrigation with broadcast method of fertilizer application.

Irrigation at 20% depletion of available water content (AWC) combined with fertilizer treatment of 500 g N + 140 g P + 70 g K tree^{-1}year^{-1} produced a significantly higher fruit yield and canopy volume in addition to higher nutrient status and fruit quality compared to other treatments with 10% depletion or 30% depletion of AWC with 600 g N + 200 g P + 100 g K^{-1}tree^{-1}year in 14-year-old Nagpur mandarin (*Citrus reticulata cv.* Blanco) on an alkaline calcareous Lithic Ustochrept soil type [66, 67, 88]. Irrigation at 30% depletion of AWC combined with fertilizer treatment of 500 g N + 140 g P + 70 g K tree^{-1}year^{-1} produced a significantly higher fruit yield canopy volume in addition to higher nutrient status and fruit quality compared to other treatments involving irrigation either 10% depletion or 20% depletion of AWC with 600 g N + 200 g P + 100 g K^{-1}tree^{-1}year in 10-year-old acid lime (*Citrus* aurantifolia Swingle) on an alkaline calcareous Lithic Ustochrept soil type [74, 77]

5.4 NUTRIENT USE EFFICIENCY AND FERTIGATION

The purpose of fertilization is to improve the nutritional status of a crop. Citrus trees need high-amounts of fertilizers, unfortunately, farmers have applied excessive dosages of nutrients because of poor fertilizing criteria and slight increase in fruit yield with increased dosages. This has resulted in poor quality of the fruit [21], a reduction in the profitability of the citrus crops [97] and a NO_{3-} displacement, mainly, to deeper soil layers. In this case, many studies have shown direct relationships between the addition of N in areas of intensive agriculture and the alarming increase of NO_{3-} concentration in groundwater [9, 12, 14, 23, 66, 86].

At present, efforts are being directed to understand the large number of processes in which nutrients are involved in the tree-soil system, like: irrigation management, application frequency, timing of application, as well as soil processes, in order to reduce rates and losses, which may result in surface and ground water pollution, maintaining crop productivity. This section reviews several research results carried out by different authors with the aim of reevaluating current fertilization programs. This information is necessary to understand nutrient use efficiency and thus advance towards Best Management Practices (BMP) for citrus crops.

5.4.1 NITROGEN FERTIGATION

In citrus orchards, irrigation systems directly affects the N absorbed from fertilizer (Naff) by the entire tree and the amount retained in soil or leached in drainage. Quiñones et al. (2005) obtained higher N recovery percentages in Navelina using micro irrigation (73%) than in flood irrigation (63%). This data are similar to those of Syvertsen and Smith (1996) who found a nitrogen use efficiency (NUE) value for lysimeter – grown citrus trees of 61 to 68%. Further improvement of NUE by citrus with fertigation compared with dry granular fertilizer was reported by Dasberg et al. [24], Alva and Paramasivam [2], Alva et al. [3] and Alva et al. [4]. Li et al. [44] studied the influence of fertigation strategies on N distribution in soil profile with micro irrigation. For a given volume of water applied, increasing the application rate allowed more water to distribute in the horizontal direction, as in micro irrigation, while decreasing the rate leads to more water in vertical direction and, therefore, nitrate leaching could be higher. Quiñones et al. [59] showed that the percentages retained in soil profile as NO_3^-N were significantly higher for the flood irrigated (around 38% of the N retention) than for the drip irrigated trees (8%). Nevertheless, no significant differences were observed in the amount of organic ^{15}N for both irrigation systems. Citrus trees demand high-amounts of nitrogenous compounds as nitrogen (N) has a greater influence on growth and production than other nutrients [87].

Frequency of N application also affects N distribution in tree-soil-leaching system. More frequent application of dilute N solutions double NUE compared with less frequent application of more concentrated N solutions [58, 61]. In another study, Alva et al. [5] demonstrated a slight increase in NUE as a result of better management practices associated with N placement, timing of application, and optimal irrigation scheduling when comparing fertigation (FRT – 15 N applications) versus water soluble granular (WSG – 4 N applications). Also increases in NUE were obtained by other authors expressed as increment in fruit yield. Bowman [13] reported a greater NUE (9% greater fruit yield) in grapefruit trees receiving a combination of one dry granular broadcast application (33% of the annual rate) and 18 fertigations at 2-week intervals compared to trees that received three applications of dry fertilizer. Alva et al. [1] evaluated different combinations of irrigation and nitrogen management. In young trees, Morgan et al. [48] found higher yields when compared controlled-released fertilizer and fertigation applied 30 times annually with dry granular fertilizer and fertigation applied four times.

Greater N recovery was observed by whole tree in trees fertilized with potassium nitrate (40.1 and 37.0% in sand and loam soil, respectively) than those under ammonium fertilization (37.9 and 33.9% in sandy and loamy soil, respectively). Use of nitrification inhibitors (NI) can also affect NUE. Nitrate-N fertilizers are absorbed more efficiently than ammonium-N by citrus trees, however, ammonium fertilizers are recommended during the rainfall period. The addition of NI to ammonium-N fertilizers increases NUE (16%), resulting in lower N-NO_{3-} content in the soil (10%) and in water drainage (36%).

5.4.2 PHOSPHORUS FERTIGATION

Research studies on phosphorus uptake are not abundance because, in general, the soils have enough phosphorus. In practice, the main important question the citrus grower can ask: whether there is enough available P in the soil solution to ensure a proper tree development [34]. Under nonirrigated conditions, phosphorus shows very low mobility into the soil profile [45], and therefore losses by leaching of this element are negligible [22]. High fertigation frequency ameliorates this situation, since there is a continuous mass flow. Increased saturation of P fixation sites in the soil due to high frequency and application rate results in higher amounts of P released to solution, which combined with the forced flow of water into the soil, facilitates the distribution and the consequent increased levels of P [26]. Therefore, P fertigation can increase the movement of this nutrient in the soil profile, compared to the conventional fertilizer application. Also, the use of phosphoric acid provides increased mobility of soil P when compared to superphosphate [95, 100]. Phosphate rapidly reacts with Ca in basic soils, and with Fe and Al in acid soils, being the distance traveled by applied P quite limited, even in sandy soils, as compared with the water [10]. The low availability of P in the bulk soil limits the tree uptake. The efficiency of absorption of P can vary up to 10% for furrow irrigation system and up to 35% for irrigation [52], because about 80% of the P becomes immobile and unavailable for tree uptake due to adsorption, precipitation, or conversion to the organic form [32].

In Florida, citrus orchards traditionally receive about 40 kg phosphorus ha^{-1} at treeing, followed by applications of up to 100 Kg per ha per year until they enter into the fruit-bearing years. From then onwards, citrus receive 20–50 kg ha^{-1} per year [93]. However, according to Obreza, [49] there is a lack of fertilizer response in newly citrus trees in sandy soils. Similarly, adult citrus trees rarely respond to P fertilizer [87], except when treed on soils with extreme P fixation capacity. Cantarella et al. [16] and Quaggio et al. [57] observed positive yield responses of Valencia oranges and lemons to annual P fertilizer rates up to 62 kg ha^{-1} on a high P-fixing Brazilian soil. On the contrary, Alva et al. [1] found negligible effects of fertilization source (granular, controlled release formulation or liquid) and rates on citrus trees P content grown in a sandy soil.

5.4.3 POTASSIUM FERTIGATION

Citrus trees remove large amounts of potassium (K) compared with other nutrients; and K enhances fruit set and thus yield, as well as affects fresh fruit qualities. Potassium deficiency reduces fruit number and size, increases fruit creasing, plugging and drop, and decreases juice soluble solids, acid and vitamin C content. Potassium is present as component of rocks and soil (fixed position) or an exchangeable cation on all clay particles. Since the rate of K release from fixed position is slower than the rate of K tree demand, additions of K in fertilizers are needed for normal tree development. This is especially important when micro irrigation is used, since the volume of soil occupied by the active root is small and not all the soil volume contributes K to the growing tree [34]. In soils containing appreciable amounts of organic matter or clay, mobility of K can be limited, because positive charge of K ion enables it to be held by the soils' negatively charged cation exchange complex. However, in sandy soils, with very low concentrations of clay or organic matter, the ability to hold K against leaching can be almost nonexistent [50]. According to this situation, and considering that citrus trees use large quantities of K, K is applied at relatively high rates in a typical citrus fertilization program. Potassium is applied at a K_2O rate equal to the N rate, however, this rate is increased by 25% when leaf K is consistently below optimum and especially in calcareous soils [50, 51].

The efficiency of absorption of K can vary up to 60% for furrow irrigation and up to 90% for fertigation [52]. The effect of K-doses on yield and fruit quality of the bearing Nagpur mandarin was studied during 2009–2012 and results showed that the highest fruit yield (26.67 tons/ha) with 50 g K_2O/tree potassium sulfate followed by in K-fertigation with 40 g K_2O/tree dose (25.52 tons/ha), as indicated by Shirgure et al. [79]. The research results in bearing Nagpur mandarin have shown the highest response of the fruit yield (31.13 t/ha) with potash fertigation using mono potassium phosphate followed by fertigation with potassium nitrate (29.4 t/ha). The total soluble solid was highest (10.49 – °Brix) in K fertigation with mono potassium phosphate followed by fertigation with potassium sulfate (10.48–°Brix). Highest juice content (38.76%) and low acidity (0.77%) was found in K fertigation with mono potassium phosphate. The highest TSS to acidity ratio (sweetness indicator) was observed in Mono potassium Phosphate (13.6) followed by Potassium sulfate (13.1) [83].

5.5 RECOMMENDATIONS FOR FUTURE RESEARCH

Nowadays, techniques and managements of agricultural production are directed towards the need to conserve resources, energy and a commitment to the environment. Therefore, fertigation is a valuable tool in recent years that has spread around the world in all agricultural areas, field and horticultural crops. This has led to an increase in fertilizer and water use efficiency. In the future, fertigation should continue to replace traditional flood irrigation. Citrus are mainly grown in arid and semiarid region with water scarcity. Furthermore in India, climatic conditions are characterized by low rainfall (400–600 mm $year^{-1}$) and irregular spatial and temporal distribution. On the other hand, the world's population has undergone an exponential growth, which has led to soaring food demand and, therefore, high natural-resource exploitation. Therefore, future trends in fertigation should be addressed to use micro irrigation with recycled sewage or/and desalination water.

In this context, improved water use efficiency (WUE), using different strategies, is also a key concept to solve this water scarcity. So nowadays, efforts are being focussed on developing not only alternative irrigation methods but also new water management methods in order to reduce water dosages while maintaining maximum tree growth, without significantly affecting yield. In micro irrigation systems, subsurface micro irrigation (SDI), where is applied below the soil surface, using buried drip tapes, is being part of modern agriculture. Current commercial and grower interest levels indicate that future use of SDI systems will continue to increase. Improvement of WUE can be also achieved by means of drip irrigation. It is possible to increase efficiency under different irrigation management methods based on regulated deficit-irrigation (RDI) programs. These RDI strategies are defined precisely, where the total water provided for the tree (irrigation plus effective rainfall) is below to the crop water needs in order to reduce ETc, and hence save water, while simultaneously minimizing or eliminating negative impacts of stress on fruit yield or quality. However, these principles of scheduling fertigation are still far from factual basis since they do not take into account the nature and properties of the root zone. In this regard, use of available water content has shown a definite edge over the other methods of scheduling fertigation.

Lastly, nutrient use efficiency can be meliorated by using nitrification inhibitors or tree growth-promoting bio-effectors. Nitrification inhibitors restrict the microbial conversion of ammonium to nitrate that it is mobile in soils and therefore leached. Thus, nitrification inhibitors have potential to reduce nitrate leaching. Bio-effectors or bio-stimulant describes microorganisms and active natural compounds involved in tree growth which, not being a tree nutrient or pesticide, but in some manner have a positive impact on tree health. The biostimulant may increase chlorophyll efficiency and production, enhance metabolism, increase antioxidants, enhance nutrient availability and increase the water holding capacity of the soil. In addition to all these factors, precise soil sampling, whether to take samples from below drippers or in between drippers or mixing soil samples from both the sites and finally, drawing a representative soil samples, find a greater intervention while evaluating nutrient-water interaction in citrus.

5.6 SUMMARY

Citrus is the main fruit group grown in tropical as well assub tropical climate of more than 150 countries throughout the world. Irrigation scheduling and water requirement of the citrus crops are one of the main concerns of the modem citrus fruit production irrespective of availability of natural resources. A large number of research findings have emerged world over the irrigation scheduling and citrus water requirements based on available soil water content, AWC depletion and on pan evaporation replenishment. The research review has revealed best promising results with irrigation scheduling based on depletion patterns of soil available water content, irrigation scheduling of various citrus crops and fertigation. Irrigation water management with proper water use is one of the prime concerns for citriculture irrespective of soil and water resource availability. A variety of recommendations have reviewed the world over on irrigation scheduling based on analysis of meteorological parameters, evapotranspiration, depletion of available water content, soil and leaf water potential. The review of the literature has revealed best promising results on irrigation scheduling based on depletion patterns of soil available water content. Similarly, irrigation scheduling has shown good responses on growth, yield, and quality compared to calender method of irrigation scheduling. The present status of irrigation scheduling, fertigation and water requirement of citrus cultivars is reviewed in this chapter.

KEYWORDS

- **acid lime**
- **acidity**
- **band fertilizer application**
- **basin irrigation**
- **black polythene mulch**
- **canopy volume**
- **Citrus**
- ***Citrus reticulate* cv. Blanco**
- **drainage**
- **micro irrigation**
- **drippers**
- **fertigation**
- **fertilizer use efficiency, FUE**
- **field drains**
- **flowering**
- **fruit quality**

- **fruits**
- **grass mulch**
- **harvesting**
- **high density treeing**
- **input use efficiency**
- **iron**
- **irrigation**
- **irrigation scheduling**
- **juice percent**
- **leaf nutrient composition**
- **lemons**
- **maturity period**
- **micro irrigation**
- **microjet irrigation**
- **microjets**
- **mulches**
- **mulching**
- **Nagpur mandarin**
- **net returns**
- **nitrogen**
- **nutrient management**
- **nutrient uptake**
- **orchard efficiency**
- **orchards**
- **organic farming**
- **organic mulches**
- **phosphorous**
- **tree growth**
- **potash**
- **production**
- **Rangapur lime**
- **rootstocks**
- **rough lemon**
- **scion girth**
- **stock girth**

- **surface irrigation method**
- **total soluble solids, TSS**
- **trickle irrigation**
- **water requirement**
- **water stress period**
- **water use**
- **water use efficiency, WUE**
- **weed population**
- **weight**
- **yield**
- **Zinc**

REFERENCES

1. Alva, A. K., Fares A., Dou, H. (2003). Managing citrus trees to optimize dry mass and nutrient partitioning. *J. Pl. Nutri.* 26(8):1541–1559.
2. Alva, A. K., Paramasivam, S. (1998). Nitrogen management for high yield and quality of citrus in sandy soils. *Soil Sci. Soc. Am. J.* 62(5):1335–1342.
3. Alva, A. K., Paramasivam, S., Graham, W. D. (1998). Impact of nitrogen management practices on nutritional status and yield of Valencia orange trees and groundwater nitrate. *J. Environ. Qual.* 27:904–910.
4. Alva, A. K., Paramasivam, S., Graham W., Wheaton, T. A. (2003). Best nitrogen and irrigation management practices for citrus production in sandy soils. *Water Air Soil Pollut.* 143:139–154.
5. Alva, A. K., Paramasivam, S., Obreza, T. A., Schumann, A. W. (2006). Nitrogen best management practice for citrus trees I. Fruit yield, quality, and leaf nutritional status. *Sci. Hort.* 107:233–244.
6. Alva, A. M., Syvertsen, J. P. (1991). Irrigation water with salinity affects soil nutrient distribution root density and leaf nutrient levels of citrus under drip fertigation. *J.Pl. Nutri.* 14:715–727.
7. Autkar, V. N., Patel, V. S., Deshpande, S. L., Bagade, T. R. (1989). Management of micro irrigation in Nagpur mandarin. *Ann. Tree Physiol.* 3:74
8. Azzena, M., Deidda, P., Dettori, S. (1988). Drip and micro-sprinkler irrigation for young Valencia orange trees. In: Proc. Sixth Intl Citrus Congr., Balaban Publishers, Tel Aviv, Israel, Vol. 2, pp.747–751
9. Babiker, I. S., Mohamed, A. A., Terao, H., Kato, K., Ohta, K. (2004). Assessment of groundwater contamination by nitrate leaching from intensive vegetable cultivation using geographical information system. *Environ Int.* 29(8):1009–1017.
10. Ben-Gal, A., Dudley, L. M. (2003). Phosphorous availability under continuous point source irrigation. *Soil Sci. Soc. Am. J.* 67:1449–1456.
11. Bester, D. H., Fouche, P.S., Veldman, G. H. (1977). Fertilising through micro irrigation systems on orange trees. In: Proc. Int. Soc. Citriculture. Grierson W. (ed.) May 1–8, Orlando, Florida, Vol. 1, pp. 46–49.
12. Bingham, F. T., Davis, S., Shade, E. (1971). Water relations, salt balance and nitrate leaching losses of a 960 acre citrus watershed. *Soil Sci.* 112:410–418.
13. Bowman, B. J. (1996). Fertigation versus conventional fertilization of flatwoods grapefruit. *Fert. Res.* 44(2):123–128.
14. Burkart, M. R., Stoner, J. D. (2002). Nitrate in aquifers beneath agricultural systems. *Water SciTechnol* 45(9):19–29.

15. Burke, J. H. (1967). The commercial citrus regions of the world. In: The Citrus Industry. Reuther, W., Webber, H. J., Batchelor, L. D. (ed.), Univ. of California, Davis, USA. Vol. 1, pp. 214–215.
16. Cantarella, H., Quaggio, J. A., Bataglia, O. C., Raij, B., Van. (1992). Response of citrus to NPK fertilization in a network of field trials in São Paulo State, Brazil. In Proc. Int Citrus Cong, Tribulato, E., Gentile, Refrigerato, G. (eds.) Mar. 8–13, Acireale, Italy Vol. 2 pp. 607–612.
17. Capra, A., Nicosia, O. U. D. (1987). Irrigation management in citrus orchards. *Irrigazine* 34(1):3–15.
18. Castel, J. R., Buj, A., Ramos, C. (1989). Comparison of drip and border irrigation of mature Salustina orange trees. *Investigation Agraria, Produciton Y Protection Vegetables* 4(3):393–412.
19. Castel, J. R. (1994). Response of ypung Clementine citrus trees to micro irrigation I. Irrigation amount and number of drippers. *J. Hort. Sci.* 69: 481–489.
20. Cevik, B., Kaplankiran, M., Yurdakul, O. (1987). Studies for determining the most efficient irrigation method for growing lemons under Cukurova conditions. *Doga, Tarumve, Ormaniciuk* 11:42–43.
21. Chapman, H. D. (1968). The mineral nutrition of citrus. In: The Citrus Industry. Reuther, W, Batchelor, L. D., Webber, M. J. (eds). 2nd edn, Univ Calif Div Agri Sci Berkeley, California.
22. Coelho, F. S. (1973). fertilidade do solo. 1 ed. Campinas. Instituto Campineiro de Ensino Agrícola: 384 p.
23. De Paz, J. M., Ramos, C. (2004). Simulation of nitrate leaching for different nitrogen fertilization rates in a region of Valencia (Spain) using a GIS-GLEAMS system. *Agri. Ecosyst. Environ.* 103(1):59–73.
24. Dasberg S., Bar-Akiva, A., Spazisky, S., Cohen, A. (1988). Fertigation versus broadcasting in an orange grove. *Fert. Res.*15:147–154.
25. Dasberg, S. (1995). Drip and spray irrigation of citrus orchards in Israel. In: Micro-irrigation for a Changing World: Conserving Resources/ Preserving the Environment. Ion: Proc. 5th Intl Micro-irrigation Congress, American Society of Agricultural Engineers, 2–6 April 1995, Orlando, Florida, U. S. A., pp. 281–287.
26. Duenhas, L. H., Villas Bôas, R. L., Souza, C. M. P., Ragozo, C. R. A., Bull, L. T. (2002). Fertirrigaçãocom diferentes doses de NPK e seusefeitos sobre a produção e qualidade de frutos de laranja (*Citrus sinensis* O.) "Valencia". *Revista Brasileira de Fruticultura, Jaboticabal*, v. 24, 20(1): 214–218.
27. Ferguson, J. J., Davies, F. S. (1989). Fertilization of young citrus. Florida Coop. Ext. Fact Sheet FC-79.
28. Fouche, P. S., Bester, D. H. (1986). The influence of water soluble fertilizer on nutrition and productivity of Navel orange trees under micro-jet irrigation. *Citrus Sub-trop. Fruit J.* 62:8–12.
29. Garcia-Petillo, M. (2000). Fertigation versus conventional nitrogen fertilization of Valencia oranges. *Agrociencia* (Montevideo). 4 (1) 23–30.
30. Germana, C. (1994). Increasing water use efficiency through irrigation management. In Proc. Int. Soc. Citriculture, Vol. 2, pp.638–642.
31. Ghosh, S. P., Singh, R. B. (1993). Citrus in South Asia. RAPA Publication, FAO, Regional Office, Bangkok, Thailand, p. 21
32. Holford, I. C. R. (1997). Soil phosphorus: its measurement, and its uptake by trees. *Aust. J. Soil Res.* 35:227–239.
33. Intrigliolo, F., Coniglione, L., Germana, C. (1992). Effect of fertigation on some physiological parameters in Orange trees. In: Proc. Int. SocCitriculture. Tribulato E., Gentile, A., Refergiato, G. (eds). Mar. 8–13, Acireale, Italy, Vol 2. pp. 584–589.
34. Kafkafi, U., Tarchitzky, J. (2011). Fertigation: A tool for efficient fertilizer and water management. First edition, IFA, Paris, France and IPI, Horgen, Switzerland, May (2011).
35. Koo, R. C. J. (1979). The influence of N,K and irrigation on tree size and Fred production of 'Valencia Late' orange. *Proc. Fla. St. Hort. Soc.* 92:10–13.
36. Koo, R. C. J. (1980). Results of citrus fertigation studies. *Proc. Fla. St. Hort. Soc.* 92:33–36.

37. Koo, R. C. J. (1984a). The importance of ground coverage by fertigation to citrus on sandy soils. *J. Fert. Issues* 1(2): 75–78.
38. Koo, R. C. J. (1984b). Fertigation for citrus on sandy soil: ground coverage. *Citrus & Veg. Mag.* 48(1): 10–11.
39. Koo, R. C. J., McCornack, A. M. (1965). Effects of irrigation and fertilization on production and quality of 'Dancy' tangerine. *Fla. Sta. Hort. Soc.* 10:10–15.
40. Kumar, A. P. A., Bhojappa, K. M. (1994). Studies on the effect of micro irrigation on yield and quality of fruit in sweet oranges and economy in water use. *Mysore J. Agric. Sci.* 28:338–344.
41. Legaz, F., Ibanez, R., DeBarreda, P. G., Millo-Primo, E. (1981). Influence of irrigation and fertilization on productivity of 'Navelate' sweet orange. In: Proc. Int. Soc. Citriculture. K. Matsumoto (ed.) Nov. 9–12, Tokyo, Japan, Vol. 2. pp. 591–595.
42. Legaz, F., Primo-Millo, E. (1988). Normas para la fertlización de los agrios. Serie Fullets Divulgació no 5–88. Conselleríad' Agricultura i Pesca. Generalitat Valenciana, 29 pp.
43. Lekchiri, A. (1983). Interactions between water and phosphate and potash fertilizers. Nutrient Balances and the Need for the Fertilizers in Semi-arid and Arid Regions. Int. Potash. Institute, Bern, Switzerland, pp. 305–314.
44. Li, J., Zhang, J., Rao, M. (2004). Wetting patterns and nitrogen distributions as affected by fertigation strategies from a surface point source. *Agri. Water Mgmt.* 67:89–104.
45. Malavolta, E., ViolanteNetto, A. (1989). Nutrição mineral, calagem, gessagem e adubação dos citros. Piracicaba: Potafós 153p.
46. Marler, T. E., Davis, F. S. (1990). Micro-sprinkler irrigation and growth of young 'Hamlin' orange trees. *J. Am. Soc. Hort. Sci.* 115:45–51.
47. Martínez-Alcántara, B., Quiñones, A., Primo-Millo, E., Legaz, F. (2011). Nitrogen remobilization response to current supply in young Citrus trees. Pl. Soil 342:433–443.
48. Morgan, K. T., Wheaton, T. A., Castle, W. S. (2009). Response of young and maturing citrus trees grown on a sandy soil to irrigation scheduling, nitrogen, fertilizer rate, and nitrogen application method. Hort Sci. 44:145–150.
49. Obreza, T. A. (1990). Young 'Hamlin' orange tree fertilizer response in southwest Florida. *Proc. Fla Sta. Hort. Soc.* 103:12–16.
50. Obreza, T. A. (2003). Importance of potassium in Florida citrus nutrition program. *Better Crops* 87:19–22.
51. Obreza, T. A., Morgan, K. T. (2011). Nutrition of Florida citrus trees. Document SL 253. Dept Soil and Water Sci. 2ed. Univer Florida 96 p.
52. Papadopoulos, I. (2001). Processo de transição da fertilização convencional para fertirrigação. In: Follegatti, M. V., Casarini, E., Blanco, F. F., Brasil, C. R. P., Resende, R. S. (Coord.) Fertirrigação: flores, frutas e hortaliças. Guaíba: Agropecuaria, cap. 1, pp. 9–69.
53. Paramasivam, S., Alva, A. K., Wheaton, T. A., Syvertsen, J. P., Tucker, D. P. H. (2000). Critical leaf nutrient status for optimal fruit production under differing fertilizer management practices. In: Proc. Int. Soc. Citriculture, pp. 437–439.
54. Peng, Young Hong, Rabe, E. (1999). Effect of irrigation methods and ground cover on the fruit quality, yield and light levels in the canopy of microwave Satsuma. *J. Fruit Sci.* 15:128–132
55. Plessis, S. F. Du. (1985). Irrigation of citrus. *Citrus & Sub trop. Fruit J.* 614:12.
56. Pyle, K. R. (1985). An appraisal of micro-irrigation for use in citrus with an emphasis on micro irrigation. *Citrus Sub-trop. Fruit J.*61: 4–7.
57. Quaggio, J. A., Cantarella, H., van Rajj, B. (1998). Phosphorous and potassium soil test and nitrogen leaf analysis as a base for citrus fertilization. *Nutrient CyclAgroexcosyst.* 52:67–74.
58. Quiñones, A., Bañuls, J., Primo-Millo, E., Legaz, F. (2005). Recovery of the ^{15}N-labelled fertilizer in citrus trees in relation with timing of application and irrigation system. *Pl. Soil* 268:367–376.
59. Quiñones, A., Martínez-Alcántara, B., Legaz, F. (2007). Influence of irrigation system and fertilization management on seasonal distribution of N in the soil profile and on N-uptake by citrus trees. *Agri. Ecosyst. Environ.* 122:399–409.

60. Rondey, D. R., Roth, R. L., Gardner, B. R. (1977). Citrus response to irrigation methods. In: Proc. Intl Soc. Citricuture, Vol.1, pp.106–110.
61. Scholberg, J. M. S., Parsons, L. R., Wheaton, T. A., McNeal, B. L., Morgan, K. T. (2002). Soil temperature, nitrogen concentration and residence time affect nitrogen uptake efficiency in citrus. *J. Environ. Qual.* 31:759–768.
62. Shirgure, P. S., Lallan, R., Marathe, R. A., Yadav, R. P. (1999). Effect of Nitrogen fertigation on vegetative growth and leaf nitrogen content of acid lime. Indian J. Soil Conserv., 27(1), 45– 49.
63. Shirgure, P. S., Marathe, R. A., Lallan Ram, Singh, S. (2000a). Irrigation scheduling in acid lime as affected by different soil moisture regimes. *Indian J. Agri. Sci.70* (3): 173–176.
64. Shirgure, P. S., Srivastava, A. K., Shyam, S. (2000b). Water management in citrus – A review. Agri. Rev., 21(4), 223–230.
65. Shirgure, P. S., Srivastava, A. K., Shyam, S. (2001a). Effect of pan evaporation based irrigation scheduling on yield and quality of drip irrigated Nagpur mandarin. Indian J. Agri. Sci., 71(4), 264–266.
66. Shirgure, P. S., Srivastava, A. K., Singh, S. (2001b). Growth, yield and quality of Nagpur mandarin (Citrusreticulata Blanco) in relation to irrigation and fertigation. Indian J. Agri. Sci., 71(8), 547–50.
67. Shirgure, P. S., Srivastava, A. K., Singh, S. (2001c). Fertigation and Micro irrigation in Nagpur mandarin (Citrus reticulata Blanco). South Indian Hort., 49(Special), 95–97.
68. Shirgure, P. S., Srivastava, A. K., Singh, S. (2001d). Effect of micro-jet irrigation system on growth, yield and fruit quality in Nagpur mandarin. South Indian Hort., 49(Special), 357–359.
69. Shirgure, P. S., Srivastava, A.K., Singh, S. (2001e). Effect of nitrogen fertigation and band placement fertilizer application on soil -leaf nutrient build-up and incremental growth of acid lime. J. Soil Water Conserv., 45(3&4), 176–181.
70. Shirgure, P. S., Srivastava, A. K., Singh, S. (2001f). Effect of drip, microjets and basin irrigation method on growth, soil and leaf nutrient change in acid lime. Indian J. Soil Cons., 29 (3), 229–234.
71. Shirgure, P. S., Srivastava, A. K., Singh, S. (2002a). Economics of drip and fertigation in acid lime orchards. J. Soil Water Conserv., 46(1), 56–60.
72. Shirgure, P. S., Srivastava, A. K., Singh, S. (2002b). Effect of micro-irrigation systems and basin irrigation on growth and yield of acid lime (Citrus aurantifolia Swingle). Indian J. Citriculture., 1(2), 154–161.
73. Shirgure, P. S., Srivastava, A. K., Singh, S. (2003a). Evaluating micro-irrigation systems in Nagpur mandarin under sub-humid tropical climate. Trop. Agr., 80(2), 91–96.
74. Shirgure, P. S., A. K., Srivastava, Shyam, S. (2003b). Irrigation scheduling and fertigation in acid lime (Citrus aurantifolia Swingle). Indian J. Agr. Sci., 73(7), 363–7.
75. Shirgure, P. S., Srivastava, A. K., Singh, S., Pimpale, A. R. (2004). Micro irrigation scheduling growth, yield and quality of acid lime (Citrus aurantifolia Swingle). Indian J. Agr. Sci., 74(2), 92- 4.
76. Shirgure, P. S., Srivastava, A. K., Singh, S. (2004a). Growth, yield and quality of acid lime under pan evaporation based micro irrigation scheduling. Indian J. Soil Conserv., 32(1), 32–35.
77. Shirgure, P. S., Srivastava, A. K., Singh, S. (2004b). Integrated water and nutrient management in acid lime. Indian J. Soil Conserv., 32(2),148–151.
78. Shirgure, P. S., Srivastava, A. K. (2012). The effect of four under tree micro-jet irrigation (180–300^0) systems on fruit yield and quality of Nagpur mandarin in Central India. Sci. J. Agr., 1(7), 177–186.
79. Shirgure, P. S., Srivastava, A. K., Singh, S. (2012). Potassium (K) fertigation using Sulphate of Potash (SOP) in Nagpur mandarin. Abstract National Dialogue on Citrus, Improvement, Production and Utilization conducted by National Research Centre for Citrus, Nagpur in the silver jubilee celebration during 27–29th February, 2012, p. 162.
80. Shirgure, P. S. (2012a). Effect of pulse irrigation scheduling with hybrid station controller on fruit yield and quality of Nagpur mandarin (Citrus reticulate Blanco). Sci. J. Crop Sci., 1(5), 76–82.

81. Shirgure, P. S. (2012b). Micro-irrigation systems, automation and fertigation in Citrus. Sci. J. Rev., 1(5), 156–169.
82. Shirgure, P. S. (2013a). Yield and fruit quality of Nagpur mandarin (Citrus reticulata Blanco) as influenced by evaporation based micro irrigation schedules. Sci. J. Crop Sci., 2(2), 28–35.
83. Shirgure, P. S., Srivastava, A. K. (2013). Tree growth, leaf nutrient status, fruit yield and quality of Nagpur mandarin (Citrus reticulate Blanco) as influenced by potassium (K) fertigation with four potash fertilizer sources. Scientific Journal of Crop Science, 2 (3): 36–42
84. Shirgure, P. S. (2013b). Citrus fertigation – a technology of water and fertilizers saving. *Scientific Journal of Crop Science*, 2(5) 56–66
85. Simpson, G. H. (1978). Developments in under tree irrigation systems in the Murray valley. In Proc. Intl Soc. Citriculture, pp. 234–235
86. Singh, B. R., Kanehiro, Y. (1969). Adsorption of nitrate in amorphous and *kaolinitic* Hawaiian soils. (Ion exchange). *Proc. Soil Sci. Soc. Am.* 33(5):681–683.
87. Smith, P. F. (1966). Citrus Nutrition In: Temperate and tropical fruit nutrition. N. F. Childers (ed.) Hort. Pub State Univ New Brunswick, New Jersey, Chap. VII:174–207.
88. Srivastava, A. K., Shirgure, P. S., Singh, S. (2003). Differential fertigation response of Nagpur mandarin (*Citrus reticulata* Blanco) on an alkaline Inceptisol under sub-humid tropical climate. *Trop. Agri.* Trinidad 80(2):97–104.
89. Srivastava, A.. K., Singh, S. (1999). Recent trends in citrus production research. In: Proc. Int. Symp. Citriculture, National Research Centre for Citrus, Nagpur Nov. 23–27, pp. 339–357.
90. Srivastava, A. K., Singh, S. (2003). Diagnosis of nutrient constraints in citrus. In: Manual No. 2. National Research Centre for Citrus, Nagpur Maharashtra, India, p. 1–70.
91. Syvertsen, J. P., Smith, M. L. (1996). Nitrogen uptake efficiency and leaching losses from lysimeter-grown citrus trees fertilized at three nitrogen rates. *J. Am. Soc. Hort. Sci.* 121(1):57–62.
92. Syvertsen, J. P., Jifon, J. L. (2001). Frequent fetigation does not affects citrus tree growth, fruit yield, nitrogen uptake, and leaching losses. *Proc. Fla. Sta. Hort. Soc.* 114:88–93.
93. Tucker, D. P. H., Davis, R. M., Wheaton, T. A., Futch, S. H. (1990). A nutritional survey of south central, southwest and east coast flat-woods citrus groves. *Proc. Fla. Sta Hort. Soc.* 103:324–327.
94. Tuzucu, O., Cevik, B., Kaplankiran, M., Yesiloglu, T., Eebeci, Z. (1997). The effect of different irrigation methods on root distribution of Kudi-ken lemon cv in Adana, Turkey. In: Proc. 5th World Congress, Int. Soc. Citrus Nurseryman, Montepillier, France, Mar. 5–8, 1997.
95. Vivancos, A. D. (1996). Fertigación 2ed. Madrid. Ediciones Mundi-Prensa. p 233.
96. Weinert, T. A., Thompson, T. A., White, S. A., Maner, M. A. (2002). Nitrogen fertigation of young Navel orange: Growth, N status and uptake of fertilizer N. *HortSci.* 37:334–337.
97. Wild, A. (1992). *Russell's soil conditions and tree growth.* Mundi Prensa (ed.) (Madrid).
98. Willis, L. E., Davis, F. S. (1990). Fertilization, nitrogen leaching and growth of young Hamlin orange trees on two rootstocks. *Proc Fla Sta Hort. Soc.* 103:30–37.
99. Willis, L. E., Davies, F. S., Graetz, D. A. (1991). Fertigation and growth of young Hamiln orange trees in Florida. *Hort Sci.* 26:106–109.
100. Zanini, J. R., Villas Bôas, R.,L., FeitosaFilho, J. C. (2002). Uso e manejo da fertirrigação e hidroponía. Jaboticabal Funep, 65 p.
101. Zhang, M., Alva, A. K., Li, Y.C., Calvert, D. V. (1996). Root distribution of grapefruit trees under dry granular broadcast vs. fertigation method. *Pl. Soil* 183:79–84.

CHAPTER 6

MICRO IRRIGATION POTENTIAL IN FRUIT CROPS: INDIA

R. K. SIVANAPPAN

CONTENTS

In this chapter, the conversion rate for Indian currency is Rs. 64.00 = US $1.00 on September 30, 2013. The prices of drip irrigation system/crops are based on 1999–2000 prices. Information in this chapter is based on my onsite visits and interviews with farmers/scientists in India. I acknowledge the consultancy opportunity by Water Technology Center at TNAU – Coimbatore; Central and State Governments in India; NGOs in India; and the International Development Enterprises.

6.1 MICRO IRRIGATION IN INDIA

Micro irrigation (MI) system consists of: pump head, main, submain, lateral, drippers, fertigation equipment, filtration system, and other accessories to deliver the required quantity of water near the root zone of crop. Filters are necessary to prevent clogging of the drippers. Fertilizers can be applied through venturi or fertilizer tanks to avoid wastage of this costly commodity. In India, micro sprinklers and micro sprayers are also available to provide the water around the root zone of the tree crops (fruit/orchards). The water consumption and crop yield for MI and conventional methods is given in Table 1.

TABLE 1 Water used and yield for various crops in drip and conventional irrigation methods in India (1990).

Crop	Yield (100 Kg per ha)			Water depth applied (cm)		
	Conventional	Drip	Increase in yield (%)	Conventional	Drip	Water Saving (%)
Banana	575.00	875.00	52	176.00	97.00	45
Grapes	264.00	325.00	23	53.20	27.80	48
Mosambi (,000)	100.00	150.00	50	166.00	64.00	61
Pomegranate (,000)	55.00	109.00	98	144.00	78.50	45
Papaya	13.40	23.48	75	228.00	73.30	68

Source: Ref. [5].

The development of the system has been slow and is adopted only in the South and North Western States of India, due to water scarcity and maintenance of sustainable agriculture. The area has increased from almost zero in 1970 to 1,000 ha in 1985; and 60,000 ha in 1995; 225,000 ha in 1998; 400,000 ha in 2002 covering about 30 crops in India. The experience of numerous farmers in micro irrigation technology has revealed that the old coconut farms are not possible to irrigate as the ground water is depleting. Numerous farmers in Coimbatore district have taken up this irrigation method for the coconut trees. Now water is not available even for these systems.

In Kerala state, the coconut and other plantation crops need water during January to May, and the farmers are introducing micro irrigation due to the shortage of water. The experiences in Karnataka and Andhra Pradesh states are also encouraging, especially for grapes, coconuts and other fruit crops. It is slowly catching up in Gujarat, Rajasthan and Madhya Pradesh states. This system is not familiar (not adopted) in Northern and North Eastern States. This method is very well suited to the undulated terrain and for plantation crops like tea, cardamom, rubber, etc., to get more yield. Therefore, the farmers should be educated to adopt this system: to get more income, to save fertilizer and labor.

Micro irrigation is thought to be an expensive venture. The social acceptability is also decision making factor. The large scale adoption involves a crucial question of economic viability. The cost of the system depends on: the crop type, row spacing, crop water requirements, location of water source etc. The cost is about Rs 15,000/ha for coconut/mango (wide spaced crops).

The benefit-cost ratio (BCR) for drip system was worked out by interviewing the farmers in Maharashtra and Tamil Nadu states. The range of BCR excluding the proposition of water saving for grape was about 13.35. If water saving is considered, the BCR range goes up to 33.00 for grapes (Table 2).

TABLE 2 Benefit-cost ratio (BCR) for tree crops and vines under drip irrigation in India.

Crop	Row spacing m × m (ft × ft)	Cost of the MI system Rs/acre	Benefit-cost ratio, BCR	
			Excluding water saving	Including water saving
Coconut	7.62 × 7.62 (25′ × 25′)	7,000	1.41	5.14
Grapes	3.04 × 3.04 (10′ × 10′)	12,000	13.35	32.32
Grapes	2.44 × 2.44 (8′ × 8′)	16,000	11.50	27.08
Banana	1.52 × 1.52 (5′ × 5′)	18,000	1.52	3.02
Orange	4.57 × 4.57 (15′ × 15′)	9,000	2.60	11.05
Acid lime (citrus Sp.)	4.57 × 4.57 (15′ × 15′)	9,000	1.76	6.01
Pomegranate	3.04 × 3.04 (10′ × 10′)	12,000	1.31	4.04
Mango				
Papaya	7.62 × 7.62 (25′ × 25′)	7,000	1.35	
	2.13 × 2.13 (6′ × 6′)	18,000	1.54	8.02
				4.01

The farmers have been compelled to opt for the advanced method of irrigation, due to limited available water resources and high demand of water from all sectors (industries and drinking). The awareness of the farmers to increase the production and income has kindled them to use the water more efficiently. It is reported that drip irrigation farmers in Maharashtra and other states are able to get a net profit of Rs 50,000 to 100,000 per ha (0.40 ha = one acre) by growing grapes, orange, pomegranate and other fruit crops. Similarly, the yield of tea crop has increased by about 30% with drip irrigation in the summer dry months. In spite of above facts, the area under drip is very meager in India. Therefore, it is projected to have 1 million ha in 2005 that is 1% of the irrigated area and about 10 million ha in 2025. The cost of MI for 1 million ha will be about Rs 40 billion (average Rs 40,000 per ha). This financial investment is not much compared to the benefits achieved by increasing the crop yield. It will also generate employment opportunity on large scale.

6.1.1 SUSTAINABLE MICRO IRRIGATION TECHNOLOGY

The net cultivated/sown area in India is about 145 Mha. The gross sown area will be about 210–220 Mha in another 10 to 15 years. The irrigation potential is about 85 Mha at the present, though the net irrigated is only about 82.0 Mha. The potential utilizable water is about 112 Mhm (surface water 69 Mhm + ground water 43 Mhm), which may be sufficient to irrigate about 135–140 Mha. Therefore, even with full exploitation, about 40–50% of the cultivated area will remain as rain-fed agriculture.

The per hectare investment cost for an irrigation project has increased enormously from about Rs 1500 during first five year plan (1951–1956) to more than Rs 100,000 during 10th plan period (2002–2007). Therefore, it is necessary to economize on the use of water for agriculture to bring more under irrigation and reduce the cost of irrigation per hectare.

Micro irrigation system is very well suited for undulated terrain, shallow soils and water scarce areas. Saline/brackish water can also be used to some extent, since water is applied daily, which keeps the salt stress at minimum and the salt will be pushed to the periphery of the moisture regime which is away from the root zone of the crop. Therefore, it does not affect the crop growth.

6.1.2 NEED FOR MICRO IRRIGATION

The total rainfall is about 400 Mhm in India, but the utilizable quantity is only about 112 Mhm both from surface and ground water. Even with full exploitation of the potential, nearly 40–50% of the cultivated area will remain under rain-fed. In addition, the ground water table in many parts of India is depleting year after year at a rate of about one meter per year. In many parts India, thousands of wells are abandoned in view of the alarming depletion of water in the wells.

In Tamil Nadu state, it is reported that more than 150,000 open wells are abandoned for lack of water. The crop yield per hectare is not comparable with other developed and even in some developing countries for all crops including paddy, vegetable, fruit, etc. Therefore, it is necessary: to economize the use of water for agriculture to bring more area under irrigation; to reduce the cost of irrigation per hectare; and to increase the productivity per unit area from unit quantity of water. This can be achieved by introducing advanced irrigation methods like micro irrigation with improved water management practices.

The experiments/ studies conducted in various research institutions in India have indicated that the water saving for any crop is about 40–70% and the crop yield has increased up to 100% (i.e., double the yield). In spite of high installation cost of MI system, the economics were worked out by the author for various crops and it is viable. The payback period varies from 6–24 months and the BCR ratio is about 2.0 to 7.0 (Table 3).

TABLE 3 Benefit-cost ratios (BCR) and payback periods for selected crops under micro irrigation in India (1996).

Variable	Crops					
	Banana 5′×5′ 3′×5′ ×6′	**Papaya 6′×6′ 8′×8′**	**Grapes 10′×6′**	**Pome-granate 14′×14′**	**Ber 15′×15′**	**Strawberry 9′×12′ ×9′**
System Cost, Rs/acre	19,000	16,000	17,000	12,000	12,000	75,000
System Cost, Rs per hectare	47,500	40,000	44,000	30,000	30,000	188,000
Water used	15–20 Lpd/ acre	15 Lpd/P	12–20 Lpd/P	50–60–60 Lpd/P 12,000–15,000 Lpd/acre		2 Lpd/p1
Yield tons/acre	30 tons	750 kg Latex 60 tons fruit	20 tons	9 tons	10 tons	3 tons
Payback period, months	12	18 one crop season	<12	<12	<12	24 or two season
B.C.R.	3.08	4.09	3.64	7.03	6.51	2.34
Extra Income* Rs	49,320	72,040	2,64,200	1,51,280	1,33,280	67,000

Note: *Due to drip irrigation over conventional method. B.C.R. = Benefit-cost ratio.

Source: Ref. [15].

Therefore, the advanced method can play an important role in improving the living condition of the farmers in the country. Further to popularize the system, the Central and State Governments in India are giving subsidy. It was up to 70–90% but now it has come down to 25%. In addition, 90% of the subsidy is borne by the Central Government and only 10% is borne by the State Government. However, it is not available for all crops, which are suitable for drip irrigation. Another constraint is that the system and its usefulness is not known among farmers and even to Government officials who are in charge of popularizing drip irrigation, that is, agriculture and irrigation department officials. But the main constraint is its cost especially to the small and marginal farmers in India who comprise about 83% of total number of farmers and get only about 35% by income.

6.1.3 ADVANCES IN MICRO IRRIGATION AND FERTIGATION IN INDIA

Table 4 indicates water productivity gains due to shifting from surface to drip irrigation in India. Drip system costs about Rs 20,000–75,000/ha depending upon the crop, topography, source of water supply etc. Most farmers think it to be expensive. However, the economic studies by author indicate that the system is economically viable (Table 5). This may be the case in many developing countries. Scientists/NGO's are researching to bring down the cost of the system so that the poor and small farmers can afford the system. It is possible to reduce the cost by about 40–50% by proper geometry of the crop planting, irrigating 2 to 4 rows by providing suitable micro tubes on both sides of the laterals and by proper design and layout of the system. For popularizing this low cost/low energy system in India, field trails are being conducted in farmer's field in large scale for orchard crops by Universities and NGO etc. and the response are very much encouraging.

TABLE 4 Water productivity gains from shifting to drip irrigation from surface irrigation in India (1994).

Crop	Change in yield	Change in water use, %	Change in water productivity
Banana	+52	–45	+173
Grapes	+23	–48	+134

Source: Sandra Postel – Pillar of sand – 1999. Adapted from Refs. [18, 13].

TABLE 5 Benefit-cost ratio (BCR) and payback period for various crops under micro irrigation (1993).

Crops	Crop spacing	Cost of the system	Water used	Yield	Payback period	BCR
	m	Rs./ha	Lpd/plant	Tons/ha	months	
Banana	0.91×1.5×1.8 pair row	47,500	15–20	75	12	3.00
Grape	3.03×1.8	44,000	15–20	45	<12	3.28

TABLE 5 *(Continued)*

Crops	Crop spacing	Cost of the system	Water used	Yield	Payback period	BCR
	m	Rs./ha	Lpd/plant	Tons/ha	months	
Pomegranate	4.3×4.3	30,000	50–60	25	<12	5.16
Ber	4.5×4.5	30,000	60	25	12	4.56
Papaya	1.81×1.81	40,000	15	60	12	4.09

Source: Case studies conducted by the author with numerous farmers in Maharashtra State, November 1993 [2].

6.1.4 FUTURE PROSPECTS

The studies conducted and information gathered from various farmers revealed that drip irrigation is technically feasible, economically viable and socially acceptable. Drip irrigation can be implemented in most of the areas irrigated by open/ tube wells, which make about 35% of the total irrigated area in the country. The drip irrigation can be extended to the following category of lands:

1. Waste lands after planting tree crops including fruit trees
2. Hilly area
3. Semi arid Zones
4. Coastal sandy belts
5. Water scarcity areas
6. Command area of the community wells
7. Wind mill (farm) areas.

At present, on an average about Rs. 100,000 is invested to bring one hectare of land under irrigation in the new irrigation project if water is available. As water is becoming increasingly scarcer, adoption of micro irrigation system offers potential for bringing nearly double the area under irrigation with the same quantity of water.

TABLE 6 Areas (Mha) sown and irrigated: Suitable for drip irrigation in India, (2000).

Crop	Present area, 2000		Expected area, 2020/25	
	Sown	Irrigated	Sown	Irrigated
Coconut/Aricanut	1.5	0.9	2.0	1.0
Fruits	4.0	1.2	4.2	2.2
Plantation crops	2.8	1.0	3.0	1.6
Total	8.3	3.1	9.2	4.6

It has been considered as a boom for wide spaced perennial crops namely, mango, coconut, banana, grapes, pomegranate, ber, citrus, tea, coffee, cardamom and the like. The details of cultivated and irrigated areas under fruits, plantations crops at present and in the year 2020/25 are given in Table 6, which gives an idea of large potential area to bring under drip irrigation system in the coming years.

6.2 MICRO IRRIGATION IN FRUIT CROPS

6.2.1 INTRODUCTION

Water is a prime natural resource to achieve number of significant functions. Unlike most other natural resources, water does not have a substitute in its main uses. It can be used more or less lavishly or efficiently, but it cannot be replaced. It is indispensable, finite and vulnerable resource. Virtually no activity in society or process in the landscape or in the environment is possible in the absence of water. Water is one of the plentiful resources. Covering more than 2/3 of the earth, water travels from the sea into the air to the land and back to the sea in a seemingly endless cycle of renewal (Hydrologic cycle).

Yet water is a finite resource, and the tiny fraction suitable for drinking or irrigating crops is distributed unevenly throughout the regions. At the same time, the human needs for water is escalating because of rapid population and industrialization, especially in the regions where water is a most scarce. Between 1940 and 1990, world population has more than doubled, from 2.3 to 5.3 billion and the per capita use of water has also doubled from 400 to 800 m^3/year. Hence, the global water use has increased by more than 4 times during this period. In many of the regions of the world, population is growing more rapidly; the needed water is simply unavailable. The critical limits are not at the global level but at regional, national and local levels. The area of irrigated land worldwide nearly doubled in the first half of the twentieth century (from 48 Mha to 94 Mha). Land under irrigation has nearly trippled (260 Mha).

Worldwide agriculture is a single biggest user of water supply, accounting for about 69% of all use. About 23% of water is used to meet the demands of industry and just 8% to domestic use. Pattern of use varies greatly from country to country, depending on factors such as: Economic development, climate and population. As an example, India and Africa consume about 90% of water for agriculture, (irrigation) while highly industrialized countries in Europe allot more than half the water for industry and energy production. Only in the recent years, the growth has slowed down. In California and some parts in India, farmers are selling their land and the accompanying water rights to the metropolitan area with huge demand. The proportion of water used for industrial purposes is often seen as an indicator of economic development.

According to the data available on the global water supply, there seems to be no lack of fresh water worldwide (Table 7). However, it must be taken into account that in individual region, because of the spatial and temporal variations in precipitation, the potential usable water supply is very small. The most frequently used criterion for assessing the availability of the renewable water supply is the per capita supply, a supply of less than 500 M^3 per capita per annum being regarded as the critical lower limit, 1000 M^3 per capita as very low, 2000 M^3as critical. According to a study carried out by the World Resource Institute it is especially the countries which are in or near semiarid regions that are in a critical situation. Based on a supply of less than 500 M^3 per capita per annum, the most endangered regions are North Africa with 3 and the Middle East with 8 countries as threat. Twenty-two of the countries in the world currently [2] have renewable water supplies of less than 1000 M^3 per capita per year. The World Bank estimates that by the year 2025 one person in three, in other words 3.25 billion people, in 52 countries will live in conditions of water shortage.

TABLE 7 Global water supply for earth (1999).

Region	Km^3/a	mm	$M^3/p/year$
Europe	3110	319	4410
Asia	13190	293	4130
Africa	4225	139	6581
North America	5960	287	13925
South America	10380	583	34949
Australia	1965	225	75577
	38830 (Total)	294 (Avg)	7337 (Avg)

Source: National Resource and Development, Focus – Water the life line of our future, Vol. 49/50, Institute of Scientific Co-operation, Tubingen, Germany.

The major potential of drip irrigation is for fruit crops where the system can provide a substantial water economy and better productivity. Further the cost of the system will be reasonable, economical, and viable. India has a total area of about 3.5 Mha of fruit crops producing about 42 metric tons per year. The major fruit crops are mango, apple, guava, pineapple, grapes, papaya etc. where good water economy can be affected if drip irrigation is used with technical and scientific recommendations these fruit crops. The area of different fruit crops and their production in India is given in Table 8.

TABLE 8 Area and production of fruit crops in India.

Crop	Area $\times 10^5$ha	Production $\times 10^5$ tons
Mango	12	110
Banana	4.45	130
Citrus	4.5	38.0
Apple	2.2	12
Guava	1.3	15
Pineapple	0.71	10.7

TABLE 8 *(Continued)*

Grapes	0.35	6.0
Papaya	No data	3
Other fruits:		
Ber, Pomagranate, Custard apple, Strawberry, Sapotaetc	5	65

Source: NCPA, Perspective plan for Drip and Sprinkler Irrigation (1990–2000) [5].

6.2.2 RESEARCH ADVANCES IN SUSTAINABLE MICRO IRRIGATION IN FRUIT CROPS

The results of the survey conducted on micro irrigation for fruit crops has revealed that in Europe, the yield increase was in the range of 10–50% as a result of switching to micro irrigation and the water saving was significant (20–25%) compared to sprinkler irrigation, and was 40–60% compared to surface methods. France reported a similar trend from mini-sprinkler to drippers in Orchards.

However, in Australia and South Africa the trend was opposite with an expanding use of microsprinkler/sprayer in fruit orchards. The major problem reported by most of the countries has been clogging of drippers which were overcome in most cases by installation of efficient filters. In some cases, injection of chemicals was necessary to overcome buildup of algae or carbonate/iron compounds in the lines and drippers. The experiments conducted in India for fruit crops at various Agricultural Universities/Research Institution have revealed that the water savings is about 40–70% and yield increase varied from 20–100%. The author has analyzed economics of micro irrigation system, BCR ratios, etc., by interviewing farmers/scientists in Maharashtra, Tamil Nadu, Karnataka at two different occasions (1990 and 1993–94). The results are summarized in Tables 9 and 10.

TABLE 9 Benefit cost ratio for various fruit crops under micro irrigation (1990).

Crops	Spacing	Benefit-cost ratio, BCR	
	m × m	Excluding water saving	Including water saving
Grapes	3×3	13.35	32.32
	8×8	11.50	27.08
Acid lime	4.57×4.57	1.76	6.01
Banana	1.52×1.52	1.52	3.02
Mango	7.62×7.62	1.35	8.02
Orange	4.57×4.57	2.60	11.05
Papaya	1.84×1.84	1.54	4.01
Pomegranate	3.04×3.04	1.31	4.04

Source: Ref. [12].

TABLE 10 Benefit – cost ratio and payback period for fruit crops under micro irrigation (1993–94).

Crop	Spacing, m × m	Payback period	Benefit – cost ratio
Banana	0.91×1.5×1.8	One year	3.00
Grapes	3.03×1.8	One year	3.28
Pomagranate	4.3×4.3	One year	5.16
Ber	4.5×4.5	One year	4.56
Papaya	1.80×1.80	One year	4.09

Source: R K. Sivanappan, Case study with number of farmers in Maharashtra, 1993–94.

At Gujarat Agricultural University in a 14 ha farm, thr research experiments were conducted having various fruit crops namely: Mango, sapota, ber, guava, pomegranate, acid lime, sonala, phalsa (or falsa, Indian word for *Grewia asiatica*), sweet orange, mandarin, coconut. The research studies indicated: Higher yield; reduction in farm labor; weed/pest; huge water saving; and a better fruit quality. Experiments conducted at Dapoli has revealed that irrigation for mango applied at 60 L/tree/week through MI produced 152.7% higher yield compared to manual watering with equivalent amount of water. For Pomegranate, MI gave comparable yield (6.84 T/Ha) and saved 45% water over check basin method of irrigation as indicated by the scientists of Agricultural University at Rahuri. Several studies carried out at the UAS, Bangalore on fertigation of fruit crops (grapes/sapota) have revealed that application of 80% of water soluble fertilizers were superior over the conventional method of application of 100% normal fertilizer. The experiment conducted at Marathwada Agricultural University – Parbhani has indicated that drip irrigation for banana crop gave better yield than surface/conventional method of irrigation.

6.2.3 FUTURE OF MICRO IRRIGATION FOR FRUIT CROPS

The irrigated area at present is about 12×10^5 ha of which only about 1.35×10^5 ha is under drip irrigation (Table 11).

About 2 Mha of fruit crops under micro irrigation by 2020/25 will require detailed and phrased plans by the Governments, NGO sand manufacturers, and determination on the part of farmers. Furthermore, micro irrigation should be supported by the suppliers and extension staff to help farmers to maintain and operate their system properly.

TABLE 11 Area under drip for fruit crops (1999).

Crop	Area ha	Crop	Area ha
Amla	220	Guava	4,930
Banana	24,565	Mango	21,863

TABLE 11 *(Continued)*

Crop	Area ha	Crop	Area ha
Ber	4,700	Papaya	2,115
Citrus	22,210	Pomegranate	19,250
Custard apple	810	Sapota	5,125
Grapes	29,630	Strawberry	170
Total for all crops		**1,34,588 Ha**	

Source: Proceedings of the All India Seminar on Micro Irrigation. Prospects and potential in India held in June 1999 at Hyderabad.

6.3 SUMMARY

Micro irrigation is well suited for fruit crops but it has not been fully exploited. In India, the area of irrigation of fruit crops is only about 30% and the average productivity is not even in sufficient quantity prescribed for the population, though there is tremendous scope in extending the area of irrigation for fruit crops. Drip irrigation can increase productivity and also the quality of fruits.

Micro irrigation is very well suited for fruit crops but it has not been fully exploited. In India, the area of irrigation of fruit crops is only about 30% and the average productivity is minimum. The fruit production is not even sufficient for the entire Indian population, though there is tremendous scope for the total demand by extending the area of irrigation for fruit crops. Drip irrigation can increase productivity and also the quality of fruits. Therefore, it is planned to bring at least 2 Mha under micro irrigation in the year 2020–2025. This will not only meet the demand of the population and at the same time, it will fetch the much-required foreign exchange by exporting the fruits.

KEYWORDS

- **accepted method**
- **action plan**
- **advanced method**
- **advantages**
- **affordable**
- **availability**
- **clogging**
- **command area**
- **commercial crops**
- **compact**
- **confidence**
- **conveyance and distribution efficiency**

- **cost benefit**
- **cost, drip system**
- **cultivable land**
- **development**
- **drippers**
- **economic development**
- **education and training**
- **emitter**
- **emitter spacing**
- **evaporation**
- **expensive**
- **exploit**
- **fertigation**
- **filter**
- **food security**
- **foreign exchange**
- **fruit crops**
- **great potential**
- **increased yield**
- **India**
- **inefficient extension**
- **investment**
- **investment cost**
- **irrigation investment**
- **Jain irrigation**
- **lay out**
- **low cost drip system**
- **maintenance**
- **overall efficiency**
- **plantation crops**
- **plastics in agriculture**
- **precision farming**
- **sustainability**
- **Tamil nadu – India**
- **target area**

- **technical feasibility**
- **technology**
- **uniformity**
- **water economy**
- **water management**
- **water saving**
- **water use efficiency**
- **weed growth**
- **yield improvements**

REFERENCES

1. Abbot, J. S., Micro Irrigation: World Wide Usage, (Personal commuication). Section 5.
2. Engelman, Robert, Pamela Levoyt, (1993). Sustaining water Population and the future of renewable water supply. Population Action International. Section 3.
3. Institute for Scientific Co-operation. Natural Resources and Development, Focus: Water – the life line of our future. Volume 49/50, Tubingen, Germany, (1999). Section 5 and 3.
4. Keller, Jack, (2000). Gardening with low-cost drip irrigation in Kenya: For health and profit. Technical Report prepared for International Development Enterprises (IDE), http://www.siminet.org/fs_start.htm.
5. National Committee on the Use of Plastics in Agriculture (NCPA), (1990). Status, Potential and Approach for Adoption of Drip and Sprinkler Irrigation Systems. Pune – India. Section 1.
6. Polak, Paul, Nanes, Bob, Adhikari, Deepak. A Low Cost Drip System for Small Farmers in Developing Countries. Journal of the American Water Resources Association, February (1997), Vol33, No. 1.
7. Polak, Polak, Sivanappan, R. K (2001). The potential contribution of low cost drip irrigation to the improvement of irrigation productivity in India Technical Report prepared for International Development Enterprises (IDE), http://www.siminet.org/fs_start.htm.
8. Postel, Sandra (1999). Pillar of Sand: Can the Irrigation Miracle Last. WW Norton & Co, New York. Section 3.
9. Proceedings of the XI International Conference on the Use of Plastics in Agriculture, New Delhi – India, (1990). Section 5.
10. Proceedings of VIII IWRA World Congress on Water Resources (1994). National water resources Center, Cairo – Egypt. Section 3.
11. Proceedings of the National Seminar on Micro Irrigation and Sprinkler Irrigation Systems, Delhi – India, April 28–30 (1998).
12. Sivanappan, R. K. (1990). Constraints and potential in popularizing drip and sprinkler irrigation system. National committee in the Use of Plastics in Agriculture, New Delhi - India. Section 5.
13. Sivanappan, R. K. (1994). Prospects of Micro Irrigation in India. Journal of irrigation and drainage System. 8:49–58. Section 5.
14. Sivanappan, R. K. (1994). Status of drip irrigation in India. Journal Irrigation & Drainage Systems, 8:49–58. Section 3.
15. Sivanappan, R. K. (1996). Strengths and weakness of growth of drip irrigation in India, Paper presented at the National seminar organized by the National Water Development agency at Bhopal. SECTION 2.
16. Sivanappan, R. K. (1999). Scope for micro irrigation in India. Proceedings of the National seminar on Problems and prospects of Micro Irrigation – A critical Approach, IE(I) Bangalore, pages 12–23. Section 3.

17. Sivanappan, R. K. (2000). Micro irrigation including low energy precision applications. International Conference on Management of Water Resources for 21st century, New Delhi – India. Section 3.
18. Sivanappan, R. K., Rao, A. S., Dikshit, N. K. (1994). Drip Irrigation in India. Indian National Committee on Irrigation and Drainage, Jolly Reprographics, New Delhi – 110 008.

CHAPTER 7

QUALITY OF MUNICIPAL WASTEWATER FOR MICRO IRRIGATION

VINOD KUMAR TRIPATHI, T. B. S. RAJPUT, NEELAM PATEL, and LATA

CONTENTS

7.1 INTRODUCTION

Worldwide, availability of good quality water for irrigation sector is expected to decline as the requirement of fresh water for all other sectors (domestic, industry, power, inland navigation, ecology) increases [29]. Therefore, it is greatly essential to reduce the fresh water consumption in irrigation sector by adopting efficient methods of irrigation and making reuse of waste water (WW) generated as by-product from other sectors for irrigation. In India, WW generation from Class I and Class II cities are 38,254 million liters per day [10]. This huge quantity of WW gives opportunity for its reuse in agricultural sector to mitigate water demand for irrigation in water scares areas. Numerous groups have described future issues that must be addressed to ensure water quantity, quality, security and controlling emerging contaminants and health risk with protection of environment. There is an urgent need to focus on integrated management of WW use to ensure sustainability of water quality and quantity for future generation [26].

It was estimated that about 20 million hectares of land worldwide was irrigated with untreated WW [21, 27]. The use of untreated WW (or polluted water) poses risks to human health since it may contain excreta related pathogens (viruses, bacteria, protozoan and multi cellular parasites), skin irritants and toxic chemicals like heavy metals and pesticide residues. When WW is used in agriculture, pathogens and certain chemicals are the primary hazards to human health. The risk for human health is mainly with consumption of WW grown produce. Outbreaks of food borne illness throughout the world are increasingly linked to consumption of contaminated fruits and vegetables [8, 17, 18]. Bacterial human pathogens such as *Escherichia coli* O157:H7, *Salmonella* and *Listeria monocytogenes* have been demonstrated to be involved in such outbreaks of food borne illness [6, 7]. But WW reuse for agricultural purposes is now considered an important resource for either regions with high demand or low supply or areas vulnerable to macronutrients in several European countries [24].

Oron et al. [25] concluded that poliovirus can penetrate into the plant through the root system. Water as a medium plays a tremendous role in differential distribution of pathogens in soil and plant tissues. Vaz da Costa Vargas et al. [32] observed that when poor quality WW (trickling filter effluent with 10^6 thermotolerant coliforms per 100 mL) was used to spray-irrigate lettuces, the initial concentrations of indicator bacteria exceeded 10^5 coliforms per 100 g fresh weight. Once the irrigation ceased, no Salmonella could be detected after five days, and after 7–12 days, thermotolerant coliform levels were similar to or just above the level seen in lettuces irrigated with fresh water. The crop quality was better than that of lettuces irrigated with surface waters on sale in the local markets (10^6 thermotolerant coliforms per 100 g), presumably because of recontamination in the market through the use of contaminated water for spray of vegetables. The lettuces irrigated in uncovered plots had high level of bacterial contamination, unless a period of cessation of irrigation occurred 7–12 days before harvest. Islam et al. [20] observed no detectable populations at harvest for onions (day 140) but detectable populations at harvest for carrots (day 126). Pre-harvest contamination of carrots and onions with E. coli O157:H7 for several months can occur through contaminated manure (compost) and irrigation water. Hence, the type of crop, its texture and type of leaves/fruits can influence the retention of coliforms and their differential distribution.

Studies on drip and furrow irrigation of radishes and lettuces by Bastos and Mara (1995) with waste stabilization pond effluent (1.7×10^3 to 5.0×10^3 coliforms per 100 mL) indicated that crop quality was better under dry weather condition with 10^3–10^4 *E. coli* per 100 g for radishes and lettuces and no *Salmonella* was present. In Israel, Armon et al. [4] undertook an study where sprinkler irrigation of vegetables and salad crops with poor quality effluent from WW storage reservoirs (up to 10^7 thermo tolerant coliforms per 100 mL) resulted in high levels of fecal indicator bacteria on crop surface (up to 10^5 thermo tolerant coliforms per 100 mL). However, when vegetables were irrigated with better quality effluent (0–200 thermo tolerant coliforms per 100 mL) from a different storage reservoir, thermo tolerant coliform levels on crops were generally less than 10^3 per 100 g and often lower. Many research workers have focused their attention on survival of pathogens in irrigation water, soil and vegetable produced in different countries and climatic conditions [9, 11, 15, 22, 25, 30].

In India, municipal WW for irrigation is being used mainly for growing vegetables in periurban areas [33], which may pose serious risk of coliform outbreak [12]. Therefore, this research study investigates the possible accumulation of coliforms bacteria in soil under placement of drip laterals at surface and subsurface; and assesses the quality of eggplant fruits in terms of coliforms.

7.2 MATERIALS AND METHODS

7.2.1 EXPERIMENTAL SITE

The experiment was conducted at research farm of Water Technology Centre, Indian Agricultural Research Institute (IARI), Pusa, New Delhi, India which is located within 28°37′22″ N and 28°39′ N latitude and 77°8′45″ E and 77°10′24″ E longitude. The mean annual rainfall is 710 mm of which 75% is received during the monsoon season (July to September). WW used for irrigation of crops were collected from the drain of IARI, which is fed by domestic effluents from individual houses, group houses, hostels and runoff from agricultural field particularly in rainy season and sewage water [31]. The groundwater (GW) was collected from the tubewell, which provides water from more than 30 m below the ground level. Location of WW collection point is shown in Fig. 1.

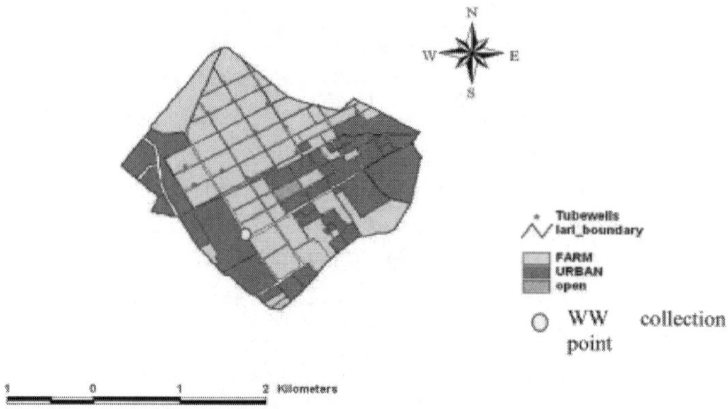

FIGURE 1 Map of IARI, New Delhi – India.

7.2.2 NURSERY RAISING AND CROP PRACTICES

Seedlings of eggplant (cv: *Supriya*) were raised in the month of September, in the plastic tray with the mixture of coco peat, vermiculite and perlite in the ratio of 3:2:1. Experiments were conducted during 2009–2010. WW was not used in the nursery. No Farm Yard Manure (FYM) was added to avoid precontamination with coliforms in the nursery. The 25 days old seedlings were transplanted in the field. Herbicides and pesticides were not applied. Water requirement of eggplant was estimated by calculating reference Evapotranspiration (ET_0) using the Penman-Monteith method and the crop coefficient (K_c) suggested by Allen et al. [2].

7.2.3 DESCRIPTION OF IRRIGATION SYSTEM

Drip irrigation system was installed for WW and GW (ground water) application separately. In-line lateral (J-Turbo Line) with 40 cm dripper spacing was laid on ground for surface and were buried at 15 cm and 30 cm depths from ground surface for subsurface drip. System included sand media filter (F1, flow rate 30 $m^3.h^{-1}$, 50 mm size, silica sand 1.0 to 2.0 mm, thickness 80 cm) with back flush mechanisms, Disc filter (F2, flow rate 30 $m^3.h^{-1}$, 20 mm size, 130 micron, disc surface 1.198 cm^2 screen surface 815 cm^2 AZUD helix system, model 2NR), and venturi injection system for chemigation. Water was passed through filter F1 and F2 alone as well as combination of both the filters (F1 and F2) to improve the quality of WW. The velocity of water was kept minimum to improve the efficiency of filters. Main lines (50 mm diameter, PVC pipe) were connected to submains (35 mm diameter, PVC pipe) for each of the plots through a gate valve.

7.2.4 SAMPLING OF WATER AND SOIL

WW samples from the drain were collected across the drain at the depth 15 cm below the surface and at three points, and then mixed. The preservation and transportation was performed according to the standard methods [3]. Soil samples were collected at time intervals of 25 and 50 days after transplanting and immediately after harvesting from each plot. Approximately 100 g of soil was aseptically collected in a sterile plastic bag from randomly selected plant at the depth of 0, 15, 30 and 45 depth for the surface and subsurface (15 cm) placed drip lateral. However, in case of subsurface drip (30 cm depth of lateral from surface), samples were collected up to 60 cm with the interval of 15 cm from ground surface. Fruit samples of eggplant were also collected randomly from each plot and transported in sterilized bags to the laboratory for analysis. All the soil samples were stored in refrigerator and analyzed within 48 h of collection.

7.2.5 DETERMINATION OF TOTAL COLIFORMS IN SOIL AND FRUIT SAMPLES

Total coliforms were analyzed by Most Probable Number (MPN) method and presented as per gram weight of dry soil/ fruit. The MPN values were determined by MPN table [1]. Soil/ water/ fruit samples were diluted tenfold. Graduated amounts of samples (10, 1 and 0.1 mL) were placed in 5–5 tubes of BCP (Bromo cresol purple) lactose broth with durhams tube. Five tubes for each dilution was incubated at 37°C for 24 h and individual tubes were checked for acid (yellow color) and gas production (Fig. 2). If no gas was present in any of the tubes, the incubation was continued for an additional 24 h.

FIGURE 2 Total coliform detection.

7.3 RESULTS AND DISCUSSIONS

7.3.1 CHEMICAL AND BIOLOGICAL PROPERTIES OF WATER

The WW and GW were analyzed for the physico-chemical and biological properties (Table 1). WW was highly turbid compared to GW but presence of total solids was higher in GW due to higher soluble salts. Higher EC, sodium and chloride content was observed in GW. Macronutrients N, P and K were found to be higher in WW. Available Mg was almost same in both the water samples. Carbonate content in WW and GW were 119 and 58 mg l^{-1} respectively. Population of total coliforms an indicator of fecal contamination was found to be in the range of 1.8×10^1 to 2.6×10^4 MPN mL^{-1}. No coliforms were detected in GW samples. Worldwide, research in 23 laboratories with 1000 strains of coliforms from various types of water has proven that only 61% of the total numbers examined were nonfecal in origin [14].

TABLE 1 Physicochemical and biological properties of water used for irrigation.

Properties	Unit	Waste water	Ground water
		Mean ±SD	Mean ± SD
EC	dS m^{-1}	1.48±0.23	2.17±0.25
pH		7.33±0.35	7.4±0.43
Total Solids	mgL^{-1}	849.8±148.5	967.4±212.6
Turbidity	NTU	46.25±10.23	1.50±0.52
$NO_{3-}N$	mgL^{-1}	4.57±1.91	5.22±0.44
P	mgL^{-1}	2.68±1.45	0.35±0.15
K	mgL^{-1}	26.83±14.78	10.3±2.98
Na	mgL^{-1}	139.91±37.4	287.8±62.4
Mg	mgL^{-1}	33.61±5.5	35.28±5.81
CO_3	mgL^{-1}	119.5±41.69	58.0±8.23
DO	mgL^{-1}	7.37±0.84	7.65±0.92
BOD_5	mgL^{-1}	95.17±19.71	0.725±0.339
COD	mgL^{-1}	139.25±30.7	16.67±4.03
Total coliforms	MPN mL^{-1}	2040± 1085	nd

nd = not detected.

7.3.2 EMITTER PERFORMANCE

Primary treatment of collected WW was done by sedimentation and filtration. WW was allowed to settle for 24 h and upper portion of settled water was used for filtration before application to plants. The coefficient of variation of emitter discharge (CV_q) for different filters and for their combination is presented in Table 2. Maximum CV_q's of 3.49% and 7.28% were observed with WW in surface and subsurface (30 cm) treatments, respectively. The performance of emitters under combination of both filters with WW and GW (ground water) was excellent with less than 5% of variation. The effect of chemical deposition in the emitters did not cause much variation in the emitter discharge.

TABLE 2 Coefficient of variation of emitter discharge under different filtrations.

Filter	Time	Lateral placement		
		Surface	15 cm	30 cm
Sand media	Beginning	1.78	1.78	1.78
for WW	End	2.36	6.08	7.28
Disk for WW	Beginning	1.29	1.29	1.29
	End	3.03	6.77	4.01
Sand media and	Beginning	1.26	1.26	1.26
disk for WW	End	3.49	4.57	3.60
Sand media and	Beginning	1.26	1.26	1.26
disk for GW	End	2.40	3.03	2.99

7.3.3 COLIFORM POPULATION IN SOIL SAMPLES

Coliforms population were evaluated for eggplant crop (Fig. 3) at three stages of crop growth (initial, middle and maturity). The variation in population of coliforms were also quantified in soil samples collected from different depth of soil (Fig. 3). The variation was not detected in the soil irrigated with GW. However, the presence of coliforms were detected in the plots irrigated with WW. In soil, total coliform count increased up to middle stage of crop (50 days after transplanting) and then stabilized up to its maturity. This may be due to availability of limited nutrients in soil to maintain the threshold population of coliforms. Maximum population of coliforms (36.10×10^5, 2.01×10^{10}–2.03×10^{10} g^{-1} soil for initial, middle and maturity stages, respectively) was observed at soil surface.

Presence of fecal contamination in soil with the subsurface placement of drip lateral at 15 and 30 cm depth was also estimated at different crop stages by analyzing the soil samples collected from surface to 60 cm depth at an interval of 15 cm (Fig. 3). Subsurface drip irrigated soil showed the vertical distribution of coliforms bacteria in bell-shaped curve with maximum population adjacent to the lateral while lower population was observed below the lateral and higher population above the lateral. This may be due to bulk density and tilth of the soil. Interestingly no coliforms were observed at the surface of soil when placement of lateral was at 30 cm below the ground level. The results in this chapter are in agreement with research studies by Ijzerman et al. [19] and WHO [34]. Hassan et al. [16] also observed that the movement of water in unsaturated soil up to the depth of 30 cm was effective in the removal of fecal coliforms.

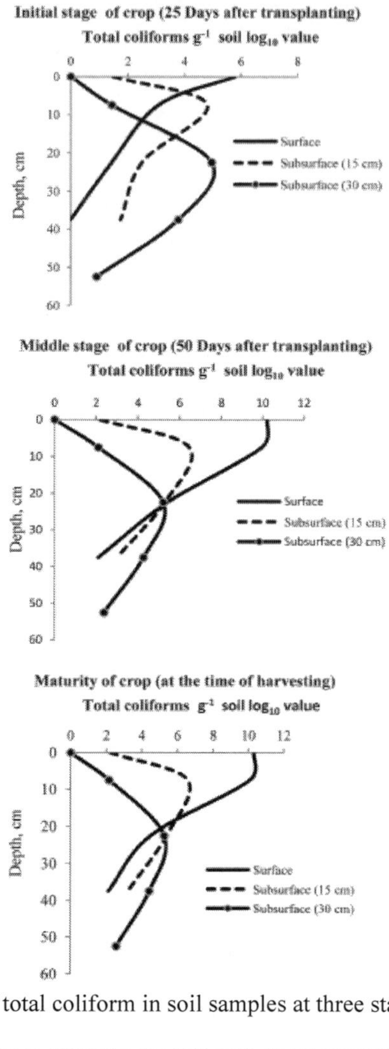

FIGURE 3 Distribution of total coliform in soil samples at three stages of crop.

7.3.4 COLIFORM POPULATION IN EGGPLANT FRUIT

The Table 3 indicates data on the presence of total coliforms in eggplants fruits after washing and crushing. Maximum concentration of total coliforms was observed in crushed eggplant fruit. In general, the fruit after crushing showed higher concentration of coliforms compared to that seen in other treatments. This could be attributed to the contamination from soil. Fruit washing with sterilized water indicated higher contamination of coliforms in the treatments having lateral pipe on ground surface. It may be due to the spread of pathogens from surface through aerosol. No coliforms were observed in fruit washing, with subsurface placement of drip lateral at 30 cm depth. It may be due to the smooth surface of eggplant, that lets only minimum attachment with the skin of the fruit. In crushed fruit, maximum concentration of total coliforms were observed with subsurface placement of drip lateral at 15 cm depth. Fardous and Jamjoum [13] found high number of coliforms on the leaves of a corn plant irrigated with treated WW. In fruit crushing with subsurface placement of drip lateral at 15 cm depth have show maximum concentration of total coliforms. No coliforms were detected after boiling the fruit crush and wash sample. Availability of moisture may have prolonged the survival of bacteria or even allowed their regrowth. Kirkham [23] reported that pathogens may survive on the surface of a plant irrigated with WW because a warm, dark, and moist place could harbor bacteria. High levels of organic matter in treated effluent can also enhance the regrowth of bacteria [28]. WHO [34] recommended a bacteriological standard of 1000 fecal coliforms per 100 g of food/vegetable.

TABLE 3 Coliform population in eggplant fruit.

S. No.	Treatment	Population of total coliforms (MPN g^{-1} log_{10} value)	
		Washing	Crushing
	WW through media filter and surface placement of drip laterals	0.12	1.21
	WW through media filter and subsurface placement of drip laterals at 15 cm depth	0.18	1.30
	WW through media filter and subsurface placement of drip laterals at 30 cm depth	nd	1.24
	WW through disk filter and surface placement of drip laterals	0.14	1.08
	WW through disk filter and subsurface placement of drip laterals at 15 cm depth	0.19	1.19
	WW through disk filter and subsurface placement of drip laterals at 30 cm depth	nd	1.23
	WW through media and disk filters and surface placement of drip laterals	0.19	1.18
	WW through media and disk filters and subsurface placement of drip laterals at 15 cm depth	0.14	1.26
	WW through media and disk filters and subsurface placement of drip laterals at 30 cm depth	nd	1.24
	GW through media and disk filters and surface placement of drip laterals	nd	nd
	GW through media and disk filters and subsurface placement of drip laterals at 15 cm depth	nd	nd
	GW through media and disk filters and subsurface placement of drip laterals at 30 cm depth	nd	nd
	After boiling	nd	nd

nd = not detected.

7.4 CONCLUSIONS

This research study examined the presence of coliform bacteria in soil and plant system of eggplant with the impact of municipal WW irrigation under surface and subsurface drip irrigation system. Maximum reduction in coliforms population was observed with sedimentation and combination of both filters. Total coliform count in soil at different depth from soil surface depends upon the placement of drip lateral. Maximum population of coliforms was observed at soil surface under surface irrigated drip system. Subsurface system with placement of drip lateral at 30 cm shows no coliform bacteria at soil surface. Higher concentration of total coliforms was observed in crushed eggplant fruit in comparison to washed fruit. No coliforms were observed in fruit wash with subsurface placement of drip lateral at 30 cm depth. Uncooked fruit grown under municipal WW is not advisable for consumption. It must be stressed that agricultural manipulation of the irrigation method described above can be used as an auxiliary means of public health protection.

Long-term study is suggested to evaluate the effects of wastewater on the survival of other beneficial soil microorganisms. Its impact on soil fertility should be assessed. There is also a need to develop suitable filtering mechanism so that transmission of harmful coliforms can be prevented to enter in drip irrigation system laterals and subsequently soil as well as plant system.

7.5 SUMMARY

Application of waste water (WW) with efficient irrigation methods is a viable option to protect the environment and mitigate the irrigation demand in arid and semi arid regions. A research study on vegetable crop eggplant (*Solanum melongena* cv. *Supriya*) with surface and subsurface drip system was conducted at PFDC field of Water Technology Centre, Indian Agricultural Research Institute, New Delhi, India. Irrigation was done with WW and it was fed to the drip system after 24 h of settlement of foreign material as primary treatment. Municipal WW was applied through sand media type filter, disk type filter and combined sand media and disk type filters, under surface and subsurface drip irrigation system.

Total coliforms population in WW was in the range of 7 $\log_{10}$ value. This population was reduced to 5 $\log_{10}$ after primary treatment (i.e., sedimentation and passing through the filters). Maximum population of coliforms 2.03×10^{10} MPN g^{-1} were observed in surface soil at maturity stage of crop growth. Presence of harmful pathogens on soil surface were not detected by placement of irrigation lateral at the depth of 30 cm under subsurface irrigation but in case of surface and subsurface (15 cm) population was 10 $\log_{10}$ value and 2 $\log_{10}$ value respectively on soil surface. In crushed fruit of eggplant, maximum concentration of total coliforms were observed with subsurface placement of drip lateral at 15 cm depth. But no coliforms were observed in fruit washing, with subsurface placement of drip lateral at 30 cm depth.

ACKNOWLEDGMENTS

The authors are thankful for the financial support by the National Committee on the Plasticulture Applications in Horticulture (NCPAH), Department of Agriculture and Cooperation, Ministry of Agriculture, Government of India.

KEYWORDS

- **Coliforms**
- **Crop growth**
- **Drip laterals**
- **Dripper**
- **Eggplant**
- **Emitter**
- **Ground water. GW**
- **Media and Disk filters**
- **Micro irrigation**
- **Most Probable Number, MPN**
- **National Committee on the Plasticulture Applications in Horticulture India, NCPAH**
- **Waste water, WW**
- **World Health Organization, WHO**

REFERENCES

1. Alexander, M. (1982). Most probable number method for microbial population. Methods of soil analysis, Part 2. Chemical and Microbiological properties-Agronomy monograph no.9 (2nd edition), 815–820.
2. Allen, R. G., Pereira, L. S., Raes, D., Smith, M. (1998). *Crop evapotranspiration. Guidelines for computing crop water requirements*. FAO Irrigation and Drainage Paper 56, FAO, Rome, Italy.
3. APHA. (2005). *Standard methods for the examination of water and wastewater* (21st ed.). American Public Health Association, Washington, DC.
4. Armon, R., Dosoretz, C. G., Azov, Y., Shelef, G. (1994). Residual contamination of crops irrigated with effluent of different qualities: a field study.*Water Science and Technology*, 30(9):239–248.
5. Bastos, R, K, X, Mara, DD. (1995). The bacteriological quality of salad crops drip and furrow irrigated with waste stabilization pond effluent: an evaluation of the WHO guidelines. *Water Science and Technology*, 31(12):425–430.
6. Beuchat, L. R. (1996). Pathogenic microorganisms associated with fresh produce. *Journal of food Protection*, 59:204–216.
7. Brandl, M. T. (2006). Fitness of human enteric pathogens on plants and implications for food safety. *Annual Review of Phytopathology, 44:367–392.*
8. Campbell, V. J., Mohle-Boetani, J., Reporter, R., Abbott, S., Farrar, J., Brandl, M. T., Mandrell, R. E., Werner, S. B. (2001). An outbreak of *Salmonella* serotype Thompson associated with fresh cilantro. *J Infect Dis*, 183, 984–987.

9. Cools, D., Merckx, R., Vlassak, C., Verhaegen, J. (2001). Survival of E. coli and Enterococcus spp. derived from pig slurry in soils of different texture. *Appl. Soil Ecol.* 17, 53–62.
10. CPCB, (2009). Status of water supply, wastewater generation and treatment in class-I cities and class –II towns of India. *Control of urban pollution, series: CUPS/ 70 / 2009–10.*Central Pollution Control Board (CPCB), Ministry of Environment and Forest, Government of India.
11. Deshmukh, S. K., Singh, A. K., Dutta, S. P., Annapurna, K. (2010). Impact of long-term wastewater application on microbiological properties of vadose zone. In: *Environmental Monitoring and Assessment.* Springer Netherlands, published online first, DOI 10.1007/s10661-010-1554-9.
12. Doyle, M. P., Erickson, M. C. (2008). The problems with fresh produce: an overview. *J. of Appl. Microbiol.*, 105:317–330.
13. Fardous, A., Jamjoum, K. (1996). Corn production and environment effects associated with the use of treated wastewater in irrigation of Khirbet al-Samra Region. Ann. Rep., NCARTT, Amman, Jordan.
14. Gavini, F., Leclerc, H., Mossel, D. A. A. (1985). Enterobacteriaceae of the coliform group in drinking water: Identification and worldwide distribution. *Syst. Appl. Microbiol.* 6, 312–318.
15. Gerba, C. P., Wallis, C., Melnick, J. L. (1975). Fate of wastewater bacteria and viruses in soil. *J. Irrig. Drain. E-ASCE IR3*, 157–174.
16. Hassan, G., Reneau Jr., R. B., Hagedorn, C. C. (2008). On-Site Waste Treatment and Disposal by Sequencing Batch Reactor – Drip Irrigation: Effluent Distribution and Solute Transport. Communication in Soil Sci. and Plant Analy., 39(1):141–157.
17. Hedberg, C. W., Angulo, F. J., White, K. E., Langkop, C. W., Schell, W. L., Stobierski, M. G., Schuchat, A., Besser, J. M., Dietrich, S., Helsel, L., GriYn, P. M., McFarland, J. W., Osterholm, M. T. (1999). Outbreaks of salmonellosis associated with eating uncooked tomatoes: implications for public health. *Epidemiol Infect*, 122, 385–393.
18. Hilborn, E.D., Mermin, J.H., Mshar, P.A., Hadler, J.L., Voetsch, A., Wojtkunski, C., Swartz, M., Mshar, R., Lambert-Fair, M.A., Farrar, J.A., Glynn, M.K., and Slutsker, L., 1999. A multistate outbreak of *Escherichia coli* O157:H7 infections associated with consumption of mesclun lettuce. *Arch Intern Med*, 159, 1758–1764.
19. Ijzerman, M. M., Falkinham, J. O., Hagedorn, C. (1993). A liquid, colorimetric presence-absence coliphage detection method. *J. Virol. Methods*, 45, 229–234.
20. Islam, M., Doyle, M. P., Phatak, S. C., Millnerc, P., Jiangd, X. (2005). Survival of Escherichia coli O157:H7 in soil and on carrotsand onions grown in fields treated with contaminated manure composts or irrigation water. *Food Microbiol.*, 22(1), 63–70.
21. Jimenez, B., Asano, T. (2008). Water reclamation and reuse around the world. In: B. Jimenez and T. Asano (eds.) *Water Reuse: An International Survey of Current Practice, Issues and Needs,* IWA Publishing, London, p 648.
22. Jimenez, B. (2003). Health risk in aquifer recharge with recycled water. In: Aertgeerts, R, Angelakis, A, eds. State of the art report: health risk in aquifer recharge using reclaimed water. Copenhagen, World health Organization Regional Office for Europe, PP. 54–190 (Report No. EUR/03/5041122).
23. Kirkham, M. B. (1986). Problems of using wastewater on vegetable crops. *Hort. Sci.* 21 (1), 24–27.
24. Kvanrstrom, E., Schonning, C., Carlsson-Reich, M., Gustafsson, M., Enocksson, E. (2003). Recycling of wastewater – derived phosphorus in Swedish agriculture - A proposal. *Water Science and Technology*, 48(1), 19–25.
25. Oron, G., Goemans, M., Manor, Y., Feyen, J. (1995). Poliovirus distribution in the soil plant system under reuse of secondary wastewater. *Water Research*, 29(4):1069–1078.
26. Rose, J. B. (2007). Water reclamation, reuse and public health. *Water Science and Technology*, 55(1–2), 275–282.
27. Scott, C. A., Faruqui, N. I., Raschid-Sally, L. (2004). Wastewater use in irrigated agriculture: Management challenges in developing countries. In: C. A. Scott, N. I. Faruqui, L. Raschid-Sally

(eds) *Wastewater Use in Irrigated Agriculture: Confronting the Livelihood and Environmental Realities*, CABI Publishing, Wallingford, UK, pp 1–10.
28. Shatanawi, M. (1994). Minimizing environmental problems associated with the use of treated wastewater forirrigation in Jordan Valley: phase I. Technical Report No. 18. Water and Environ. Res. and Study Center, University of Jordan, Amman, Jordan.
29. Singh, A. K., Minhas, P. S. (2011). Water management and environmental issues. In: *Souvenir, Seminar on water use in agriculture; challenges ahead*. Directorate of Water Management (ICAR), Bhubaneswar, India, 14–19 pages.
30. Strauss, M. (1985). Health aspects of nightsoil and sludge use in agriculture and aquaculture – Part II: Survival of excreted pathogens in excreta and faecal sludges. IRCWD News, 23, 4–9. Duebendorf, Swiss federal Institute for Environmental Science and Technology (EAWAG)/ Department of water and sanitation in developing countries (SANDEC).
31. Tripathi, V. K., Rajput, T. B. S., Patel, N. (2011). Hydraulic performance of drip irrigation system with municipal wastewater. *J. Agril. Engg.*, 48(2), 15–22.
32. Vaz da Costa Vargas, S, Bastos, R. K. X., Mara, D. D. (1996). Bacteriological aspects of wastewater irrigation. Leeds, University of Leeds, Department of Civil Engineering, Tropical Public Health Engineering (TPHE Research Monograph No. 8).
33. Water 21, (2010). UNICEF survey finds half of water sources are polluted. Water 21 Global News Digest Magazine, International Water Association, 27 April 2010.
34. WHO, 2006 *Guidelines for the safe use of wastewater, excreta and gray water*. Vol. II and IV. World Health Organization (WHO) Press, Geneva, Switzerland.

CHAPTER 8

EVALUATION OF MICRO IRRIGATION WITH MUNICIPAL WASTEWATER

VINOD KUMAR TRIPATHI, T. B. S. RAJPUT, NEELAM PATEL, and LATA

CONTENTS

8.1 INTRODUCTION

There is a great challenge ahead to produce more food for increasing population using from less and less water, because the demand of domestic and industrial water consumption is increasing. Agriculture sector is in completion with allocation of fresh water. Contrarily, increasing urbanization is resulting in increasing domestic wastewater (WW) generation. Currently, partially treated and untreated WW is discharged into rivers or lands causing various environmental concerns. On the other hand, WW is beneficial, if it is scientifically used for irrigation as it can act as an important source of water and nutrient [19]. Although WW has been used to irrigate crops, rangelands, forests, parks and golf courses in many parts of the world [1], yet unrestricted irrigation may expose the public to a variety of pathogens such as bacteria, viruses, protozoa, or helminthes and exposure to heavy metals. The factors that influence the use of WW for irrigation are: the degree of wastewater treatment, the crop type and its use (e.g., human consumption or not, consumption after cooking, animal consumption fresh or sun-dried, etc.), the degree of contact with WW, and the irrigation method. Therefore, it is preferable to have the irrigation method having specific characteristics to minimize the various risks namely plant toxicity due to direct contact between leaves and water; salt accumulation in the root zone; health hazards related to aerosol spraying and direct contact with irrigators and product consumers; water body contamination due to excessive water loss by runoff and percolation [18]. In this sense, use of WW to agricultural crops through micro irrigation system is the safest way to manage WW resource [4].

Micro irrigation system applies precise amount of water to the crop at the right time and ensure its uniform distribution in the field. Although, it is the most effective method for WW reuse, yet the suspended solids and organic matter contained in WW can lead to a high risk of system failure due to clogging of the drippers and inadequate filtering systems. These risks depend on the level of treatment, the WW has undergone. Tertiary treatment and chlorination have been found to be effective to reduce clogging caused by bacteria and algae, but, in most arid and semiarid developing countries and in small communities, extremely stringent quality standards would lead to unsustainable costs [8]. In micro irrigation system, quality of water, emitter characteristics and filter efficacy would play a key role in minimize clogging but other factors being same, most important feature for success with WW is filtration [15, 17].

In India mostly gravel media filter, screen filter and disk filters are used to clean the water for micro irrigation system. Capra and Scicolone [7] indicate that screen filters are not suitable for use with WW, with the exception of diluted and settled WW. They also observed almost similar performance in disk and gravel media filter with treated municipal WW. Besides, many researchers have conducted studies on WW using micro irrigation mostly by surface placement of lateral and mostly in laboratories [7, 9, 13, 22]. In India, subsurface micro irrigation has not been evaluated using WW. Therefore, there is a need to develop the methodology for using untreated WW through micro irrigation on sustainable basis. Therefore, this chapter discusses the research studies in realistic field situations using surface and subsurface drip systems with three kinds of filters to develop guidelines for using wastewater in micro irrigation.

8.2 MATERIALS AND METHODS

8.2.1 WATER RESOURCES

The field experiments were conducted at Precision Farming Development Centre of Water Technology Centre, IARI, Pusa, New Delhi during 2008–09 and 2009–10. Randomized block statistical design was used in field experiments. WW was collected from the drain passing through Indian Agricultural Research Institute (IARI). Water samples were analyzed for pH, electrical conductivity (EC), total Solids (dissolved and undissolved), turbidity, calcium, magnesium, carbonate, bicarbonate, total Coliform and *E. coli* according to the standard methods [2].

8.2.2 EXPERIMENTAL SET-UP

Micro irrigation system was installed for WW and GW separately (Fig. 1). Inline lateral (J-Turbo Line) with 40 cm emitter spacing was laid on the ground for surface drip and was buried at a depth of 15 cm from ground surface for subsurface drip irrigation. System included sand media filter (F1), disc filter (F2), and screen filter (F3). WW was allowed to pass through filter F1 and F2 singly as well as in combination of both the filters (F3). GW was also passed through combination of filters for comparison. Main line was connected to submains for each of the plots through a gate valve.

8.2.3 OPERATIONAL PROCEDURE

WW collected from the drain was stored in the tank one for settlement for 24 h so that all the suspended foreign particles were settled. After that the settled WW was transferred to tank two. This step was important to improve the quality of WW by reducing the suspended particles and to avoid frequent chocking of filters. Water from tank two was fed to the filtration system and then allowed to pass through emitters. The pump was turned on, and emitters were allowed to operate for approximately 2 min to allow air to escape. The water collection period was set at 5 min. Quantity of flow of water from drip emitter was collected in containers at 98.06 kPa pressure and was repeated for three times. The flow rate was estimated by dividing total volume collected to the time of collection. The measurement was taken from randomly located sampling emitters to evaluate the performance evaluation of micro irrigation. Discharge from SDI laterals was measured by excavating the soil around the buried drip laterals so that an emitter is visible with sufficient space below it for placement of the container to collect discharged water from it as suggested by Camp et al. [6]; and Magwenzi, [14]. Performance of system was evaluated at normal operating pressure to discharge sufficient water for infiltration and to avoid ponding near the emitter. As per manufactures recommendation, operating pressure of 98.06 kPa was considered appropriate. To achieve accurate pressure, emitter level measurement was done at the lateral with digital pressure gauge having the least count of 0.01 kPa.

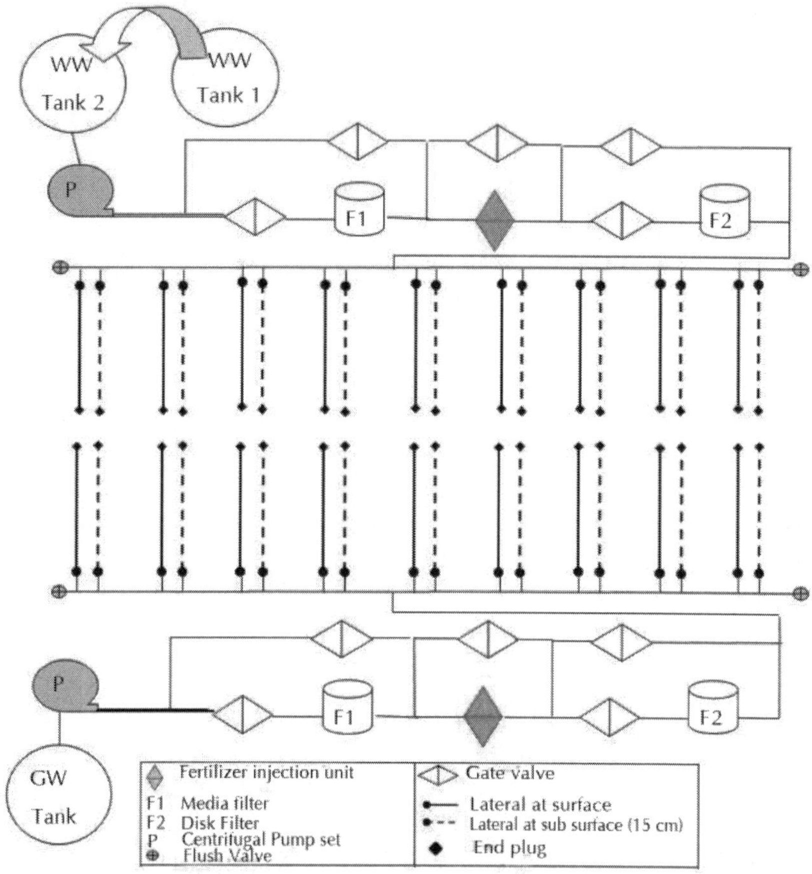

FIGURE 1 Experimental layout.

8.2.4 PARAMETERS FOR EVALUATION OF PERFORMANCE OF MICRO IRRIGATION

The parameters to evaluate the performance of the sustainable micro irrigation system were: Head-discharge relationship of emitters; irrigation uniformity, discharge variation, coefficient of variation and uniformity coefficient.

8.2.4.1 HEAD-DISCHARGE RELATIONSHIP OF EMITTERS

A numerical description of pressure flow characteristics for a given emitter device is based on flow rate versus pressure curve described below:

$$q = CH^x \tag{1}$$

where, q = emitter flow rate ($m^3\ s^{-1}$); C = dimensional emitter coefficient that accounts effects of real discharge ($l\ s^{-1}$); H = pressure head in the lateral at the location of emitters (m); and x = exponent characteristic of the emitter (dimensionless). The exponent x indicates the flow regime and emitter type and typically ranges between 0.0 and 1.0. This exponent is a measure of flow rate sensitivity to pressure change. A higher value of x indicates higher sensitivity. The emitter exponent x and constant value C were derived using a linear regression equation: (Log q) = (Log $C + x$ Log H), or $Y = mx + C$.

8.2.4.2 COEFFICIENT OF VARIATION

The coefficient of variation (CV_q) of the emitter discharge in the lateral was calculated [5, 23] using the following relation:

$$CV_q = \frac{SD}{q} 100 \tag{2}$$

where, SD = standard deviation of emitter discharge (lph); and q = mean discharge in the same lateral (lph). Minimum CV_q was observed at 98.06 kPa pressure. Therefore, it was selected as the operating pressure for evaluation of clogging.

8.2.4.3 EMITTER FLOW RATE (% OF INITIAL)

The emitter flow rate (% of initial) (R) is defined in equation (3):

$$R = \frac{q}{q_{ini}} 100 \tag{3}$$

where, q = the mean emitter discharges of each lateral (lph); and q_{ini} = corresponding mean discharge (lph) of new emitters at the same operating pressure of 98.06 kPa.

8.2.4.4 UNIFORMITY COEFFICIENT

Uniformity coefficient, UC, is defined by Christiansen [10] as follows:

$$UC = 100\left[1 - \frac{\frac{1}{n}\sum_{i=1}^{n}|q_i - q|}{q}\right] \tag{4}$$

where, q_i = the measured discharge of emitter i (lph); q = the mean discharge at drip lateral (lph); and n = the total number of emitters to be evaluated.

8.2.4.5 VARIATION IN FLOW RATE (Q_{VAR})

Emitter flow rate variation, q_{var} [23] was with the following equation:

$$q_{var} = \frac{q_{max} - q_{min}}{q_{max}} \quad (5)$$

where, q_{max} = maximum flow rate (lph); and q_{min} = minimum flow rate (lph).

8.2.5 STATISTICAL ANALYSIS

Statistical analysis was carried out using the GLM procedure of the SAS statistical package (SAS Institute, Cary, NC, USA). The model used for analysis of variance (ANOVA) included water from different filters and placement of lateral as fixed effect and interaction between filtered water and depth of emitter. The ANOVA was performed at probabilities of 0.05 or less level of significance to determine whether significant differences existed among treatment means.

8.3 RESULTS AND DISCUSSIONS

8.3.1 CHARACTERIZATION OF THE WASTE WATER

The physical, chemical and biological characteristics of WW and GW are presented in Table 1. It was observed that EC values for WW were lower than those for groundwater (GW). The EC values for GW varied from 1.89 to 2.58 dS m^{-1} with an average of 2.16 dS m^{-1}, and for WW it was in the range of 1.63 to 1.90 dS m^{-1} with a mean of 1.74 dS m^{-1}. Lower EC values indicate that salt content in the WW did not contribute much in chemically induced emitter clogging. Variation in pH values for WW was 6.60 to 7.32 with an average of 6.87 that was lower than GW (mean value 7.40 with range 6.95 to 8.57) indicating slight acidic nature of WW in comparison to GW. The pH may not have direct impact on clogging but it can accelerate the chemical reactions or biological growth involved in clogging [12, 16]. Variation in values of total solids for WW was 733 to 1297 $mg.l^{-1}$ with an average value of 989 $mg.l^{-1}$ but for GW it was in the range of 800 to 1533 $mg.l^{-1}$ with a mean of 967 $mg.l^{-1}$. Total solids (1533 $mg.l^{-1}$) were highest for GW in the month of May.

Turbidity for WW was always high and in the range of 33 to 68 NTU with a mean value of 55 NTU but GW had negligible turbidity levels with maximum of only 2 NTU. The WW contained surface runoff and foreign particles from anthropogenic pollution. Variation in calcium content of WW was from 68 to 136 mgl^{-1} with a mean value of 94.6 $mg.l^{-1}$ but for GW it was in the range of 36 to 66 $mg.l^{-1}$ with an average of 45 $mg.l^{-1}$. Variation in magnesium content of WW was 25 to 38 $mg.l^{-1}$ with mean value of 32 $mg.l^{-1}$ but for GW it was in the range of 23 to 42 $mg.l^{-1}$ with mean value of 36 $mg.l^{-1}$. Carbonate content of GW was in the range of 48 to 78 $mg.l^{-1}$ with mean value of 58 $mg.l^{-1}$ but for WW it was in the range of 12 to 78 $mg.l^{-1}$ with an average value of 43.5 $mg.l^{-1}$. The variation in bicarbonate content of WW was 440 to 610 $mg.l^{-1}$ with an average of 516 $mg.l^{-1}$ but for GW it was in the range of 264 to 496 $mg.l^{-1}$ with a mean of 364 $mg.l^{-1}$. Presence of carbonate for WW was less than GW but the range for WW was higher. It may be due to the reason that carbonate get converted into bicarbonate with the availability of other ions and variation in temperature. This also gives an indication of the presence of magnesium carbonate in GW and calcium carbonate in WW. Microbial contamination as indicated by total coliforms (mean value 2.0×10^7) was observed for WW only. Based on these quality parameters, it was concluded that clogging problem can be encountered more in WW than GW.

TABLE 1 Physicochemical and biological properties of water used for irrigation.

Properties	Units	Wastewater	Groundwater
		Mean ± SD	Mean ± SD
EC	dS m^{-1}	1.48±0.23	2.17±0.25
pH	—	7.33±0.35	7.4±0.43
Total Solids	mgL^{-1}	849.8±148.5	967.4±212.6
Turbidity	NTU	46.25±10.23	1.50±0.52
Ca	mgL^{-1}	82.16±19.99	44.58±8.27
Mg	mgL^{-1}	33.61±5.5	35.28±5.81
CO_3	mgL^{-1}	119.5±41.69	58.0±8.23
HCO_3	mgL^{-1}	415.27±69.7	364.33±70.7
Total coliforms	MPN mL^{-1}	41787± 172437	nd

nd = not detected.

8.3.2 HYDRAULIC CHARACTERISTICS OF EMITTER

Coefficient for *Q-H* equation (Eq. (1)) decreased with the time of operation of emitters in all filtration systems as a result of partial clogging (Table 2). Theoretically the exponent for the emitter was 0.5, which comes under category of completely turbulent hydraulic regime [11]. In normal pressure range, exponent was more than 0.5 for gravel media filtered WW and less than 0.5 in case of disk filter. Performance of exponent was close to 0.5 in combination filter for both WW and groundwater. The coefficient of regression (R^2) was 0.99 in most of the situations indicating that the *Q-H* equation described the flow-pressure relationship precisely.

8.3.3 COEFFICIENT OF VARIATION OF EMITTER DISCHARGE (CV_Q)

The CV_q of the discharge for different filters and for the combination of both filters are presented in Fig. 2. After one year, maximum CV_q of 3.49 and 4.57% was observed with WW in surface and subsurface drip irrigation systems, respectively. As shown in Fig. 2 after one year in all filter condition, CV_q was less than 5%. Hence the performance can be rated as excellent [3]. After two years of experimentation, maximum variation of 10.16% was observed with disk filter in subsurface drip system. The performance of combination with filter for WW and GW was excellent with only 4% of variation in surface drip system. Maximum deviation of 6.46% was observed in subsurface drip system with both filter combination in GW. The results indicate that one-year operation of the emitters did not cause much variation but continuous two years of operation caused significant variation in emitter discharge. This is also supported by the computation of the standard error, which was lower in first year but was significantly higher in second year under all filters situation with both types of water. Coefficient of variation in subsurface condition was always poor than surface.

TABLE 2 *Q-H* relationships for emitter under different filtration system with wastewater and groundwater.

Filter	Placement of lateral	Stage	Coefficient, c	Exponent, x	R^2
Gravel media (F1) (Waste water)	Surface	Beginning	3.768	0.521	0.99
		Middle	3.599	0.511	0.99
		End	3.484	0.534	0.99
	Subsurface	Beginning	3.768	0.521	0.98
		Middle	3.520	0.533	0.98
		End	3.455	0.533	0.99
Disk (F2) (Wastewater)	Surface	Beginning	3.538	0.485	0.99
		Middle	3.435	0.489	0.99
		End	3.368	0.486	0.99
	Subsurface	Beginning	3.538	0.485	0.99
		Middle	3.417	0.488	0.99
		End	3.345	0.485	0.99
Combination of F1 and F2 (with Wastewater)	Surface	Beginning	3.548	0.494	0.99
		Middle	3.476	0.492	0.99
		End	3.403	0.494	0.99
	Subsurface	Beginning	3.548	0.494	0.99
		Middle	3.465	0.494	0.99
		End	3.388	0.497	0.99

Filter	Placement of lateral	Stage	Coefficient, c	Exponent, x	R^2
Combination of F1 and F2 (with Groundwater)	Surface	Beginning	3.548	0.494	0.99
		Middle	3.446	0.494	0.99
		End	3.350	0.493	0.99
	Subsurface	Beginning	3.548	0.494	0.99
		Middle	3.431	0.494	0.99
		End	3.350	0.493	0.99

8.3.4 EMITTER FLOW RATE VARIATION

The study on flow rate variation was carried out with pressure variation so that flow rate reduction can be explained by clogging of emitters alone. Maximum reduction in flow rate was observed with gravel media filter (F1) and minimum with combination of both filters under WW. The clogging due to disk filter (F2) remained in between these two values (Fig. 3). The results of the statistical analysis revealed that after 2 years of experiment, there was significant effect of filter, emitter placement and their interaction on the discharge of drip emitters (Table 3). In the beginning of experiment there was no significant effect of emitter placement and their interaction with filtered water because emitters were new and there was no clogging. After continuous use, clogging takes place and effect of different filtration system start showing up in the discharge of emitters. At the end of one-year effect of filtration system was significant but effect of emitter placement was not significant. Both were significant after two-year use. These results prove that clogging is a dynamic phenomenon over time [21].

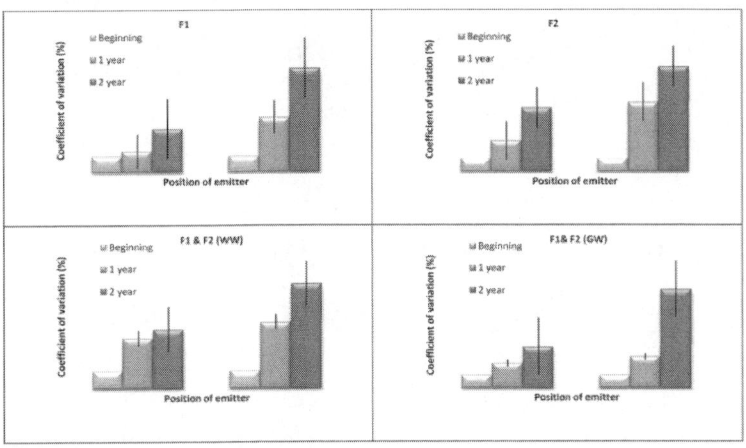

FIGURE 2 Coefficient of variation in emitter discharge under different filtration systems for wastewater and groundwater at 98.06 kPa pressure.

TABLE 3 Significance level (P-value) of the statistical model and of each factor and interaction for emitter flow rate.

Parameter	Time		
	Beginning	1 year	2 year
Model	*** (R^2=0.97)	** (R^2=0.87)	*** (R^2=0.93)
Filter (F)	n.s.	**	***
Emitter placement (EP)	n.s.	n.s.	*
F × EP	n.s.	**	***

n.s.: not significant, $P > 0.05$; *: $P < 0.05$; **: $P < 0.01$; ***: $P < 0.001$.

8.3.5 UNIFORMITY OF WATER APPLICATION

Variations in uniformity coefficient and flow rate are presented in Table 4. Least variation in flow rate with maximum uniformity was observed at the beginning of experiment. The variation in flow rate increased with the operation of drip system and maximum variation with minimum uniformity coefficient were reached at the end of two years of experimentation. Performance of filter combination for both types of water could be rated as good [20]. After two years minimum uniformity coefficient under both filter combinations was 94.17 and 93.50 with WW and GW, respectively, for subsurface drip. As per general criteria for q_{var} values of 0.10 or less are desirable and 0.1 to 0.2 is acceptable and greater than 0.2 being unacceptable. Two out of three filtration systems (F2 and F3) gave variation in flow rate under acceptable limit.

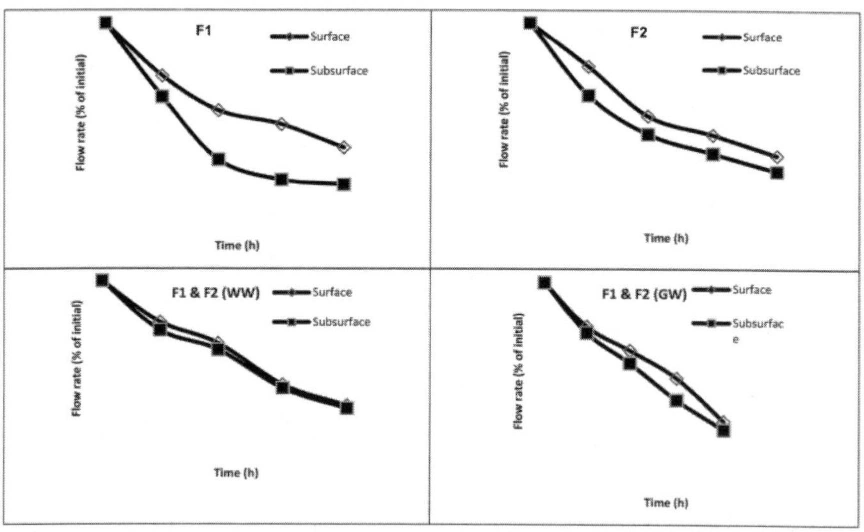

FIGURE 3 Emitter flow rate (% of initial flow rate) under different filtration system for wastewater and groundwater at 98.06 kPa pressure.

TABLE 4 Uniformity coefficient and variation in flow rate (q_{var}) resulting from the performance evaluation of micro irrigation system.

Filter	Depth of placement of lateral	Uniformity Coefficient			Variation in flow rate (q_{var})		
		Beginning	1 year	2 year	Beginning	1 year	2 year
Gravel media (F1)	Surface	98.59	96.26	94.84	0.049	0.106	0.171
	15 cm	98.56	95.24	92.05	0.049	0.120	0.219
Disk (F2)	Surface	98.89	96.52	94.73	0.032	0.090	0.117
	15 cm	98.91	95.00	93.15	0.032	0.106	0.181
Combination of F1 & F2	Surface	99.01	96.67	95.27	0.048	0.092	0.158
	15 cm	99.05	95.70	94.17	0.048	0.108	0.189
Combination of F1 and F2 with GW	Surface	99.07	97.29	95.55	0.048	0.102	0.131
	15 cm	99.02	96.02	93.50	0.048	0.100	0.160

8.4 CONCLUSIONS

The hydraulic performance of the drip emitters revealed that for continuous use of WW, filtration with a combination of gravel and disk filter would be most appropriate strategy against emitter clogging. It resulted in a better emitter discharge exponent, a reasonably good coefficient of variation and uniformity coefficient.

8.5 SUMMARY

Generation of WW in huge amounts is putting a lot of pressure to irrigation engineers for its safe reuse in agriculture. Though WW supports major and minor nutritional requirements of crops, but the presence of microbial contaminants and toxic elements in WW, limits its use. Utilization of WW for irrigation through micro irrigation system is the best choice to reduce the chances of contamination due to restricted quantity of application. Since clogging is the main problem associated with WW utilization through micro irrigation system, its remediation is required for enhanced utilization of WW through micro irrigation system. Physical and chemical characteristics of WW were determined and compared with GW.

While higher EC, pH, Mg, and carbonate were observed in GW but WW contained higher turbidity, total solids, HCO_3, and Ca. The population of total coliforms (2.72×10^4 to 5.2×10^7) and *E. coli.* (1.8×10^3 to 2.64×10^6) were detected in WW. The hydraulic performance of drip emitters was studied for two years (2009 and 2010) with municipal WW and groundwater (GW) using gravel media (F1), disk filter (F2) and combination of gravel and disk filters (F3). Filtration using F3 gave emitter discharge exponent close to 0.5 with R^2 value of 0.99. Emitter flow rate decreased with the increase in time of operation of the system. Coefficient of variation less than 4% with WW and GW showed excellent performance in surface placed drip lateral after two years of operation. After filtration with F3, coefficient of variation (CV_q) of 4.0% with WW and 6.46% with GW was observed under subsurface (15 cm deep) placement of lateral.

ACKNOWLEDGMENTS

Authors are thankful to the National Committee on Plasticulture Applications in Horticulture (NCPAH), Department of Agriculture and Cooperation, Ministry of Agriculture, Government of India for providing the necessary funds to conduct this research.

KEYWORDS

- **coefficient of variation**
- **coliforms**
- **crop growth**
- **dripper**
- **emitter**
- **emitter discharge**
- **emitter discharge exponent**
- **emitter flow rate**
- **emitter hydraulics**
- **emitter placement**
- **filter**
- **ground water, GW**
- **head-discharge relationship**
- **hydraulic coefficient**
- **micro irrigation**
- **National Committee on the Plasticulture Applications in Horticulture India, NCPAH**
- **uniformity coefficient**
- **waste water, WW**
- **water characterization**
- **World Health Organization, WHO**

REFERENCES

1. Al-Jamal, M. S., Sammis, T. W., Mexal, J. G., Picchioni, G. A., Zachritz, W. H. (2002). A growth-irrigation scheduling model for wastewater use in forest production. *Agricultural Water Management*, 56, 57–79.

2. APHA, (2005). Standard methods for the examination of water and wastewater (21st ed.). Washington, DC American Public Health Association.
3. ASABE, (2003). Design and installation of micro irrigation systems, EP405.1, *ASABE standards, 50th edition*. American Society of Agricultural and Biological Engineers: St. Joseph, Michigan.
4. Ayers, R. S., Westcot, D. W. (1991). Water quality for agriculture. *FAO Irrigation and Drainage paper* 29, Rome: FAO.
5. Bralts, F. V., Kesner, D. C. (1983). Micro irrigation field uniformity estimation. *Trans. ASAE*, 26, 1369–1374.
6. Camp, C. R., Sadler, E. J., Busscher, W. J. (1997). A comparison of uniformity measures for micro irrigation systems. *Trans. ASAE*, 40(4), 1013–1020.
7. Capra, A., Scicolone, B. (2004). Emitter and filter test for wastewater reuse by micro irrigation. *Agricultural Water Management*, 68(2), 135–149.
8. Capra, A., Scicolone, B. (2007). Recycling of poor quality urban wastewater by micro irrigation systems. *Journal of Cleaner Production*, 15, 1529–1534.
9. Cararo, D. C., Botrel, T. A., Hills, D. J., Leverenz, H. L. (2006). Analysis of clogging in drip emitters during wastewater irrigation. *Applied Engineering in Agriculture*, ASABE, 22 (2), 251–257.
10. Christiansen, J. E. (1942). Hydraulics of sprinkler systems for irrigation. *Trans ASCE*, 107, 221–239.
11. Cuenca, R. H. (1989). Irrigation system design: an engineering approach; Prentice-Hall: Englewood Cliffs, New Jersey, 317–350.
12. Dehghanisanij, H., Yamamoto, T., Rasiah, V., Utsunomiya, J., Inoue, M. (2004). Impact of biological clogging agents on filter and emitter discharge characteristics of micro irrigation system. *Irrigation Drainage*, 53, 363–373.
13. Liu, H., Huang, G. (2009). Laboratory experiment on drip emitter clogging with fresh water and treated sewage effluent. *Agricultural Water Management*, 96, 745–756.
14. Magwenzi, O. (2001). Efficiency of subsurface micro irrigation in commercial sugarcane field in Swaziland. http://www.sasa.org.za/sasex/about/agronomy/aapdfs/ magwenzi.pdf. 1–4.
15. McDonald, D. R., Lau, L. S., Wu, I. P., Gee, H. K., Young, S. C. H. (1984). Improved emitter and network system design for reuse of wastewater in micro irrigation. *Technical Report no 163*, Water Resources Research Centre, University of Hawaii at Manoa, Honolulu.
16. Nakayama, F. S., Bucks, D. A. (1991). Water quality for drip/trickle irrigation: a review. *Irrigation Science*, 12, 187–192.
17. Oron, G., Shelef, G., Turzynski, B. (1979). Trickle irrigation using treated wastewaters. *J. Irrig. Drain. Div.*, 105 (IR2), 175–186.
18. Pereira, L. S., Oweis, T., Zairi, A. (2002). Irrigation management under water scarcity. *Agricultural Water Management*, 57, 175–206.
19. Pescod, M. D. (1992). Wastewater treatment and use in agriculture. *FAO Irrigation and Drainage paper 47*, Rome: FAO.
20. Puig-Bargues, J., Arbat, G., Barragan, J., Ramirez de Cartagena, F. (2005). Hydraulic performance of micro irrigation subunits using WWTP effluents. *Agricultural Water Management*, 77 (1–3), 249–262.
21. Ravina, I., Paz, E., Sofer, Z., Marcu, A., Shisha, A., Sagi, G. (1992). Control of emitter clogging in micro irrigation with reclaimed wastewater. *Irrigation Science*, 13 (3), 129–139.
22. Rowan, M., Manci, K., Tuovinen, O. H. (2004). Clogging incidence of micro irrigation emitters distributing effluents of different levels of treatments. Conference proceeding on On-Site wastewater Treatment, Sacramento, California, USA, 21–24 March, 2004, pp. 84–91.
23. Wu, I. P., Howell, T. A., Hiler, E. A. (1979). Hydraulic design of micro irrigation systems. *Hawaii Agric. Exp. Stn. Tech. Bull. 105*, Honolulu.

CHAPTER 9

EFFECTS OF IRRIGATION METHODS ON FRUIT PERFORMANCE OF ACID LIME

P. S. SHIRGURE, A. K. SRIVASTAVA, and SHYAM SINGH

CONTENTS

9.1 INTRODUCTION

We must increase production and productivity of fruit crop per unit use of inputs. Water is a valuable resource that must be used wisely. Popularity of micro irrigation systems is popular among citrus growing farmers. The farmers have to decide the best system among various micro irrigation systems to suit the requirement and to increase productivity. The major constraints in fruit production system include inadequate rainfall and its distribution, depleting ground water table, adaption of surface irrigation methods and excess use of available water resources.

Acid lime occupies 25% of the total area under citrus cultivation, and is one of the important citrus fruit grown in different states of India. Low yield is due to inadequate irrigation management as the plants are sensitive to availability of soil moisture status. The irrigation methods in the acid lime orchards affect the distribution and availability of soil water to the plants and ultimately affect nutrient uptake and growth.

The surface irrigation is most common in acid lime orchards. However, in recent years, micro irrigation is being adopted due to many advantages of this system. Besides the higher cost, micro irrigation causes saving in labor/water/energy, greater irrigation uniformity and immediate response to crop need, better soil-water-plant relationship, favorable rooting environment, and better yield and quality [1, 6, 12]. In developed countries, the under tree sprinkler irrigation system for efficient use of water resource is practiced in citrus orchards [10, 11, 13]. The research studies are being conducted to determine most efficient irrigation method for growing lemon and it has been found that under tree sprinkler and micro irrigation gave the best pomological results. Micro irrigation was most efficient and water use efficiency was highest [5]. The response of lemon trees to micro irrigation, microjet irrigation has also been studied under different locations [8]. Drip and microsprinkler irrigation trials have been conducted in 'Valencia' orange to evaluate water use efficiency, growth and yield by Grieve [2]. The conventional method of basin irrigation was compared with micro irrigation and microjet irrigation in oranges by several investigators [7, 9]. The studies on efficacy of micro and mini sprinkler irrigation on growth, water use and yield of 'Hamlin' orange [3] and 'Shamouti' orange [4] have also conducted and it was concluded that the microsprinkler produced the best results over the flood irrigation method. This chapter evaluates effects of irrigation methods in acid lime (*Citrus aurantifolia* cv. Swingle) orchards on water use, crop performance, and soil-leaf fertility changes, during the 1995–1997.

9.2 MATERIALS AND METHODS

During 1995–1998, the irrigation trials in acid lime were conducted in a 0.5 hectare block located in the research farm of National Research Centre for Citrus, Nagpur (79°22'E longitude, 21°09′ N latitude, 311 m above msl. The experimental site is considered as a subtropical climate with no summer rains. The average annual precipitation (June through October) in the area is about 900 mm. The soil is medium deep with average depth of 50 cm underlain by parent material. The field capacity and permanent wilting point was 29.65% and 18.48%, respectively. The bulk density and available water content of the field under study was 1.19 g.cm^{-3} and 11.17% [=29.65–18.48], respectively. The acid lime plants were planted in August 1993 at 5 × 5 m spacing. The orchard was maintained for one-year establishment. The irrigation systems were installed in early January 1994, and irrigation treatments were initiated in April 1994.

The irrigation treatments consisted of dripper (T1, 8 lph, pressure compensating, 3/plant), microjet 300°(T2, Rayjet, 1/plant), microjet 180°(T3, Ejet, 2/plant) and basin (ring) irrigation method (*T4*) as a control in a randomized block design with six replications. The number of plants per treatments was seven. The micro irrigation system was laid out in the field as per the statistical design and water meter control valves were installed to monitor irrigation water in each treatment. The aluminum access tubes (50 mm) were also installed and the soil moisture status was recorded continuously. The rainfall and evaporation were recorded from the agrometeorology station at the research farm. The dripper and microjet treatments were scheduled based on class A pan evaporation (0.7 Epan). The quantity of irrigation water was calculated using the spread area (150 cm during 1996) and depth of irrigation was equal to 0.7xEpan.The irrigation frequencies were daily in dripper and microjet 300° treatments and at 2 days interval in microjet 180° treatment. The average discharge per tree was 23.1 lph in drippers, 26.7 in microjet 300° and 64.2 lph in microjet 180°, respectively. The equal amount of water was allowed to each tree during each month. The irrigation in basin method was scheduled at 50%depletion of available water content.

Soil water content was monitored using subsurface Newton moisture probe (model Troxler 4300, Soil moisture Santa Barbara, California). Aluminum access tubes were inserted at 70 cm depth within the tree basin and 70 cm away from the trunk in between two emitters. Soil-water measurements were recorded daily during December 1996 through June 1997.The neutron probe was calibrated against gravimetric sampling for the site using calibration equation: $Y= 0.0228x + 0.123 (r = 0.95)$, where, y is ratio of actual counts to standard counts and x is soil moisture content (volume basis).

The depth of irrigation water, quantity and number of irrigations (in case of basin irrigation) were recorded. The vegetative growth parameters of acid lime plants (tree height, tree girth, and canopy volume) were recorded in January 1997. The plant girth was taken 15 cm above the soil surface. The canopy volume of the tree was calculated using the Castle's formula. The growth parameters of the acid lime trees were statistically analyzed using computer program.

9.3 RESULTS AND DISCUSSION

9.3.1 RAINFALL AND EVAPORATION

The average annual rainfall ranged from 850 mm to 1050 mm. During 1995–1999, average annual rainfall was 875 mm. The effective rainfall occurred during June through September. The average monthly rainfall during rainy season was 190 mm. The rainfall recedes after October onwards. The daily pan evaporation ranged from 2.85 to 13.87 mm at the site. The maximum daily evaporation was 13.87 mm in May and the minimum was during November to January. Figure 1 shows monthly rainfall and evaporation pattern during 1995–1999. The irrigation systems were not operated during July to October as the rainfall exceeded evaporation. Irrigation was started in November and continued till the onset of monsoon season in July.

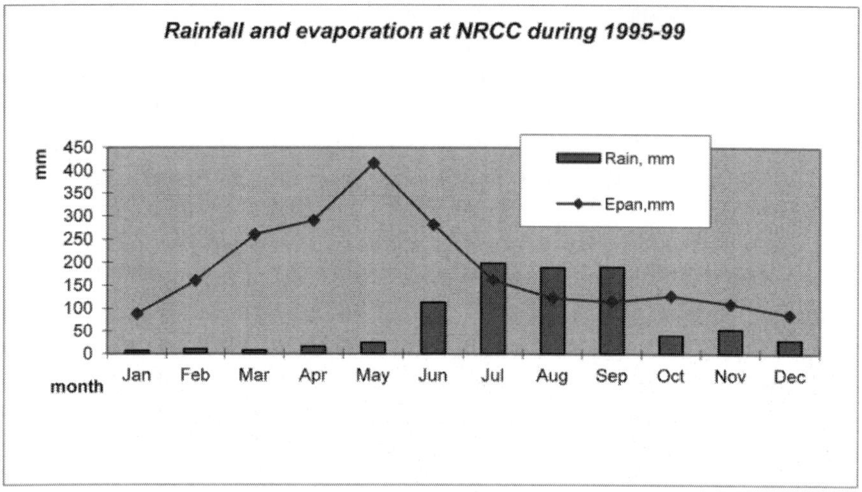

FIGURE 9.1 Rainfall and evaporation pattern at experimental site during 1995–1999.

9.3.2 WATER CONSUMPTION OF ACID LIME

The acid lime trees were irrigated uniformly with different irrigation systems. Daily irrigations were scheduled based on pan evaporation during 1995–1997. Water use of the acid lime was estimated using canopy area and evaporation. Table 1 indicates that the water requirement of acid lime during 1997–1998 was higher than during 1996–1997. Increase in canopy volume increased the water requirement. Water applied through micro irrigation systems varied from 22 to 90 L/day/tree during 1996–1997 and 12 to 96 L/day/tree during 1997–1998. In basin irrigation, the water was applied after 50% depletion of available water. Total amount of water required during a month is given in Table 1. Number of irrigations in basin irrigation was more in summer due to more evaporation. The water requirement in basin irrigation varied from 23 to 175 L/day/tree. Basin method of irrigation required 40 to 50%more water compared to micro irrigation systems. Water saving was more in summer in all the irrigation treatments. Increasing evapo-transpiration during summer required larger amount of water in all the irrigation systems. The water application during May was 90 to 96 L/day/tree for micro irrigation systems and 160 to 175 L/day/tree for basin method of irrigation, respectively.

TABLE 1 Monthly water applied to Acid lime (Liters/day/plant).

Month	Micro irrigation systems[1]		Basin irrigation	
	1996–1997	1997–1998	1996–1997	1997–1998
October	24	36	38	62
November	27	28	42	46
December	28	12	49	23
January	22	21	34	35
February	41	37	72	64
March	62	47	120	91
April	60	77	105	140
May	90	96	160	175
June	55	64	85	112

[1]Dripper (8 lph); Microjet (300°); and Microjet (180°).

9.3.3 GROWTH PERFORMANCE OF ACID LIME

The highest increase in plant height was 66.6 cm in microjet 300° treatment, and was 66.18 cm in dripper, and 62.72 in microjet 180° irrigation system (Table 2). Similarly, the increase in stock girth was 14.59 cm in microjet 300°, 14.49 cm in dripper, and 14.17 cm in microjet 180° irrigation system. The increase in canopy volume was 7.75 m^3 in microjet 300°, 7.21 m^3 in dripper and 6.69 m^3 microjet 180° irrigation system. The increase in plant height, stock girth and canopy volume in basin method of irrigation was 52.07 cm, 12.27 cm and 6.07 m^3, respectively.

TABLE 2 Acid lime performance under different micro irrigation systems in 1995 and 1997.

Treatments	Plantheight (cm)			Stock girth (cm)			Canopy volume (m^3)		
	1995	1997	Increase	1995	1997	Increase	1995	1997	Increase
Dripper, 8 lph (3/plant)	179.0	245.2	66.2	20.1	34.6	14.5	1.95	9.16	7.21
Microjet,300° (1/ plant)	193.2	259.8	66.6	21.4	36.0	14.6	2.35	10.1	7.75
Microjet,180° (2/plant)	180.6	243.4	62.8	20.6	34.8	14.2	2.20	8.99	6.79
Basin (ring) irrigation	172.9	225.0	52.1	19.7	32.0	12.3	1.85	7.92	6.07

9.3.4 CHANGES IN AVAILABLE SOIL NUTRIENT STATUS

The observations on year-wise changes in patterns of available N showed a reduction for all the treatments. However, the basin method of irrigation recorded 42.2 mg/kg compared to 6.9 mg/kg to 11.4 mg/kg using micro irrigation systems. The available soil P improved in dripper (3.13 mg/kg) and microjet 300° (1.27 mg/kg), but indicated depletion pattern in microjet 180° (0.92 mg/kg) and basin irrigation (5.42 mg/kg) method. The available soil K was improved with dripper (23.0 mg/kg) and microjet 300° (23.5 mg/kg), but was decreased with microjet 180° (2.3 mg/kg) and basin irrigation (23.8 mg/kg). Details of soil N, P and K are presented in Table 3.

TABLE 3 Effects of micro irrigation systems on soil nutrient status of acid lime plants.

Treatments	Nitrogen (mg/Kg)			Phosphorous (mg/Kg)			Potassium (mg/Kg)		
	1995	1997	change	1995	1997	change	1995	1997	change
Dripper, 8 lph (3/plant)	144.8	137.9	-6.9	20.98	24.11	+3.13	144.0	167.0	+23.0
Microjet, 300° (1/plant)	134.8	123.4	–11.4	19.37	20.64	+1.27	152.1	175.6	+23.5
Microjet, 180° (2/plant)	122.2	112.7	–9.5	20.73	19.81	-0.92	165.8	163.5	-2.3
Basin (ring) irrigation	147.4	105.2	–42.2	24.13	18.71	-5.42	179.1	155.3	-23.8

9.3.5 CHANGES IN AVAILABLE LEAF N, P AND K NUTRIENT STATUS

The leaf nitrogen content decreased (a total of 1.2–1.63%) in micro irrigation systems compared to the basin irrigation (1.83%). The leaf P content improved marginally in all the treatments (Table 4). The leaf K content decreased in different micro irrigation systems. The magnitude of reduction in leaf K varied from 0.08 to 0.62%compared to 0.21% in basin irrigation method. The results revealed that the growth and soil-leaf nutrient status of acid lime in micro irrigation systems was superior over the conventional method of basin irrigation.

TABLE 4 Effects of micro irrigation systems on leaf nutrient status of acid lime during 1996 and 1997.

Treatments	Nitrogen %			Phosphorous %			Potassium %		
	1995	1997	change	1995	1997	change	1995	1997	change
Dripper, 8 lph (3/Plant)	3.00	1.37	–1.63	0.13	0.14	+0.01	1.48	1.01	–0.47

TABLE 4 *(Continued)*

Treatments	Nitrogen %			Phosphorous %			Potassium %		
	1995	1997	change	1995	1997	change	1995	1997	change
Microjet, 300° (1/plant)	2.42	1.33	–1.09	0.10	0.11	+0.01	1.38	1.30	–0.08
Microjet, 180° (2/plant)	2.40	1.20	–1.20	0.12	0.12	0.00	1.82	1.20	–0.62
Basin (ring) irrigation	3.08	1.25	–1.83	0.11	0.13	+0.02	1.65	1.44	–0.21

9.4 SUMMARY

A field experiment was conducted on acid lime (*Citrus aurantifolia Swingle*) during 1995–1998 in an Inceptisol to evaluate the comparative efficiency of micro irrigation systems versus basin method of irrigation. The irrigation treatments were dripper 8 lph (3/plant), microjet 300° (1/plant), microjet 180° (2/plant) and double ring method of basin irrigation. The highest increase in plant height, stock girth and canopy volume was recorded in microjet 300° followed by dripper and microjet 180° irrigation system and basin method of irrigation. The observations on year-wise changing pattern of available N showed a reduction with all the treatments. However, the basin method of irrigation recorded 42.2 mg/kg compared to 6.9 mg/kg to 11.4 mg/kg using micro irrigation systems. The available soil P improved in dripper and microjet 300°, but indicated depletion pattern in microjet 180° and basin irrigation method. The available soil K improved with dripper and microjet 300°, but decreased with microjet 180° and basin irrigation. The leaf nitrogen content decreased in micro irrigation systems compared to the basin irrigation. The leaf P content improved marginally in all the treatments. The leaf K content decreased in different micro irrigation systems. The magnitude of reduction in leaf K varied from 0.08 to 0.62% compared to 0.21% in basin irrigation method. The results revealed that the growth and soil-leaf nutrient status of acid lime in micro irrigation systems was superior over the conventional method of basin irrigation.

KEYWORDS

- **Acid lime**
- **basin irrigation**
- **drip irrigation**
- **irrigation methods**
- **leaf K**
- **micro irrigation**
- **microjet irrigation**
- **microsprinkler irrigation**
- **soil-leaf nutrient change**
- **'Valencia' orange**
- **water use**

ACKNOWLEDGMENTS

The authors are thankful to *Mr. A.R. Pimple*, Technician and *Mr. S.P. Kadam*, Research Associate for the technical assistance in this research.

REFERENCES

1. Capra, A., Nicosia, O. U. D. (1987). Irrigation management in citrus orchards. *Irrigazine,* 34(1), 3–15.
2. Grieve, A. M. (1988). Water use efficiency of micro-irrigated citrus. *Proc. 4th International micro-irrigation congress,* Oct 23–28, volume 2. Albury-Nodonga, Australia.
3. Marler, T. E., Davies, R. S. (1990). Microsprinkler irrigation and growth of young 'Hamlin' orange trees. *J. Amer. Soc. Hort. Sci.,* 115(1), 45–51.
4. Moreshet, S., Cohen, Y Fuchs, M. (1988). Water use and yield of a mature Shamouti orange orchard submitted to root volume restriction and intensive canopy pruning. *Proceedings of the Sixth International Citrus Congress,* March 6–11, Tel Aviv, Israel, 2:739–46.
5. Ozsan, M., Tekinel, O., Tuzcu, O., Cevik, B. (1983). Studies on determining the most efficient irrigation method for growing lemons under cukurova conditions. *Doga BilimDergisiDz, Tarim Ve Ormancilik,* 7(1):63–69.
6. Pyle, K. R. (1985). An appraisal of micro-irrigation for use in citrus with an emphasis on micro irrigation. *Citrus and Subtropical Fruit Jr.,* 612:4–7.
7. Raciti, G., Scuderi, A. (1977). Micro irrigation trial in citrus orchard. *Proc. Int. Soci. Citriculture,* 3: 1040–1045.
8. Richards, S. J., Warnke, J. E. (1968). Lemon irrigation management under coastal conditions. *Calif. Citrogr.* 53:378–384.
9. Scuderi, A., Raciti, G. (1978). Citrus trickle irrigation trials. *Proc. of Int. Soc. of Citriculture,* 244 pages.
10. Simpson, G. H. (1978). Developments in under tree irrigation systems in the Murray valley. *Proc. of Int. Soc. of Citriculture*, 234–235 pages.
11. Smajstrla, A. G., Koo, R. C. J. (1984). Effects of trickle irrigation methods and amounts of water applied on citrus yields. *Proc. of Flor. State Hort. Soc.,* 97:3–7.
12. Smajstrla, A. G. (1993). Micro-irrigation for citrus production in Florida. *Hort. Sci.,* 28(4):295–298.
13. Zekri, M., Parsons, L. R. (1989). Grapefruit leaf and fruit growth in response to drip, micro-sprinkler and overhead sprinkler irrigation. *J. Am. Soc. of Horti. Sci.,* 114:25–29.

CHAPTER 10

PERFORMANCE OF *CITRUS RETICULATA CV.* BLANCO WITH MICROJET IRRIGATION

PARAMESHWAR S. SHIRGURE and ANOOP K. SRIVASTAVA

CONTENTS

10.1 INTRODUCTION

Most important citrus cultivar commercially grown in central India is Nagpur mandarin Citrus (*Citrus reticulata Blanco*), and is third largest fruit crop cultivated in India. The production is 0.875 Mha on 0.148 Mha cultivated area. The factors that contribute to low productivity (11 tons per ha) are: scarcity of water resources; poor soil; low application efficiency of conventional irrigation methods; inadequate maintenance of pressurized irrigation systems; insufficient technical services to the grower; and need for constant soil moisture throughout growth and fruit development stages. Gravity method (Basin irrigation) is commonly used in Nagpur mandarin orchards, however, it has high irrigation losses (conveyance, percolation, evaporation, and distribution losses) and the performance of citrus crop is poor with this system [20, 21].

Due to the scarcity of irrigation water, micro irrigation is becoming increasingly popular with mandarin growers. However, many growers are still unsure about the efficacy of drip irrigation, especially where soil moisture deficit stress is adopted for regulating stress and flowering; and lack of uniformity of moisture distribution within the root zone. The sustainability of Nagpur mandarin orchards is ensured by any irrigation method capable of replenishing the citrus evapotranspiration demand, and simultaneously keeping the soil moisture within the desired limit during the citrus growth [2, 16].

Micro irrigation systems have been commonly used in trees throughout the world. There is now a gradual shifts from furrow irrigation and overhead sprinkler irrigation systems to under-tree micro sprinkler systems like microjets [6, 26, 28]. Micro irrigation systems (e.g., drip irrigation, under-tree sprinklers, microsprinklers, and microjets, etc.) have been highly effective in commercial citrus cultivars like Valencia orange [1], Navel orange [7], Hamlin orange [14], Satsuma mandarin [15], Clementine [3], and lemon [4].

Earlier research in India has showed better performance using drip irrigation compared to flood irrigation in Nagpur mandarin [24, 31], sweet orange [12], and acid lime [22–24]. This chapter discusses research results to evaluate: Under-tree microjet irrigation systems with automatic irrigation scheduling using controller; and performance of bearing Nagpur mandarin (*Citrusreticulata cv. Blanco*) in central India with microjet irrigation systems. Authors studied tree growth, yield, nutritional status, optimum water use, uniform soil moisture distribution and availability.

10.2 MATERIALS AND METHODS

The 12–14 years old Nagpur mandarin trees (*Citrus reticulata cv. Blanco*) were used to study the effects of under-tree microjets irrigation system on the growth and productivity. The field experiment was conducted in a block of 50 × 50 m with 6 × 6 m tree spacing at experimental farm of NRCC during 2008–2011. The irrigation treatments consisted of: M1–180° microjet (2/tree) Fanjet; M2–180° microjet

(2/tree) Rayjet; M3–270° microjet (2/tree) Rayjet; and M4–300° microjet (2/tree) Rayjet.

There were six replications per treatments in a randomized block design Fig. 1. The soil in the 0–15 cm depth had pH of 7.5, $CaCO_3$ of 2.6%, sand of 31.5%, silt of 23.7%, and clay of 45.2%. The soil type is classified as fine, alkaline, hyperthermic, calcareous family of Vertic Ustochrept. Volumetric soil moisture content at field capacity (FC) and the permanent wilting point (PWP) soil moisture content at the site were 29.86% and 20.38%, respectively. The available water content of the soil was 9.48% (= 29.86–20.38%). The soil bulk density of the field was 1.34 g/cm^3. The microjet irrigation system was installed in January 2008 and irrigation treatments were initiated in April, 2009. The flow of water to the irrigation treatment was automatic with solenoid valves and was recorded with water meters.

FIGURE 1 *(Continued)*

FIGURE 1 Sustainable irrigation systems in Nagpur mandarin trees.

The average daily pan evaporation varied from 3.12 mm in November to 11.64 mm in May. The average discharge from microjet 1800 Fanjet, microjet 180° Rayet, microjet 270° Rayet and microjet 300° Rayet was 22, 18, 32 and 24 L per hour per tree, respectively. Irrigation was accordingly regulated daily by adjusting the duration of irrigation. The Hybrid Station Controller (E-6, Rain Bird, USA) and Solenoid valve (Hunter, USA) were installed in field. The easy Extra Simple Programmable (ESP) hybrid station controller (four stations) automatically operated the electronic solenoid valves for the specified programmed duration. It has three independent programs having six start times and four control stations. Each station runs for 4 h at the most. It has a setting for irrigation frequency. The water budgeting is also possible from 10 to 200% of the time set. Aluminum access tubes (50-mm diameter) were laid at the soil depth of 0.70 m within the tree basin and 0.90 m from the trunk considering the zone of maximum feeder root distribution. Soil moisture status in the tree basin was monitored regularly using a Neutron moisture probe at 15-cm, 30-cm, 45-cm, and 60-cm soil depth. The total monthly quantity of irrigation water in each treatment was recorded automatically. The increase in biometric growth parameters (tree height, and girth and canopy volume) was recorded in October 2009 and 2010. The stock girth was taken 25 cm above the soil surface. The vegetative growth parameters (tree height and tree spread) were expressed as canopy volume using the formula: canopy volume = 0.54 HD^2 where, H and D indicate tree height and trunk diameter, respectively. The leaf canopy temperature was measured using Infrared thermometer (AG4-Telatemp, USA), at the interval of 15 days between 13.00 to 14.00 time clock. The infrared thermometer was held at 50° inclined on the south facing the tree and 2 m away from the tree. The emissivity of the thermometer was fixed at 0.97.

Fifty mandarin fruits per treatment were randomly harvested for quality analysis. The total soluble solids (TSS) were determined using hand refractometer (0–32 °Brix). Titratable acidity was determined by titrating the juice against 0.1N NaOH. Percent juice content was determined by extracting the fresh juice and weighing. Five- to seven-month-old leaf samples from non fruiting terminals at 1.5–1.8 m from the ground were collected [30]. Leaf samples were later thoroughly washed, ground using a Willey grinding machine to obtain homogenous samples, and subsequently digested in tri-acid mixture of two parts $HClO_4$ + 5 parts HNO_3 + 1 part H_2SO_4 [5]. Leaf analysis consisted of N by auto-nitrogen analyzer (Model Perkin Elmer-2410), P using vanadomolybdophosphoric acid method, and K by flame photometry.

Fruits from each tree were harvested to evaluate the fruit yield per tree and various fruit quality parameters: Total soluble solids (TSS) using hand refractometer, acidity Ranganna [18]. Data for all parameters were statistically analyzed by Least Significant Difference (LSD) according to the method described by Rao [17].

10.3 RESULTS AND DISCUSSION

10.3.1 WATER MANAGEMENT

The automatic controlled microjet irrigations were scheduled based on tree water requirements each month using Class A pan evaporation and by setting the time clock for each treatment. The daily maximum open pan evaporation ranged from minimum 3.4 mm per day in December to a maximum 12.7 mm per day in May.

The minimum quantity of water requirement was 70.5 to 97.6 L per day per tree during November–December, 2009; and the maximum was 124.8 to 151.1 L per day per tree during May 2009. The quantity of daily water need using automatic microjet irrigation was minimum (83.2 to 93.2 L per day per tree) during October and maximum (151.1 to 178.3 L/day/tree) during May, 2010 (Table 1). The total quantity of daily irrigation water need was within 10–15% variation. The variation was not significantly different. The volumetric soil moisture at 15, 30, 45 and 60 cm depth was measured at the interval of 4–5 days from 1st March, 2009 to 22nd June during 2009 and 2010. The soil moisture was higher level (above 25% wet basis) in the automatic micro irrigation systems. The Fig. 2 shows the monthly soil moisture status and its distribution at 0–30 cm depth and 1.2 m spread for the automatic microjet irrigation systems. The soil moisture above 30% (w/w) was higher in the microjet 180° Fanjet compared to values for other jet systems. Similar results in Nagpur mandarin [23] and acid lime [19, 26] have been reported.

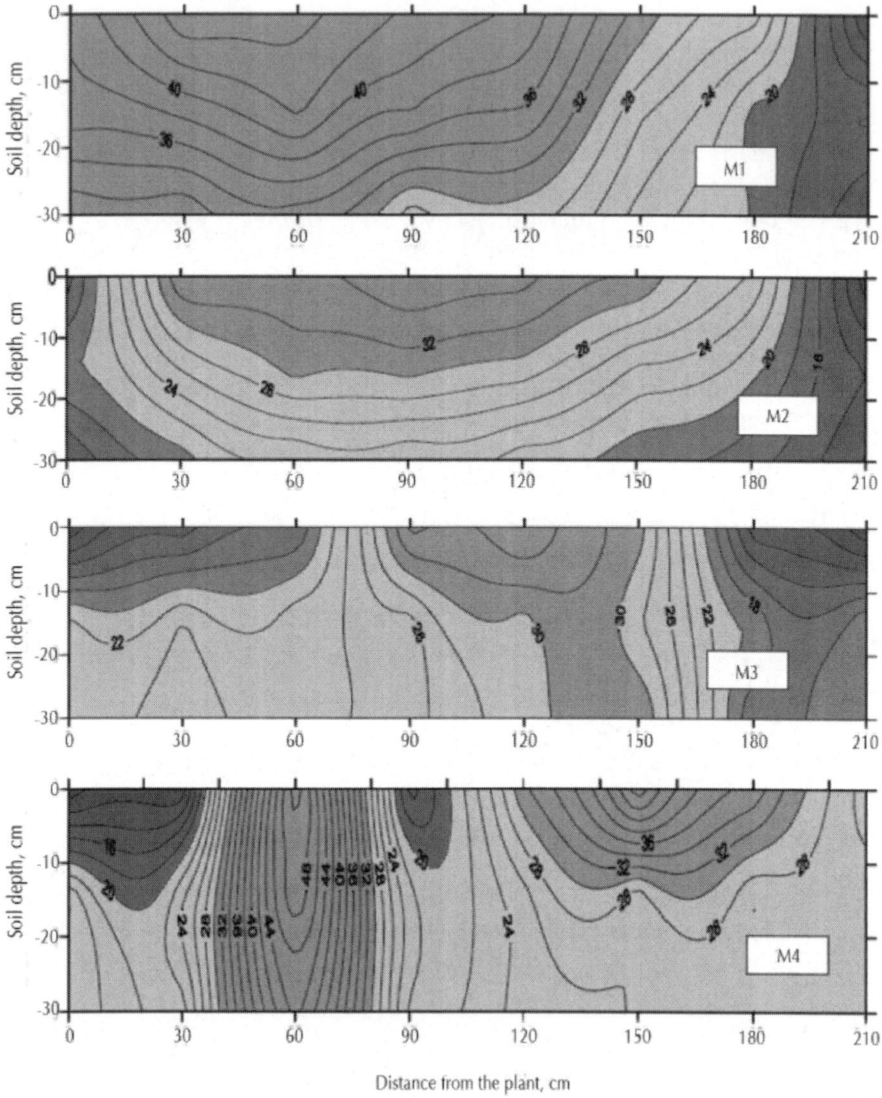

FIGURE 2 Soil moisture distribution pattern in tree root zone with micro-jet systems.

TABLE 1 Sustainable microjet irrigation requirements (liters/day/tree) during March 2009 to February 2010 and March 2010 to February 2011.

Treatments	Sustainable microjet irrigation requirements (liters/day/tree)							
	Crop season 2009–2010							
	Mar 9	Apr 9	May 9	Jun 9	Oct 9	Nov 9	Jan 10	Feb 10
M1	90.4	142.1	151.1	121.4	98.8	97.6	111.0	99.7
M2	104.7	116.6	131.2	123.9	91.3	93.9	105.4	92.9
M3	99.9	139.2	127.7	102.1	82.2	76.9	92.9	79.7
M4	94.5	116.3	124.8	101.2	79.2	70.5	86.4	76.6
Treatments	Crop season 2010–2011							
	Mar 10	Apr 10	May 10	Jun 10	Oct 10	Nov 10	Jan 11	Feb 11
M1	110.3	134.4	178.3	160.6	93.2	90.3	98.6	103.4
M2	103.4	121.7	158.2	139.3	90.7	95.4	94.5	98.8
M3	95.9	117.7	155.6	162.3	91.2	82.3	97.3	81.3
M4	93.3	112.1	151.1	149.9	83.6	86.0	93.2	88.7

M1–180° microjet (2/tree) Fanjet; M2–180°microjet (2/tree) Rayjet;

M3–270° microjet (2/tree) Rayjet; M4–300°microjet (2/tree) Rayjet.

10.3.2 GROWTH AND CANOPY VOLUME OF NAGPUR MANDARIN TREES

The efficacy of a micro irrigation system is adjudged by the extent to which evapotranspiration demand of the tree is met at critical growth stages to maintain a constant sap flow and its partitioning within the mandarin tree. The canopy volume of trees was significantly affected by the various micro irrigation systems during the year 2008–2011 Table 2.

TABLE 2 Growth parameters of Nagpur mandarin with four microjet irrigation treatments, during October 2008–2010.

Treatment	Crop season			
	2008–09	2009–10	2010–11	Mean
Tree height, m				
M1	**5.39**	**5.82**	**5.91**	**5.71**
M2	**5.24**	**5.63**	**5.7**	**5.52**
M3	**5.43**	**5.74**	**5.85**	**5.67**
M4	**5.32**	**5.53**	**5.6**	**5.48**
LSD	**NS**	**NS**	**NS**	
(P=0.05)				
Stockgirth, cm				
M1	**75.42**	**79.25**	**81.00**	**78.56**

TABLE 2 *(Continued)*

Treatment	Crop season			
	2008–09	2009–10	2010–11	Mean
M2	**74.08**	**77.92**	**81.02**	**77.67**
M3	**73.75**	**77.6**	**79.87**	**77.07**
M4	**70.00**	**77.3**	**79.50**	**75.60**
LSD (P=0.05)	**NS**	**NS**	**NS**	
Canopyvolume, m^3				
M1	**84.08**	**97.01**	**99.96**	**93.68**
M2	**85.06**	**92.71**	**94.17**	**90.65**
M3	**77.29**	**90.08**	**98.04**	**88.47**
M4	**61.01**	**74.99**	**78.44**	**71.48**
LSD (P=0.05)	**8.05**	**10.02**	**1.08**	

M1–180° microjet (2/tree) Fanjet; M2–180° microjet (2/tree) Rayjet;
M3–270° microjet (2/tree) Rayjet; M4–300° microjet (2/tree) Rayjet.

The values for tree height and stock girth were not significantly different. Maximum cumulative increase in tree canopy volume was 93.68 m^3 with microjet 180° Fanjet type irrigation system followed by 90.65 m^3 with microjet 180° Rayjet type irrigation system. The microjet 300° Rayjet type irrigation system gave lowest increase in canopy volume (71.48 m^3) due to large variation in soil moisture availability from field capacity (50–70% of AWC), thus providing a nonuniform microclimate for growth. Similar results have been reported with Hamlin orange [14] and acid lime [22, 25].

10.3.3 LEAF NUTRIENT STATUS OF NAGPUR MANDARIN

The different microjet irrigation treatments showed a significant response for the leaf nutrient composition Table 3. Optimum soil moisture distribution during the entire growth period maintained the regulated influx of macro and micronutrients within the active root zone of trees, and showed that the Fanjet micro irrigation system was highly effective compared to other three Rayjet micro irrigation systems. The leaf N, P, and K concentration increased from lower values of 2.03, 0.084, and 1.05% with microjet 300° Rayjet type irrigation system to as high as 2.17, 0.084, and 2.38% with microjet 180° Fanjet, respectively. These values were significantly higher than the other microjet irrigation systems including microjet 180° Rayjet and microjet 270° Rayjet. Similar trends in the uptake of the micronutrients (Fe, Mn, Cu and Zn) were observed with various microjet systems during 2009–2011 (Table 3). Earlier studies on Nagpur mandarin [22, 31] and acid lime [22, 27] as a test crop indicated higher leaf N, P, and K concentration under drip and micro irrigation systems compared to gravity methods of irrigation.

TABLE 3 Leaf nutrient status of Napur mandar in under different microjet irrigation treatments during 2008–2011.

Leaf nutrient status							
Treatments	**Macronutrients (%)**			**Micronutrients (ppm)**			
	N	**P**	**K**	**Fe**	**Mn**	**Cu**	**Zn**
2009–10							
M1	2.17	0.084	0.93	81.0	52.4	10.0	17.1
M2	2.17	0.078	1.04	86.6	48.0	10.4	21.5
M3	2.11	0.075	0.90	110.0	58.9	9.8	19.5
M4	2.03	0.084	1.05	140.7	33.0	8.9	14.7
CD (*P*=0.05)	NS	NS	NS	NS	NS	NS	NS
2010–11							
M1	2.34	0.12	1.42	85.4	47.6	9.8	24.8
M2	2.14	0.08	1.13	90.9	56.2	8.8	16.6
M3	2.12	0.07	0.98	94.5	52.8	9.6	15.8
M4	2.00	0.09	1.10	91.7	55.1	7.8	19.4
CD (*P*=0.05)	0.18	0.02	0.21	NS	NS	NS	NS

10.3.4 FRUIT YIELD AND QUALITY OF NAGPUR MANDARIN

The yield and fruit quality were highly influenced by the different microjet irrigation treatments Table 4. However, the response of Fanjet microjet was more pronounced than the other three microjet Ray type irrigation treatments. Higher yield under microjet irrigation was attributed to consistently and regulated supply of soil moisture within the tree rhizosphere. As shown in Table 5, the highest mandarin fruit yield was recorded with microjet 180° Fanjet (29.4 tons/ha). The moderate yield was observed with microjet 300° Rayjet (26.2 tons/ha) followed by microjet 270° Rayjet (23.6 tons/ha). The lowest fruit yield was 21.9 tons/ha was with microjet 1800 Rayjet scheduled daily. This clearly indicated that the microjet irrigation systems maintained higher as well as continuous soil moisture pattern influenced by the water and nutrient uptake resulting into good quality fruits besides enhancing the yield. The highest average fruit weight (159.8 g) and lowest acidity (0.77) was observed with microjet 180° Fanjet. The TSS (9.47 °Brix) and juice percent (39.1%) were higher with the microjet 300° Rayjet. The TSS/acidity ratio is an indicator of sweetness of the fruit. High TSS to acidity ratio implies that the fruits have more TSS (total soluble solids) and less acidity. This ratio was determined for all the treatments Table 5. The highest TSS/acidity ratio was 12.3 with microjet 300° Rayjet followed by microjet 270° Rayjet (10.9). The lowest TSS/acidity ratio was 10.8 with microjet 180° Fanjet. An improvement in fruit yield in response to irrigation systems has been reported in Navel orange [7], sweet orange [12], Satsuma mandarin [15], and Valencia orange [29].

TABLE 4 Yield and fruit performance of Nagpur mandarin in four microjet irrigation treatments during 2008–2011.

Treatments	Fruits/tree			Yield, tons/ha			Avg.weightoffruit, g		
	2008–2009	2009–2010	2010–2011	2008–2009	2009–2010	2010–2011	2008–2009	2009–2010	2010–2011
M1	500	633	763	18.9	28.7	40.3	136.6	164.4	142.6
M2	442	278	977	18.8	11.6	35.4	153.9	154.6	149.4
M3	650	88	963	26.7	4.3	39.9	148.5	171.9	150.8
M4	821	228	1015	31.5	12.0	35.1	138.7	189.4	151.5
LSD (*P*=0.05)	NS	140	121	NS	7.6	2.31	NS	23.3	1.24
	TSS, °Brix			Juice content, %			Acidity, %		
M1	8.33	8.57	10.12	33.30	40.42	42.36	0.64	1.12	0.75
M2	9.00	8.20	9.73	35.40	40.86	41.99	0.70	1.03	0.77
M3	9.20	7.57	9.63	30.10	38.41	40.60	0.70	1.00	0.72
M4	10.10	8.35	9.95	35.90	38.50	43.05	0.64	0.92	0.74
LSD (*P*=0.05)	NS	NS	NS	NS	NS	42.36	NS	NS	NS

TABLE 5 Average fruit yield and quality of the Nagpur mandarin during 2008–2011(pooled data of 3 years).

Treatments	No. of fruits	Yield	TSS	wt. of fruit	Juice content	Acidity	TSS/acid ratio
	No.	Tons/ha	°Brix	g	%	%	ratio
M1	632	29.4	9.00	147.9	38.7	0.84	10.8
M2	566	21.9	8.98	152.6	39.4	0.83	10.8
M3	567	23.6	8.80	157.1	36.4	0.81	10.9
M4	688	26.2	9.47	159.81	39.1	0.77	12.3
LSD (*P*=0.05)	NS	2.17	0.19	4.81	1.21	0.02	NS

Grieve [9] reported 12% increase in fruit yield of Valencia orange and 22% increase in water use efficiency (WUE) under micro irrigation systems compared to basin method of irrigation. Koo and Smajstrala [11] observed that fruit quality of Valencia orange was superior with trickle irrigation system. Research studies comparing drip systems with flood irrigation also demonstrated comparatively higher fruit weight, rind thickness, and juice content in sweet orange [12, 13].

10.4 SUMMARY

The sustainable higher yield and better fruit quality of Nagpur mandarin is possible with different with tree microjet irrigation systems using automatic daily scheduling. The automatic irrigation scheduling using hybrid station controller maintained higher water application in the mandarin orchard. The Nagpur mandarin yield was highest with microjet 180° Fanjet irrigation system. The fruit quality was also affected with automatic microjet irrigation systems. Highest fruit weight, TSS, juice percent and TSS/acidity ratio were obtained with microjet 300° Rayjet. The automatic microjet irrigation can be a better substitute for micro irrigation for enhancing the yield, fruit quality, water and fertilizer use efficiency.

The irrigation requirement in Ray type microjet irrigation systems was substantially optimum compared to Fanjet type microjet irrigation system. Depleting water resources in Central India and other citrus growing areas need more precise management of water in lieu of growing conditions where flowering is regulated by imposing soil-water deficit stress. The microjet 180° Fanjet or Rayjet can be used in commercial production of Nagpur mandarin in Central India. The yield and fruit quality of Nagpur mandarin can be substantially improved by adopting microjet irrigation systems and may also be used for improved soil moisture pattern, which is mainly required during fruit development stages.

The quantity of water with automatic controller in microjet irrigation systems varied from 70.5 to 142.1 L/day/tree and 82.3 to 134.4 L/day/tree during 2009–10 and 2010–11, respectively. The soil moisture distribution was higher and uniform under irrigation with 180° Fan type microjet followed by irrigation with 180°Ray type microjet. The highest average increase in canopy volume was recorded in microjet 180° Fanjet. The highest fruit yield was 29.4 tons/ha with 180° Fan type microjet followed by 26.2 tons/ha with 300° microjet (2/tree). The lowest yield was 21.9 tons/ha with 1800 Rayjet type microjet. The analysis of fruit quality revealed that total soluble solids was highest (9.47 °Brix) with 300°Ray type microjet followed by 9.0 °Brix with 180° Fan type microjet treatment. The highest juice content was 39.4% with 180° Ray type microjet irrigation compared to 180° Fan type microjet. The TSS to acidity ratio was highest with 300° Rayjet type microjet irrigation system.

KEYWORDS

- **Central India**
- **citrus**
- **citrus grove**
- **citrus orchard**
- **drip irrigation**
- **fertigation**
- **fertilizer use efficiency**
- **fruitquality**
- **irrigation**
- **leafnutrientstatus**
- **micro irrigation**
- **microjet**
- **microsprinkler irrigation**
- **Nagpurmandarin**
- **panevaporation**
- **rhizosphere**
- **soilmoisturepattern**
- **water use efficiency, WUE**
- **yield**

REFERENCES

1. Azzena, M., Deidda, P., Dettori, (1988). Drip and micro-sprinkler irrigation for young Valencia orange trees. Proc. Sixth Int. Citrus Congress, Vol. 2, Tel Aviv, Israel, pp. 747–751.
2. Capra, A., Nicosia, O. U. D. (1987). Irrigation management in citrus orchards, Irrigazine 34:3–15.
3. Castel, J. R. (1994). Response of young Clementine citrus trees to drip irrigation, Part I - Irrigation amount and number of drippers. J. Hort. Sci. 69:481– 489.
4. Cevik, B., Kaplankiran, M., Yurdakul, O. (1987). Studies for determining the most efficient irrigation method for growing lemons under Cukurovaconditions. Doga.Tarumve.Ormaniciuk, 11:42– 43.
5. Chapman, H. D., Pratt, P. F. (1961). *Methods of Analysis of Soil, Trees and Waters*. University of California, Division of Agricultural Science, USA, 182–186 pages.
6. Dasberg, S. (1995). Drip and spray irrigation of citrus orchards in Israel. In: *Micro-irrigation for a Changing World: Conserving Resources/Preserving the Environment*. Proc. Fifth Int. Micro-irrigation Congress, Orlando, Florida, U.S.A., 2–6 April, pages 281–287.
7. Fouche, P. S., Bester, D. H. (1986). The influence of water soluble fertilizer on nutrition and productivity of Navel orange trees under micro-jet irrigation. Citrus Sub-Tropical Fruit J., 62:8–12.
8. Germana, C. (1994). Increasing water use efficiency through irrigation management. Proc. Int. Soc. Citric. 2:638–642.

9. Grieve, A. M. (1988). Water use efficiency of microirrigated citrus. Proc. 4th Int. Micro-irrigation Congress, 23–28 October, Albury-Nodonga, Australia. And Trop. Agric. (Trinidad) 80(2):April 2003.
10. Kanber, R., Koksal, H., Ouder, S., Eyten, M. (1996). Effect of different irrigation methods of yield, evapo-transpiration and root development of young orange trees. Turkish J. Agric. Forestry 20:163–172.
11. Koo, R. C. J., Smjstrala, A. G. (1984). Effect of trickle irrigation and fertigation on fruit production and fruit quality of Valencia orange. Proc. Fla. State Hort. Sci. 97:8–10.
12. Kumar, A. P. A., Bojappa, K. M. (1994). Studies on the effect of drip irrigation on yield and quality of fruit in sweet oranges and economy in water use. Mysore J. Agric. Sci., 28:338–344.
13. Madrid, R., Canovas, J., Lacomba, R. F., Cano, J. A., Lorente, J., Bernal, P. (1989). Relationship between physical parameters during fruit development of oranges (cv. Valencia late), influence of irrigation method. ITFA, Production Vegetable, 20:23–26.
14. Marler, T. E., Davies, F. S. (1990). Micro-sprinkler irrigation and growth of young 'Hamlin' orange trees, J. Am. Soc. Hort. Sci. 115:45–51.
15. Peng, Y., Rabe, E. (1999). Effect of irrigation methods and ground cover on the fruit quality, yield and light levels in the canopy of microwave Satsuma. J. Fruit Sci., 15:128–132.
16. Pyle, K. R. (1985). An appraisal of micro-irrigation for use in citrus with an emphasis on drip irrigation. Citrus Sub-tropical Fruit J., 61:4–7.
17. Raghava, R. D. (1983). *Statistical Techniques in Agricultural and Biological Research*: *Designing of Experiments*, New Delhi, India, Oxford &IBH Publishing Co. Pvt. Ltd., pages 190–271.
18. Ranganna, S. (1986). *Handbook of Analysis and Quality Control for Fruit and Vegetable Products.* New Delhi, India, Tata McGraw Hill Publication Company Ltd.
19. Shirgure, P. S., Ram, L., Singh, S., Marathe, R. A., Yadav, R. P. (2000a). Water use and growth of acid lime under different irrigation systems. Indian J. Agric. Sci., 70(2):125–127.
20. Shirgure, P. S., Srivastava, A. K., Singh, S. (2001a). Effect of pan evaporation based irrigation scheduling on yield and quality of drip irrigated Nagpur mandarin. Indian J. Agric. Sci., 71(4): 264–266.
21. Shirgure, P. S., Srivastava, A. K., Singh, S. (2001b). Growth, yield and quality of Nagpur mandarin (*Citrus reticulate* Blanco) in relation to irrigation and fertigation. Indian J. Agric. Sci., 71(8):547–550.
22. Shirgure, P. S., Srivastava, A. K., Singh, S. (2001c). Effect of drip, micro-jets and basin irrigation method on growth, soil and leaf nutrient change in acid lime. Indian J. Soil Cons., 29(3):229–234.
23. Shirgure, P. S., Srivastava, A. K., Singh, S. (2001d). Effect of micro-jet irrigation system on growth, yield and fruit quality in Nagpur mandarin. South Indian Hort., 49 (Special):357–359.
24. Shirgure, P. S., Srivastava, A. K., Singh, S. (2003a). Evaluating micro-irrigation systems in Nagpur mandarin under sub- humid tropical climate. Tropical Agriculture, 80(2):91–96.
25. Shirgure, P. S., Srivastava, A. K., Singh, S. (2003b). Irrigation scheduling and fertigation in acid lime (*Citrus aurantifolia Swingle*). Indian Journal of Agricultural Sciences, 73(7):363–367.
26. Shirgure, P. S., Srivastava, A. K., Singh, S., Pimpale, A. R. (2004a). Drip irrigation scheduling growth, yield and quality of acid lime (*Citrus aurantifolia Swingle*). Indian J. of Agricultural Sciences, 74(2):92 –94.
27. Shirgure, P. S., Srivastava, A. K., Singh, S. (2004b). Integrated water and nutrient management in acid lime. Indian Journal of Soil Conservation, 32(2):148–151.
28. Simpson, G. H. (1978). Developments in under tree irrigation systems in the Murray valley. Proc. Int. Soc. Citric., 234–235
29. Smajstrla, A. G. (1993). Micro-irrigation for citrus production in Florida. Hort. Sci. 28:295–298.
30. Srivastava, A. K., Kohli, R. R., Dass, H. C., Huchche, A. D., Ram, L., Singh, S. (1999). Evaluation of the nutritional status of Nagpur mandarin (*Citrus reticulata* Blanco) by foliar sampling. Trop. Agric. (Trinidad), 76(2):93–98.

31. Srivastava, A. K., Shirgure, P. S., Singh, S. (2003). Differential fertigation response of Nagpur mandarin (*Citrus reticulata* Blanco) on an alkaline Inceptisol under sub-humid tropical climate. Tropical Agriculture (Trinidad), 80(2):97–104.

CHAPTER 11

MICRO IRRIGATION SCHEDULING IN NAGPUR MANDARIN

P. S. SHIRGURE

CONTENTS

11.1 INTRODUCTION

Citrus reticulata cv. Blanco (Nagpur mandarin) production in India is 0.875×10^6 tons that is grown on 0.148×10^6 ha (bearing area is 86,200 ha). The average citrus yield is 10–11 tons/ha that is too low compared to other citrus cultivars, due to crop water stress, inadequate method of scheduling, poor irrigation efficiency, lack of modern irrigation technology, and inadequate soil moisture during the critical plant growth and fruit developmental stages. In Nagpur mandarin orchards, the conventional irrigation methods are being replaced with micro irrigation systems, due to increasing scarcity of water. However, the micro irrigation system is not operated regularly to maintain the correct irrigation intervals and irrigation uniformity, which may be due to inadequate maintenance and the manual operation. The crop yield can be increased from 10 to 15 tons/ha with adoption of the modern fully automated micro irrigation system. The adoption of automated microjet irrigation systems in combination with fertigation enhanced the production and yield of the Nagpur mandarin [11], and fruit quality [9, 10, 12].

The computerized method of irrigation scheduling allows adequate irrigation management and the growers can know when critical moisture levels are expected to occur. The simple, reliable and accurate automated system enables rapid analysis of a number of variables. The inputs by the farmer are minimal namely: monitor in grain fall and irrigation levels; and determining oil moisture depletion level, etc. Citrus growers are provided with a projected data when crop stress will start, so that the technician can start irrigation [4] with a computer based feedback system. Information on soil moisture and fertilized levels is registered by sensors and is fed into a microcomputer, which initiates, controls and terminates the irrigation or fertigation [1]. Eight to ten year old trees cv. *Valancia* on citrus *aurantium* rootstock planted at 8 × 4 m in red loamy soil in Cuba were irrigated at 65, 75, 85, the conventional 80% of field capacity, and zero irrigation. The irrigation at 85%field capacity gave highest yield and improved fruit quality [6]. Cavazza [2] assessed the value of automatic irrigation systems to reduce labor and water consumption costs with three automation methods: Local automation control, Cyclic automation and Central programming automation.

A simplified method of irrigation scheduling for citrus orchards in the Mediterranean was developed in Sicily, Italy. The automation system was implemented using data from a meteorological station, a personal computer, field units and solenoid valves. Daily gross requirements were calculated with net water requirement and irrigation application efficiency. Reference evapotranspiration was estimated based on Hargreaves-Samani model and the class A pan evaporation data [7]. A survey of the Florida citrus industry revealed that larger operations are more likely to use a computer than smaller operations. If the computers are available, then largest citrus operations are more likely to use more specialized software applications for irrigation automations, citrus production, decision aids, and accessing local weather information [3].

The existing micro irrigation system in Nagpur mandarin orchard was converted to automatic micro irrigation system using ESP-4 Hybrid Station Controller, and 100 PGA solenoid valve. The soil moisture content at 30 cm depth in the root zone ranged from 27.2–29.8%, which is 10–15% below the field capacity. The canopy volume of the mandarin enhanced from 65.2 to 81.4 m^3 due to micro irrigation modernization. Fruit yield also increased from 17.6 to 25.1 tons/ha. The fruit quality was also improved due to automatic controller based micro irrigation system. The optimum quantity of water required for the bearing Nagpur mandarin plants was minimum (47 to 70 L/day/plant) during October–December and was maximum (118 to 129 L/day/plant) during April–June. The study indicated that the automatic micro irrigation scheduling using controller has potential for water saving and sustaining the Nagpur mandarin yield and quality [8, 13]. Pulse irrigation system is shown in Fig. 1.

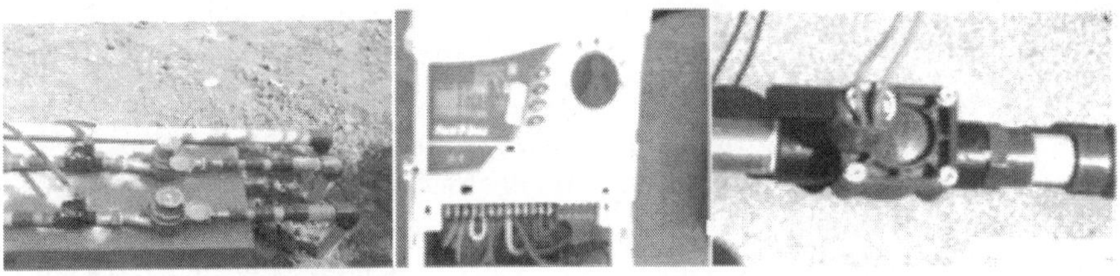

FIGURE 1 Pulse irrigation.

This chapter discusses: (1) The conversion of the existing micro irrigation system in Nagpur mandarin orchard to the automatic micro irrigation scheduling system using ESP-6 Hybrid Station Controller and solenoid valves; (2) The water was pumped automatically from the subsurface ground water according to the irrigation needs (depth, duration and frequency) indicated by the controller to maintain irrigation uniformity and continuous soil moisture in the root zone; effects on the yield and quality of Nagpur mandarin. The author studied automatic micro irrigation scheduling daily as well as alternate-day to evaluate the effects on tree performance, yield and quality of 12–14 years old Nagpur mandarin trees. He also considered potential evapotranspiration.

11.2 MATERIALS AND METHODS

The automatic irrigation scheduling experiment was conducted at experimental farm of Nagpur Research Citrus Center of ICAR (NRCC) during 2008–2011. The experimental site was 0.25 ha with six rows of citrus trees. The treatments were automatic irrigation daily with 60min interval three times (*I1*), automatic irrigation daily with 90-min interval two times (*I2*), automatic irrigation on alternate day with 120 min three times (*I3*) and automatic irrigation on alternate day with 180 min two times (*14*) with six replication sin ran domize d block design. The soil texture was clay loam at a soil depth of 41 cm. The composites oil samples were collected for determination of field capacity (FC) and permanent wilting point (PWP). The FC and PWP were 30.44% and 19.56%, respectively. The available water content of the soil was 10.89% [= 30.44–19.56]. The soil bulk density was 1.34 g/cc that was determined using core sample of 100 cm^3 and oven drying method. Hydraulic conductivity was 14.6 cm/m of soil depth.

The Extra Simple Programming (ESP) with self-display made the electronic controller easy to program, read and work. The easy programmable hybrid station controller (4–6 stations) automatically was installed to operate the electronic solenoid valve forth specified program d duration. The automatic controller had three program options (A, B and C) with six independent start times and four controlstations. Each station was run for 4 h. The controller also had a feature for setting the frequency of irrigation. The water budgeting is also possible from 10 to 200% of the time set. The Hybrid Station Controller (E-6, Rain Bird, USA) and Solenoid valve (Hunter, USA) are installed in field irrigated with micro irrigation.

The electrical control panel consisted of power supply, main switch, pump control relays and hybrid station controller. The ground station consisting of valves and water meters was also installed to operate the system according to the controller settings for each treatment. The irrigations were based on open class A pan evaporation and by setting the time in each treatment according to the water need of plant each month. The micro irrigation system consisted of 16 mm dia. lateral and 8 lph drippers (4/plant) and other accessories. The plant growth parameters were recorded during October 2008. Increase in plant height, girth and canopy volume were recorded in October 2009–2010. The stock girth was taken at 15 cm and scion girth at 25 cm above the soil surface. The canopy volume of the mandarin tree was calculated using spread and canopy height using Castle's formula. The total fruits harvested from each tree were weighed for evaluating the fruit performance. The total soluble solid was determined using hand refractometer (0–32 °Brix). Titratable acidity was determined by titrating the juice against 0.1N NaOH. Percent juice content was determined by weighing the fresh juice. The data was analyzed with standard procedure by SAS.

11.3 RESULTS AND DISCUSSION

11.3.1 WATER USE WITH AUTOMATION

The irrigation scheduling was based on class A pan evaporation and by setting the time for each treatment according to the water need of tree every month. The total quantity of irrigation water scheduled on daily as well as on alternate day basis was nearly same. The daily class A pan evaporation ranged from a minimum of 3.4 mm per day in December to a maximum of 12.7 mm per day in May. The minimum quantity of water was 46.9 to 55.4 L per day per tree during November–December, 2009 and the maximum was 118.4 to 129.1 L per day per tree during May 2011. The quantity of water for Nagpur mandarin scheduled using automatic micro irrigation, and daily and alternate day basis was minimum (65.00 to 72.4 L per day per tree) during October and was maximum (133.04 to 147.7 L/day/plant) during May, 2010. The variation in monthly water quantity of water per tree was not significant at P = 0.05. The in situ soil moisture was monitored from 1st March, 2009 to June 22, 2010. The volumetric soil moisture at 15, 30, 45 and 60 cm soil depth was measured at an interval of 4–5 days. The soil moisture was monitored at higher level (above 25% wet basis) in the automatic irrigation scheduled daily with 90 min two times and automatic irrigation scheduled on alternate day with 180 min two times. The soil moisture was maintained between 15–25% in automatic irrigation scheduled daily with 60 min three times and irrigation scheduled on alternate day with 120 min three times. This research study indicates that the automatic irrigation scheduling affected the soil moisture and it was higher during the critical summer months from March through June. This clearly indicates that soil moisture was maintained higher in automatic irrigation scheduled automatic irrigation daily with 90 min interval two times and automatic irrigation daily with 180 min interval two times, which had higher and continuous flow rates. The fluctuations over the period were not observed. The study concludes that higher soil moisture during the year 2009 was maintained in the automatic irrigation scheduling having 90 min two times daily and 180 min two times on alternate days. During the year 2010, the soil moisture was maintained higher and uniform in automatic irrigation scheduling daily with 90 min interval two times and automatic irrigation on alternate day with 180 min two times, which have higher and continuous flow rates.

11.3.2 PERFORMANCE OF NAGPUR MANDARIN TREES

The growth of Nagpur mandarin was affected by automatic irrigation scheduling treatments based on daily and on alternate day. The growth of mandarin was recorded during October of 2008–2010. Data on tree height and tree spread were used to estimate the canopy volume. The tree height and stock girth were not significantly different at P = 0.05. The average tree height of the Nagpur mandarin ranged from 5.10–5.42 m, and stock girth varied from 71.75–76.03 cm. The significant differences in canopy volume were observed at P = 0.05, ranging from 64.56–87.81 m^3 (Table 1). The average tree height and stock girth were higher in automatic irrigation on alternate day with 120 min three times followed by automatic irrigation daily with 90 min interval two times. The canopy volume was significantly affected due to the automatic irrigation scheduling. The average canopy volume was higher (87.81 m^3) in automatic irrigation scheduling on alternate day with 120 min interval three times followed by automatic irrigation scheduling daily with 180 min interval two times (84.83 m^3) compared to the automatic irrigation scheduling daily 60 min interval three times (66.6 m^3) and automatic irrigation scheduling daily with 90 min interval two times (64.56 m^3)

during 2008–2011. The differences were mainly due to availability of constant and continuous soil moisture in root zone. The similar observations have been also recorded in the studies on Nagpur mandarin [9].

TABLE 1 Effects of four irrigation scheduling treatments on growth of Nagpur mandarin during October 2008–2010.

Treatment	Plant height (m)				Stock girth (cm)				Canopy volume (m^3)			
	2008	2009	2010	Mean	2008	2009	2010	Mean	2008	2009	2010	Mean
I1	4.96	5.11	5.23	5.10	67.42	72.92	74.90	71.75	61.94	67.99	69.87	66.60
I2	5.11	5.22	5.27	5.20	69.5	76.25	78.25	74.67	56.18	67.11	70.38	64.56
I3	5.26	5.45	5.54	5.42	72.75	76.33	79.00	76.03	81.43	89.64	92.37	87.81
I4	5.27	5.29	5.45	5.34	72.75	77.13	78.20	76.03	78.99	86.87	88.62	84.83
LSD (P = 0.05)	NS	NS	NS	NS	NS	NS	NS	NS	13.61			1.34

I1 – Automatic irrigation daily with 60 min interval three times;

I2 – Automatic irrigation daily with 90 min interval two times;

I3 – Automatic irrigation at alternate day with 120 min three times; and

I4 – Automatic irrigation at alternate day with 180 min two times.

11.3.3 FRUIT YIELD AND QUALITY

The average number of fruits per tree, yield, TSS, Juice content, acidity and TSS to acidity ratio were analyzed during the study period and pooled data are presented in Table 2. The Nagpur mandarin fruits were harvested during first two weeks of November, and the samples were randomly selected to evaluate the fruit performance. During 2008–2011, Nagpur mandarin yield and fruit quality of the Nagpur mandarin were significantly affected by the automatic controller based irrigation scheduling on daily and alternate day basis having duration of 1–3 h and two to three pulses a day. The fruits per tree, fruit yield per ha, TSS, and juice content were significantly different at $P = 0.05$ among the irrigation scheduling treatments during 2008–2011. The average fruit weight and acidity were not significantly different at $P = 0.05$ that may be attributed to the internal fruit quality and micro irrigation uniformity.

TABLE 2 The average fruit yield and quality of the Nagpur mandarin during 2008–2011 (pooled data for 3 years).

Treatment	No. of fruits per tree	Yield (tons/ ha)	TSS (°Brix)	Average weight offruit(g)	Juice (%)	Acidity (%)	TSS/acidity ratio
I1	606	24.50	9.71	146.54	38.64	0.83	11.7
I2	747	30.11	9.49	151.96	37.68	0.81	11.7
I3	726	30.91	10.22	153.67	40.77	0.78	13.2
I4	638	27.04	9.92	152.23	37.93	0.80	12.4
LSD (P=0.05)	29	0.54	0.37	0.81	1.21	0.04	0.72

The treatments are described in Table 1.

The yield and fruit quality values were significant at P = 0.05 for all irrigation treatments. The average number of fruits per tree varied from 606 to 726 in all treatments. The number of fruits per tree was highest in the automatic irrigation on alternate day with 120 min three times followed by automatic irrigation daily with 90 min interval two times. The yield of the Nagpur mandarin was significantly influenced by various micro irrigation scheduling. The yield increased from 24.5 to 30.91 tons/ha. The highest mandarin fruit yield was recorded in the automatic micro irrigation on alternate day with 120 min three times (30.91 tons/ha). The moderate yield was observed in automatic micro irrigation daily with 90 min interval two times (30.11 tons/ha) followed by automatic micro irrigation daily with 180 min interval two times (27.04 tons/ha). The lowest fruit yield was seen in irrigation scheduled daily having 60 min interval three times. This concludes that the automatic micro irrigation scheduling on daily and alternate days maintained higher as well as continuous soil moisture influenced by water and nutrient uptake resulting into good quality fruits besides enhancing the yield. The mandarin fruit diameter ranged from 1.51 to 6.87 cm during the study period. High fruit growth rate was seen in automatic irrigation on alternate day with 180 min two times in 2009 and 2010.

The highest average fruit weight (153.67 g.) and lowest acidity (0.78) were observed in the automatic micro irrigation on alternate day with 120 min three times. The TSS (10.22 °Brix) and juice percent (40.77%) were higher in the automatic irrigation on alternate day with 120 min three times. The TSS/acidity ratio is an indicator of sweetness of the fruit. The high TSS to acidity ratio implies that the fruits have more TSS (total soluble solids) and less acidity. This ratio was analyzed for all the treatments. The highest TSS/acidity ratio was found in the automatic micro irrigation on alternate day with 120 min three times (13.2) followed by automatic micro irrigation on alternate day with 180 min two times (12.4). The lowest TSS/acidity ratio (11.7) was observed in automatic micro irrigation with 60 min three times daily.

11.4 CONCLUSIONS

The sustainable and higher production of Nagpur mandarin is possible with automatic micro irrigation scheduling daily or on alternate days, which maintained higher water application to the mandarin trees. Automated micro irrigation maintained the soil moisture status above 25% (wet basis) throughout the fruit growing period. The automatic irrigation on alternate day with 120 min three times gave highest values of yield, fruit weight, TSS, juice percent and TSS/acidity ratio. The automatic micro irrigation scheduling is a good substitute for manual micro irrigation operation to enhance the yield, fruit quality, water and fertilizer use efficiency.

11.5 SUMMARY

During 2008–2011, the hybrid station controller based automatic pulse irrigation scheduling field experiment was conducted on 10–12 years old bearing Nagpur mandarin (*Citrus reticulata* Blanco) at National Research Center for Citrus, Nagpur. The objective was to study the automatic daily micro irrigation scheduling daily as well as alternate day based on time schedules and potential evapotranspiration. The treatments consisted of automatic daily irrigation daily with 60 min interval three times (I_1); automatic irrigation daily with 90 min interval two times (I_2); automatic irrigation on alternate day with 120 min three times (I_3); and automatic irrigation on alternate day with 180 min two times (I_4) with six replications in randomized block design. The automatic hybrid station controller E-6 (Rain Bird, USA) was used for micro irrigation scheduling setting the time for each treatment based on the tree water need and class A pan evaporation. The various scheduling treatment timings were programmed in A, B and C options of the hybrid station controller. The sustainable production of Nagpur mandarin is possible with automated micro irrigation scheduling daily or on alternate days.

The water use in October varied from 65.0–72.4 L/day/plant and during May–June it was 133.0–147.7 L/day/plant. The leaf nutrient status was high with automatic alternate day micro irrigation scheduling. The canopy temperature was positively influenced with automatic micro irrigation scheduling. The Nagpur mandarin fruit yield was highest (30.91 tones/ha) with irrigation on alternate day 120 min three times, followed by irrigation scheduled with 90 min interval two times daily (30.11 tones/ha). Fruit weight (154.7 g), TSS (10.22 °Brix) and juice percent (40.77%) were significantly different with automatic irrigation at alternate day with 120 min three times. The automatic micro irrigation scheduling may be better option than the manual micro irrigation scheduling to enhance the water use efficiency of Nagpur mandarin.

KEYWORDS

- **automatic pulse irrigation scheduling**
- **Brix**
- **citrus**
- **citrus grove**
- **class A pan evaporation**
- **Extra Simple Programming, ESP**
- **fertigation**
- **Florida**
- **fruit quality**
- **fruit yield**
- **Hargreaves-Samani**
- **Hybrid station controller**
- **hybrid station controller**
- **irrigation**
- **juice percent**
- **mandarin**
- **micro irrigation**
- **Nagpur**
- **tensiometer**
- **TSS**
- **water use efficiency**

REFERENCES

1. Bolden, J. L., Fay, K. B., Miller, R. N. (1985). Computer based feedback system for the care and management of citrus groves. Proc. of American Society of Agriculture Engineers conference held at Chicago during 25–28th February, 90–95 pages.
2. Cavazza, D. (1991). Farm automation of irrigation systems in open–field greenhouse crops. Irrigazione Drenaggio, 38(4):21–26.
3. Ferguson, J. J., Israel, G. D. (1998). Computer use by Florida citrus growers. Proc. of 7th International Conference on Computers in Agriculture, Orlando, Florida, during 26–30th October.
4. Jackson, J. L., Ferguson, F. J. (1981). Computerized irrigation scheduling. Proc. of the Florida State Hort. Sci., 93:5–6.
5. Marsh, A. W. (1968). Automatic tensiometer signaled irrigation systems for orchards. California Citrograph, 54:2–12
6. Peres, E. M. T. (1987). Effort of irrigation on the yield and quality of orange fruits. Rosteniev dni Nauki, 24(2):71–76.
7. Sardo, V. (1992). Simplified automatic management of irrigation. Proc. of 7th International Citrus Congress, Acireale – Italy, 8–13 March.
8. Shirgure, P. S., Srivastava, A. K. (2012). Automatic pulse irrigation scheduling for enhancing the productivity and quality of drip irrigated Nagpur mandarin (*Citrus reticulata* cv. Blanco). Paper presented in 5th Indian Horticultural Congress 2012 at PAU, Ludhiana held during 6–9th November. 283 pages.

9. Shirgure, P. S., Srivastava, A. K., Singh, S. (2000). Automatic micro irrigation system in Nagpur mandarin. Abstract 9th International Society of Citriculture Congress held at Florida (USA) on 3–7 December, 141 pages.
10. Shirgure, P. S., Srivastava, A. K., Singh, S. (2001). Growth, yield and quality of Nagpur mandarin (Citrus reticulata Blanco) in relation to irrigation and fertigation. Indian J. Agri. Sci., 71(8):547–550
11. Shirgure, P. S., Srivastava, A. K., Singh, S. (2003). Evaluating micro-irrigation systems in Nagpur mandarin under sub-humid tropical climate. Trop. Agri. (Trinidad), 80(2):91–96.
12. Shirgure, P. S., Srivastava, A. K., Singh, S. (2004). Automation of micro irrigation system improved yield and quality of Nagpur mandarin (*Citrus reticulata* Blanco). Paper presented at First Indian Horticultural Congress at IARI, New Delhi held during 6–9 November. 205 pages.
13. Shirgure, P. S., Srivastava, A. K., Singh, S., Panchariya, G. S. (2005). Yield and quality of Nagpur mandarin under different automatic micro irrigation schedules. Paper presented at the International Conference on Plasticulture and Precision Farming at Hotel Ashoka, New Delhi during 17–21 November. 228 pages.

CHAPTER 12

PERFORMANCE OF NAGPUR MANDARIN WITH PRACTICES

P. S. SHIRGURE

CONTENTS

12.1 INTRODUCTION

In majority of commercially grown citrus orchards, citrus is usually cultivated as monoculture and any onset of epidemic may lead to crop destruction. In major citrus grown orchards in India, cotton is one of the commercial crops besides pulse and oilseed crops [1, 3, 12]. The intercropping in citrus is reported by Krishnamurti [7] and Gill [4]. The research on sustainable intercropping practices in Nagpur mandarin orchards with single inter strip crops is being conducted [5, 6, 8]. The integrated citrus based cropping system: (1) Optimizes land use, maximum return per unit area, soil conservation and fertility build-up, waste recycling and insurance against failure of individual crop; and (2) Provides year round employment and reduces total cost of production.

In India, the citrus production is about 0.86 million tons per annum on an area of 0.923 million hectares. The important commercial citrus cultivars of citrus are mandarin (*Citrus reticulata* Blanco), sweet orange (*Citrus sinensis* Osbeck), and *acid lime* (*Citrus aurantifolia* Swingle) with a total production of 1.634, 3.567 and 2.571 million tons, respectively. Nagpur mandarin (*Citrus reticulata* Blanco) is an important commercial citrus crop in the Vidarbha region of Maharashtra – India [13].

The declining of the citrus orchards in India is a major concern due to lack of disease-free planting material, inadequate sustainable soil and water practices, poor health management [2, 9], lack of irrigation water resources, and conventional irrigation practices. The Nagpur mandarin orchards are also declining due to shortage of irrigation water during the critical tree growth stage and poor drainage system during the rainy season [10]; inadequate best management practices (BMP). The adoption of intercropping practices provides an effective strategy to obtain additional income during off-season without inducing soil moisture stress and soil infertility. The adoption of strip cropping with pulse and cotton in prebearing citrus orchards is highly remunerative per unit area both qualitatively and quantitatively in irrigated agro ecosystem. The fruit bearing of the mandarin starts from fourth year onwards [11]. Therefore, during the prebearing Nagpur mandarin orchard establishment, unused row spaces can be effectively used for sustainable intercropping practices. This helps citrus growers economically. The innovative intercropping system in Nagpur mandarin grove has not studied in central Indian conditions, with cotton as a main intercrop and the leguminous/ oilseed crops (e.g., soybean, groundnut and gram) as sub intercrops between the cotton and mandarin trees.

In this chapter, the author presents research results on performance of Nagpur mandarin with sustainable practices to recommend suitable intercropping system consisting of cotton and leguminous crops. He also discusses the relationship and interactions between Nagpur mandarin trees and intercropping system consisting of cotton, soybean, black gram, groundnut, gram and mung beans (also known as green gram or golden gram in India: *Vigna radiata*). According to the author, the complementary, competitive and supplementary nature of such practices near the mandarin root zone improve the soil moisture and soil fertility status in the long-term with emphasis on the health and yield performance of citrus trees with regard to plant growth, yield, quality, nutrient and moisture conservation of main and intercrops.

12.2 MATERIALS AND METHODS

The research, on sustainable intercropping practices in prebearing and bearing Nagpur mandarin during 2009–2012, consisted of seven intercrops in the farmer's field at Sawandri and Brhamni villages of Nagpur District in India. The following intercropping treatments were evaluated using randomized block design with three replications and six trees per plot:

T1: Nagpur mandarin + no intercrop;
T2: Nagpur mandarin + cotton;
T3: Nagpur mandarin + cotton + soybean;
T4: Nagpur mandarin + cotton + black gram;
T5: Nagpur mandarin + cotton + groundnut;
T6: Nagpur mandarin + soybean followed by gram;
T7: Nagpur mandarin + black gram followed by gram; and
T8: Nagpur mandarin + groundnut followed by summer mung.

The soil was moderately deep (49 cm), well drained, calcareous, clayey, gently sloping with land capability class III. The soil field capacity and permanent wilting percentage were determined using pressure plate apparatus (Soil moisture Inc., Santa Barbara, USA). Soil bulk density was estimated with core method and oven drying methods. The prebearing and bearing mandarin trees were spaced at 6 m with an average canopy area of 10.75–13.85 m^2.

In prebearing stage of Nagpur mandarin, the cotton was sown at a plant spacing of 5 m between two rows of mandarin trees at 6 m spacing. In bearing mandarin grove, inter space was 3 m for intercropping. In case of Cotton with other intercrops (T2, T3, and T4), the cotton was spaced at 3 m and soybean/black gram/groundnut at one meter spacing on either side of cotton. In other than cotton treatments (T5, T6, T7 and T8), the intercrop was sown at 5 m spacing between two rows of main crop in kharif crop (Indian word: refers to the planting, cultivation and harvesting of any domesticated plant sown in the monsoon rainy season, and these crops are usually sown with the beginning of the first rains towards the end of May) and then followed with rabi crop (Indian word: refers to agricultural crops sown in winter and harvested in the spring; these crops are sown after the departure of monsoon rains and harvested in the beginning of April/May) such as gram/summer mung in the residual moisture with light irrigation if required. In bearing mandarin stage, light to medium pruning was done.

In bearing mandarin orchard, plant spacing of cotton was 3 m space between two rows of mandarin spaced at 6 m (T2). Cotton with other intercrops (T3, T4, T5) was sown at 3 m spacing with the central row at 1 m for cotton and soybean/

black gram/groundnut on both sides at one meter. In treatments without cotton crop (T6, T7, T8), the kharif intercrop was sown at 3 m spacing between the tree rows and then followed with rabi intercrop. In the kharif season, cotton (var. LRK-516), soybean (var. JS-335), black gram (var. TAU-1) and groundnut (var. JL-24) were sown as intercrops. After the monsoon season land attained field capacity (1/3 bar soil moisture), the kharif intercrops were sown. In rabi season, gram (var. Chaffa) and summer mung (var. K-581) were intercropped..

In prebearing mandarin orchard, the soil field capacity was 31.4% at 102 cm soil depth and 29.8%, at 78 cm. The soil bulk density was 1.35 g/cc in prebearing and 1.54 g/cc in bearing Nagpur mandarin orchards, respectively. From June through September, sufficient soil moisture in the citrus grove as well as in intercrop regions was maintained due to effective rainfall. From October through May, conventional method of gravity irrigation was used. Surface flooding was used for intercrops using basin method of irrigation, based on the calendar method of irrigation scheduling. Irrigation was initiated when 50% of available water content was depleted. The soil moisture at 30 cm depth was monitored every 15 days with the soil moisture monitoring probe (Profile Probe PR1, Delta T, UK) and soil moisture monitoring meter (HH2 Delta T, UK). The fiberglass-reinforced plastic (FRP) tubes were installed in each treatment for monitoring the soil moisture with the profile probe.

The experiment was initiated and initial growth parameters were recorded during October 2009 subsequently. Increase in tree growth parameters (tree height, stock and scion girth and canopy volume) were recorded in October 2011 and 2012. The stock girth was measured 15 cm and scion girth at 25 cm above the ground surface. The canopy volume of the mandarin tree was calculated using spread and canopy height with a Castle's formula. The total fruits from each tree were harvested and weighed to calculate the yield. A total of 50 fruits per treatment were randomly taken for analyzing the fruit quality. The total soluble solids (TSS) were determined using hand refractometer (0–32 °Brix). Titratable acidity was determined by titrating the juice against 0.1N NaOH. Percent juice content was determined by weighing the extracted juice.

12.3 RESULTS AND DISCUSSION

12.3.1 EFFECTS OF SUSTAINABLE INTERCROPPING PRACTICES ON SOIL MOISTURE CONSERVATION IN NAGPUR MANDARIN

The soil moisture status in Nagpur mandarin with intercrop cotton was lower than the other intercropping treatments (Table 1) due to higher soil moisture extraction by the cotton crop. It can be attributed to narrow row spacing and larger foliage coverage. This also reveals that cotton crop required more moisture compared to soybean, black gram and groundnut intercrops. Comparatively higher soil moisture was observed in soybean and groundnut intercrops. Moderately high soil moisture content was recorded in rabi season in treatments with soybean followed by gram, and groundnut followed by mung bean. Lower soil moisture was observed during April-May due to high temperature and evaporation during summer months. The advantage of conserved soil moisture was more in prebearing Nagpur mandarin orchard than the bearing grove. It is concluded that the sustainable intercropping practices can best use the available soil moisture during the rainy season and irrigation water (Table 1).

TABLE 1 Effect of sustainable intercropping practices on soil moisture status in Nagpur mandarin orchard during 2010–2012.

Treatment	Soil moisture at 30 cm depth,% (wet basis)							
	2010–2011				2011–2012			
	Jun–Sep	Oct-Dec	Jan-Mar	Apr–May	Jun–Sep	Oct-Dec	Jan-Mar	Apr–May
Pre-bearing Nagpur mandarin orchard								
T1	25.49	25.12	31.87	28.99	26.44	21.12	23.38	23.58
T2	21.81	28.87	30.82	27.50	25.87	16.13	24.42	11.05
T3	19.40	29.74	31.23	28.89	27.41	24.82	22.80	18.65
T4	25.82	29.93	33.38	26.11	25.34	18.43	22.70	13.94
T5	27.59	27.08	32.21	28.75	25.34	18.09	24.10	22.62
T6	30.65	28.82	31.72	34.19	27.40	19.69	23.18	22.67
T7	25.44	29.90	32.19	29.87	30.53	22.46	22.05	22.85
T8	28.37	28.21	25.61	26.91	28.42	19.87	24.87	22.75
CD (P=0.05)	NS	NS	NS	2.4	NS	NS	NS	1.8
Bearing Nagpur mandarin orchard								
T1	18.62	28.04	20.18	24.30	20.80	19.80	20.77	24.10
T2	20.04	27.82	23.05	27.57	18.86	20.70	20.52	25.02
T3	21.44	27.08	22.42	22.54	18.25	20.40	19.88	27.34
T4	27.94	27.82	22.03	18.11	22.78	20.90	20.87	25.75
T5	29.75	27.05	20.92	17.54	20.56	20.05	23.67	25.83
T6	24.33	27.36	24.41	22.64	18.91	21.50	25.12	28.75
T7	31.87	26.82	22.17	20.64	20.27	20.70	20.16	28.54
T8	24.55	29.26	24.06	21.35	24.03	20.60	21.42	26.31

TABLE 1 *(Continued)*

Treatment	Soil moisture at 30 cm depth,% (wet basis)							
	2010–2011				2011–2012			
	Jun–Sep	Oct-Dec	Jan-Mar	Apr–May	Jun–Sep	Oct-Dec	Jan-Mar	Apr–May
CD (P=0.05)	NS	NS	NS	1.9	NS	NS	NS	1.5

Sustainable practices: Intercropping treatments

T1 – Nagpur mandarin;
T2 – Nagpur mandarin + cotton;
T3 – Nagpur mandarin + cotton + soybean;
T4 – Nagpur mandarin + cotton + black gram;
T5 – Nagpur mandarin + cotton + Groundnut;
T6 – Nagpur mandarin + soybean followed by gram;
T7 – Nagpur mandarin + black gram followed by Gram; and
T8 – Nagpur mandarin + groundnut followed by summer mung beans.

Increase in soil moisture content during April-May months of summer was significant at P = 0.05. Higher soil moisture content at 30 cm depth was observed: 34.2% for prebearing in T6 and 29.6% for bearing in T5 during 2010–2011; and 30.53% for prebearing in T7 and 28.75% for bearing in T6 during 2011–2012 (Table 1). Amongst all treatments, the treatment with soybean, black gram and groundnut intercrops resulted in significantly higher increase in soil moisture at 30 cm during both the years. Chadha et al. [1] reported similar results in their research with intercropping practices in young citrus orchards. The higher soil moisture content below the crop canopy of the intercropping treatments can be due to reduction in soil surface evaporation and weed growth. The Nagpur mandarin with cotton intercrop also conserved soil moisture in the tree root zone compared to Nagpur mandarin alone.

12.3.2 EFFECTS OF SUSTAINABLE INTERCROPPING PRACTICES ON PERFORMANCE OF NAGPUR MANDARIN

Sustainable intercropping practices in this research affected tree height, stock girth, scion girth and canopy volume of Nagpur mandarin (Table 2). The increase in tree height, stock girth and scion girth was not significant. However, the canopy volume of the plant was significantly influenced by the various intercropping treatments during 2009–2010. The highest increase in tree height of 1.04 m in T5 and stock/scion ratio of 0.61 in T6 were observed in bearing Nagpur mandarin orchard. The highest increase in tree height of 0.53 m in T3 and stock/scion ratio of 0.67 in T2 were observed in prebearing Nagpur mandarin orchard. The tree height, canopy volume and stock/scion ratio were comparatively higher in the intercrops with Nagpur mandarin along with soybean, black gram and groundnut followed by gram and summer mung bean. The increase in tree height, canopy volume and stock/scion ratio were 0.28 m, 8.5 m3 and 0.64 for pre bearing Nagpur mandarin in treatment without cotton and intercrops (T1). The increase in tree height, canopy volume and stock/scion ratio were 0.92 m, 19.52 m3 and 0.59 for bearing Nagpur mandarin in treatment without cotton and intercrops (T1).

In bearing mandarin grove, there was no effect of various intercropping treatments on growth parameters (Table 2). The maximum canopy volume was 28.81 m3 in treatment T6, followed by T5, followed by T8, followed by T7.Maximum yield was 20.0 tons per ha in treatment T6. However, the effect of various treatments on yield was nonsignificant. In bearing orchard, tree height increments in various intercropping combinations were nonsignificant. However, the values of increase in tree height and canopy volume were much lower in prebearing orchard (Table 2).

Plant canopy volume showed significantly more increase in treatment Nagpur mandarin + black gram followed by gram (10.66 m^3) followed by Nagpur mandarin + cotton + soybean during 2010–2012. There was no significant effect on stock/scion ratio under various intercropping combinations in all treatments. The stock/scion ratio was higher in prebearing mandarin orchard than bearing mandarin plants. This is due to the establishment of the new plants in the prebearing orchard of mandarin.

TABLE 2 Effect of sustainable intercropping practices on tree performance and yield of prebearing and bearing Nagpur mandarin orchard during 2009–2012.

Treatment	Increase in tree height	Canopy volume	Stock / scion ratio	Yield	Treatment	Increase in tree height	Canopy volume	Stock / scion ratio	Yield
	m	m^3	ratio	tons/ha		m	m^3	ratio	tons/ha
Pre-bearing Nagpur mandarin orchard					**Bearing Nagpur mandarin orchard**				
T1	0.28	8.49	0.64	—	T1	0.92	19.52	0.59	12.20
T2	0.44	7.73	0.67	—	T2	0.58	17.15	0.66	12.75
T3	0.53	10.53	0.66	—	T3	0.64	16.10	0.56	14.56
T4	0.35	8.73	0.65	—	T4	0.62	10.99	0.55	15.11
T5	0.37	9.29	0.63	—	T5	1.04	27.06	0.57	14.50

TABLE 2 *(Continued)*

Treatment	Increase in tree height	Canopy volume	Stock / scion ratio	Yield	Treatment	Increase in tree height	Canopy volume	Stock / scion ratio	Yield
	m	m^3	ratio	tons/ha		m	m^3	ratio	tons/ha
Pre-bearing Nagpur mandarin orchard					**Bearing Nagpur mandarin orchard**				
T6	0.26	8.19	0.63	—	T6	1.01	28.81	0.61	20.00
T7	0.24	10.66	0.61	—	T7	0.86	22.95	0.56	12.85
T8	0.43	10.60	0.64	—	T8	0.63	23.72	0.55	13.51
C D (P=0.05)	NS	1.24	NS	—	CD (P=0.05)	NS	2.45	NS	4.71

Note: Treatments are defined in Table 1.

12.3.3 GROWTH AND YIELD OF INTERCROPS IN NAGPUR MANDARIN ORCHARDS

In prebearing mandarin orchard, Table 3 indicates that the plant volume of cotton was not significantly different among all intercropping combinations. The number of bolls/plant and cotton yield were maximum in the treatment T4 compared to cotton alone and cotton in combination with soybean, black gram and groundnut intercrops. The plant volume of soybean did not show any distinct trend. However, number of pods per plants and yield were 92.60 and 2837.33 Kg/ha for treatment T3 compared to soybean alone followed by gram (T6). Length of pod was higher in cotton + soybean (3.39 cm) but number of seeds per pod was more in soybean alone followed by gram (2.68). Yield of soybean was more in combination with cotton compared to that when grown alone followed by gram. The volume of plants in black gram did not show significant differences in the two treatments (Nagpur mandarin + cotton + black gram, and Nagpur mandarin + black gram followed by gram). The volume of plants during 2011–2012 was less due to high incidence of mosaic disease. This directly affected number of pods, length of pods, and number of seeds per pod and yield of black gram. The plant volume and number of pods per plants in groundnut were not significantly different in treatments T5 and T8. However, the length of pods, number of seeds and yield were significantly more when groundnut was grown with cotton. The plant volume was similar in all treatments combination with gram. Yield of gram was more in treatment involving soybean (Table 3). The yield of intercrops per tree during 2010–2011 and 2011–2012 was influenced by different intercropping systems. Significant differences in yield of intercrops were observed in different intercropping treatments during 2010–2012. Similar results of increased yield with intercropping practices have been reported in mandarin grove and other citrus cultivars [7, 9, 12].

TABLE 3 Performance of intercrops during 2009–2012.

Treatment	Volume of plants m^3	No. of pods per plant	Yield kg per ha	Treatment	Volume of plants m^3	No. of pods per plant	Yield kg per ha
Pre-bearing Nagpur mandarin orchard				**Bearing Nagpur mandarin orchard**			
Cotton					**Cotton**		
T2	0.027	22.44	898.58	T2	0.018	21.47	286.54
T3	0.027	25.80	975.23	T3	0.013	18.93	275.48
T4	0.026	24.08	1008.25	T4	0.014	18.93	252.93
T5	0.025	22.57	893.44	T5	0.012	18.28	211.40
Soybean					**Soybean**		
T3	0.031	92.60	2837.33	T3	0.011	24.16	390.99
T6	0.029	67.07	1358.77	T6	0.009	22.48	340.35
Black gram					**Black gram**		
T4	0.017	21.80	823.91	T4	0.008	10.11	147.63
T7	0.014	18.90	545.67	T7	0.007	8.74	67.22
Groundnut					**Groundnut**		
T5	0.021	23.18	898.37	T5	0.007	17.66	128.50
T8	0.019	25.31	814.54	T8	0.008	23.27	139.29
Gram					**Gram**		
T6	0.009	24.23	453.04	T6	0.002	3.85	123.50
T7	0.011	25.80	526.94	T7	0.002	3.18	105.84
CD (P=0.05)	0.005	2.88	27.3	CD (P=0.05)	0.002	1.73	18.4

Note: The treatments are defined in Table 11.

12.3.4 SOIL FERTILITY STATUS AND NUTRIENT UPTAKE IN NAGPUR MANDARIN ORCHARDS

In all the treatments, the soil organic carbon increased significantly in both pre-bearing and bearing orchards. Considerable higher increase was observed in treatments T6 and T7 after two years. Whereas, in no intercrop treatment (T1), it did not change. The increase is attributed to addition of organic residues by the intercrops. The soil pH value in both orchards was unaffected in all treatments. In bearing orchard, decrease in electrical conductivity (EC) values was observed in T5, T6 and T7 treatments while in prebearing orchard it decreased in treatment T4

(0.34 ds/m). In prebearing mandarin orchards soil nutrient variation was greater compared to bearing orchards due to large space available for growing different intercrops. The nitrogen content among different treatments varied significantly. The decrease was observed in treatment T2 while the highest increase was observed in T6.

The phosphorus content decreased after two years in all treatments except in T4. The potassium content also varied significantly. It decreased treatments in T2 and T4, whereas it increased in other treatments. The highest increase was observed in T3 (328.7 kg/ha). The soil iron status varied significantly among all treatments. The increase was observed in T1 and T6 treatments, where it decreased in treatment T2, from the initial value after two years. The manganese content decreased in all treatments after first year, and afterwards it increased. Considerable decrease in manganese was observed in T1 treatment. Differences in Zinc content among all treatments were not significant.

In prebearing orchards, the nitrogen uptake in all the treatment increased during 2010–2012. The highest increase was 2.24% T6 (Nagpur mandarin + soybean followed by gram). The P uptake was 0.13% in T1 (Nagpur mandarin alone), 0.12% in T2 (Nagpur mandarin + cotton var. LRK516), 0.16% in T5 (Nagpur mandarin + cotton + groundnut var. JL24), and 0.16% in T7 (Nagpur mandarin + black gram followed by gram). The potassium uptake was not significantly different among T1, T2 and T3, after two initial years. In T4 to T8, it increased over the initial content. The percentage values of Fe, Mn, Cu and Zn varied significantly among different treatments. The Fe content increased during initial two years, and then decreased during third year in T1, T2 and T3. The manganese content varied significantly among different treatments. The highest increase in Mn content was 75.8 ppm in T6 after two years. The copper content among different treatments was significant during initial two years, whereas it was nonsignificant during third year. The highest increase in Zinc uptake was 22.8 ppm in T8.

In bearing orchard, the nitrogen uptake varied significantly (from 1.62 to 2.14%) during 2011 and 2012 among all treatments. It was highest in the treatment T7 (2.08% in 2011 and 2.14% in 2012). In all the intercropping treatments except in T2, it increased. The variation in phosphorus uptake was not significantly in all treatments. In T4 (Nagpur mandarin + cotton + black gram var. TAU1), the P content was decreased from 0.19 to 0.12% after three years. In most of the intercropping treatments, the K content increased after first and second year, while it decreased during third year. The black soils of central India are rich on potassium and luxury consumption by plants may be the reason for these differences.

12.3.5 FRUIT QUALITY OF BEARING NAGPUR MANDARIN

The fruit quality of Nagpur mandarin was affected in all treatments as shown in Table 4. Better fruit weight, TSS, acidity, juice content and yield were due the in-situ soil moisture conservation by different intercropping practices. The differences were significant in juice content, acidity and TSS. The values of fruit size, peel weight and thickness, number of seeds and segments were non significant. Highest fruit juice content (3.12%) was observed in the Nagpur mandarin without any intercrops (T1).

The highest TSS (13.6 °Brix) was in T4. The lowest acidity (0.74) was in T7. The TSS to acidity ratio was highest in T7 (15.95), indicating the sweet mandarin. The TSS to acidity ratio was 14.51 in T5 and 13.72 in T8. The fruit size was 6.36 cm in T8. The mandarin fruit size was lowest in T1.This implies that sustainable intercropping practices were beneficial. Similar results in Nagpur mandarin have been reported by other scientists [12, 8]. Total soluble solids (TSS) was 13.61% in T4 and was significant. Vitamin C of the fruit was unaffected with intercropping systems in Nagpur mandarin orchards.

TABLE 4 Effect of sustainable intercropping practices on quality and performance of Nagpur mandarin fruits during 2009–2012.

Treat-ment	Fruit Diameter cm	Peel		No. of		Juice			
		Weight g	thickness mm	Seeds	Segments	Content %	T.S.S °Brix	Acidity %	TSS/ acidity ratio
T1	5.87	57.3	1.87	10.29	8.03	43.12	11.55	0.97	11.90
T2	6.04	62.3	1.76	9.95	8.73	39.64	11.62	0.97	11.97
T3	5.76	48.7	1.76	8.94	7.97	41.06	12.80	0.96	13.34
T4	5.84	55.4	1.77	10.02	7.61	43.03	13.61	1.03	13.21
T5	5.74	58.0	1.75	10.06	9.01	40.92	12.63	0.87	14.51
T6	6.14	57.1	1.89	10.18	8.56	41.74	11.41	0.84	13.58
T7	6.13	68.5	1.77	9.96	9.01	39.65	11.81	0.74	15.95
T8	6.36	63.0	1.93	10.06	7.67	37.41	11.53	0.84	13.72
CD (P=0.05)	NS	NS	NS	NS		NS	1.65	1.61	0.41

Note: The treatments are defined in Table 1.

12.4 CONCLUSIONS

In bearing orchard, tree growth was not affected by the intercrops. In prebearing orchard, tree height and stock-scion ratio of Nagpur mandarin trees was not affected by various intercrops, whereas canopy volume was maximum in the treatment Nagpur mandarin + black gram followed by gram. Cotton can be grown at a row spacing of 3 m and 5 m between the mandarin rows without affecting the yield. Cotton with the other intercrops (soybean, black gram and groundnut) can be sown between centrally grown cotton rows and in strips between Nagpur mandarin. The plant canopy volume of Nagpur mandarin was maximum in Nagpur mandarin + soybean followed by gram. The highest Nagpur mandarin yield of 20.0 tons/ha (72.3 kg/tree) was recorded in the intercropping of Nagpur mandarin + soybean followed by gram. Growth and yields of the intercrops in the bearing orchard were lower than those under prebearing orchard. In bearing orchard, combination of Nagpur mandarin + soybean followed by gram resulted in maximum yield of 20.0 tons/ha of Nagpur mandarin. The total soluble solids (TSS) to acidity ratio was more in Nagpur mandarin + black gram followed by gram than for the Nagpur mandarin + cotton + groundnut. The fruit acidity and juice percent were also significantly affected.

High soil moisture status was observed in intercrops of soybean and groundnut. The maximum cotton yield was intercropping with soybean, black gram and groundnut compared to cotton alone. The number of pods per plants was more in soybean grown with cotton as compared to soybean alone. This research revealed that the intercropping system with cotton + soybean or cotton + black gram in the interspaces of prebearing and bearing mandarin improved the yield as well as sustainability of mandarin, and the production of intercrops.

Net price-return of the intercrops was higher in prebearing orchard compared to bearing orchard. The economical return of mandarin crop was higher in the intercropping with soybean followed by gram compared to other treatments.

12.5 SUMMARY

A field experiment was conducted during 2009–2012 to evaluate the sustainable intercropping practices with cotton as the main intercrop and black gram/soybean/groundnut/gram/mung bean as intermediate intercrops in prebearing and bearing Nagpur mandarin orchards. The treatment consisted of Nagpur mandarin without intercrop, Nagpur mandarin + cotton, Nagpur mandarin + cotton + soybean, Nagpur mandarin + cotton + black gram, Nagpur mandarin + cotton + groundnut, Nagpur mandarin + soybean followed by gram, Nagpur + black gram followed by gram, Nagpur mandarin + groundnut followed by summer mung. The tree canopy volume indicated maximum increase in Nagpur mandarin + soybean followed by gram. The highest Nagpur mandarin yield was 20.0 tons/ha (72.3 kg/tree) with Nagpur mandarin + soybean followed by gram. The prebearing orchard, the Nagpur mandarin + black gram followed by gram gave significant increase compared to Nagpur mandarin + cotton + soybean. The total soluble solids (TSS) to acidity ratio was higher in Nagpur mandarin + black gram followed by gram compared to Nagpur mandarin + cotton + groundnut. The fruit acidity and juice percent were also significantly affected. Higher soil moisture was observed in intercrops of soybean and groundnut. The maximum cotton yield was in intercropping treatments with soybean, black gram and groundnut compared to cotton alone. The number of pods per plants was more in soybean grown with cotton compared to soybean alone. This study revealed that the intercropping system with cotton + soybean or cotton + black gram in the interspaces of prebearing and bearing mandarin improved the yield, sustainability of mandarin, and the production of intercrops.

KEYWORDS

- **black gram**
- **brix**
- **bulk density**
- **citrus**
- **citrus grove**
- **citrus reticulata**
- **cotton**
- **crop stage, bearing**
- **crop stage, prebearing**
- **field capacity**
- **fruit acidity**
- **fruit production**
- **fruit quality**
- **gram**
- **groundnut**
- **intercrop**
- **intercropping**
- **kharif crop**
- **micro irrigation**
- **mung beans**
- **Nagpur mandarin**
- **peel thickness**
- **permanent wilting percentage**
- **profile probe**
- **rabi crop**
- **refractometer**
- **soil moisture conservation**
- **soybean**
- **sustainability**
- **sustainable practices**
- **total soluble solids, TSS**
- **yield**

REFERENCES

1. Chadda, H. P., Randhawa, J. S., Bakshi, J. C., and Parihar, S. S. (1969). Effects of intercropping characteristics in young citrus orchards on soil properties. Paper presented at Seminar on Advances in Fruit Research, Ludhiana – India, April.
2. Chadda, K. L., Randhawa, N. S., Bindra, O. S., Chohan, J. S., and Knorr, L. C. (1970). *Citrus decline in India*. Punjab Agricultural University, Ludhiana – India, 50–55 pages.
3. Chundawat, B. S. (1993). Intercropping in orchards. In: Chanda, K L. (Ed). *Advances in Horticulture Vol. 2*. Malhotra Publishing House, New Delhi, 763–775 pages.
4. Gill, A. S. (1999). Agri-silvi-horticultural studies with mandarin orchards. Proceedings of International symposium on Citriculture. November 23–27. NRCC, Nagpur, India, 680–684 pages.
5. Gonge, V. S., and Kale, P. B. (1997). Effect of intercropping on growth of citrus during prebearing stage. Abstracts: National Symposium on Citriculture, NRCC, Nagpur, 17–19 November, 30 page.
6. Huchche, A. D., Shivankar, V. J., Shirgure, P. S., Marathe, R. A., Singh, S., Mane, P., Kadao, S., and Patil, S. (2006). Legume based intercropping systems for Nagpur mandarin (*Citrus reticulata* cv. Blanco). Paper presented at the National Symposium on Citriculture: A road map at ICAR research complex for North East Himalaya region, Umiam – Meghalaya, 22–24 February. 55 pages.
7. Krishnamurti, S. (1959). Intercropping, cover cropping and mulching of orchards. Indian J. Hort., 16:221–228.
8. Paslawar, A. N., Dalal, S. R., Golliwara, V. J., and Khobragade, R. I. (1999). Intercropping in Nagpur mandarin (*Citrus reticulata* cv. Blanco). Proceeding of International symposium on Citriculture, November 23–27, NRCC, Nagpur – India, 676–679 pages.
9. Randhawa, N. S., Bhumbla, D. R., and Dhingra, D. P. (1996). Citrus decline in the Punjab –A review. Punjab Hort. J. 6:35–44.
10. Shirgure, P. S., Srivastava, A. K., and Singh, S. (2000). Water management in citrus – A review. Agricultural Reviews, 21(4):223–230.
11. Shivankar, V. J., Singh, S., Huchche, A. D., Shirgure, P. S., Das, A. K., Mane, P. N., and Patil, S. P. (2004). Effect of intercropping systems on insect pests complex of Nagpur mandarin (*Citrus reticulata* Blanco). Abstract: International Citrus Congress held at Agadir – Morocco, 15–20 February.
12. Singh, K. K., and Bakshi, J. C. (1961). Intercropping of citrus orchards. I. Effect of intercropping of cotton on growth of young sweet orange trees. Indian J. Hort. 18:130–134.
13. Singh, S. (1999). Citrus in India. Proceedings of International Symposium on Citriculture, 23–27 November, National Research Centre for Citrus, Nagpur – India, 278–303 pages.

CHAPTER 13

POTASSIUM FERTIGATION IN NAGPUR MANDARIN

P. S. SHIRGURE and A. K. SRIVASTAVA

CONTENTS

13.1 INTRODUCTION

Nagpur mandarin occupies an area of 0.0286 Mha in Maharashtra, Madhya Pradesh, Rajasthan and Chattisgarh States of Central India. In the current citriculture crops of India, the major concerns are efficient use of water and fertilizer to increase the crop yield and fruit quality [24]. To maximize yield and fruit quality and nutrient uptake in citrus groves with limited amounts of water and fertilizer, the essential factors are: Maximization of water use; minimization of input cost of irrigation and fertilizers; the adoption of sustainable micro irrigation systems [16]; irrigation scheduling based on class A pan evaporation [13, 18, 19, 25] ; and fertigation [17]. For higher yield and good fruit quality of Nagpur mandarin, the use of optimum quantity of Potassium fertilizer along with other N and P fertilizers at appropriate time of the fruit growth stages is among the various factors. To enhance the growth and productivity, amount of NPK fertilizer either in the form of organic or inorganic and its application method play an important role [1].

The current practice dose is to split into three the fertilizer doses applied in June, October and February. These doses of fertilizers are applied as banded causing nutrient pollution and waste of fertilizer due to leaching, evaporation and soil fixation. The fertilizer application efficiency with conventional methods is low due to its lateral movement away from the active root zone. Sustainable micro irrigation in combination with fertigation is the most efficient method of pressurized irrigation because of saving in water and fertilizer use [7, 10, 25]. Fertigation is most effective, economical and convenient means of maintaining optimum fertility level according to the specific requirement of each crop and resulting in higher yield and better fruit quality [26, 29]. In areas with inadequate rainfall, fertigation offers the best option to ensure that nutrients reach the root zone. The Nitrogen fertigation have been evaluated on Shamouti sweet orange [2], Valencia orange [8], Naval orange [5, 9], Sunburst mandarin [4], Nagpur mandarin [14, 22, 23] and acid lime [12, 15]. The Potassium (K) fertigation during flower initiation to fruit growth and development is a latest technology and not enough literature is available on Nagpur mandarin under Central Indian agro-climatic conditions.

In this chapter, the authors discuss the effects of fertigation with various sources of potash (K) fertilizers on tree vegetative growth, leaf nutrient status and uptake, yield and fruit quality of Nagpur mandarin (*Citrus reticulate* cv. Blanco) in Central India.

13.2 MATERIALS AND METHODS

At experimental farm of National Research Centre for Citrus (NRCC) – ICAR-Nagpur – India, the field experiment was set up on a 0.25 ha with 6 × 6 m tree spacing to evaluate effects of four different Potassium fertilizers on: Nutrient uptake due to K fertigation; growth and productivity of 12–14 years old bearing Nagpur mandarin during 2009 to 2012. The treatments consisted of fertigation with:

T1 Potassium Chloride (KCL)(150 g K_2O/tree) at 15 days interval,

T2 Potassium Nitrate (KNO_3)(150 g K_2O/tree) at 15 days interval,

T3 Potassium Sulfate of Potash (K_2SO_4)(150 g K_2O/tree) at 15 days interval, and

T4 Mono Potassium Phosphate (KH_2PO_4)(150 g K_2O/tree) at 15 days interval.

Randomized block design with six replications was used to setup the field experiment. The soil texture at the site is clay loam and the soil depth is 45 cm. The composite soil samples were collected for determination of field capacity and permanent wilting point. Volumetric soil moisture content at field capacity (FC) and the permanent wilting point (PWP) were determined using pressure plate method. The FC and PWP was 28.14% and 19.1%, respectively. The available water content of the soil is 9.04% (= 28.14–19.10). The soil bulk density was determined using core sampler having 100 cm^3 volume and oven drying method. The soil bulk density was 1.5 g/cm^3. The soil water holding capacity was 12.23 cm/m of soil depth.

The micro irrigation system consisted of 4 lph drippers @ 4 per tree at 4 locations on the lateral (16 liters per hour per tree) and the liquid dispenser (DOSTRAN, France). For studying suitability of different Potassium fertilizers for K fertigation, the authors used Potassium Chloride (KCL, 0:0:60), Potassium Nitrate (KNO_3, 13:0:46), Potassium Sulfate (K_2SO_4, 0:0:50) and mono Potassium Phosphate (KH_2PO_4, 0:52:34). The recommended fertigation dose for the Nagpur mandarin grove is 500:150:150 of N:P:K, respectively. Fertigation was started in October at an interval of 15 days, on 2nd and 16th day each month. Nitrogen was fertigated from October to January, and all N, P and K were fertigated from February to June. From October to January months, Nitrogen (N) was fertigated with urea (46% N) @ 11.60 Kg of urea in all treatments. From February to June, different Potassium fertilizers were fertigated. The various fertilizer combinations along with the quantity of fertilizers for 48 trees in each fertigation treatment were:

1. **For Treatment T1**, KCL: In this treatment, Urea Phosphate, Urea and KCL were used in following quantities: KCL, 1.20 Kg + Urea Phosphate, 1.632 Kg + Urea, 2.256 Kg.
2. **For Treatment T2** , KNO3: Urea Phosphate, Urea and KNO_3 were used in following quantities: KNO_3, 1.6 Kg + Urea Phosphate, 1.637 Kg + Urea, 1.806 Kg.
3. **For Treatment T3**, K2SO4: Urea Phosphate (UP), Urea and K2SO4 were used in following quantities: K2SO4, 1.44 kg + UP, 1.632 Kg + Urea, 2.256 Kg.
4. **For Treatment T4**, KH2PO4: KH_2PO_4, Phosphoric acid (86 %) and Urea, were used in following quantities: KH2PO4, 1.385 Kg + P2O5 acid, 0.290 + Urea, 2.898 Kg.

The biometric growth parameters of Nagpur mandarin trees (height, girth, spread) were recorded in October during 2009, 2010 and 2011. The tree stock girth was taken 15 cm above the soil surface. The canopy volume of the mandarin tree was calculated with Castle formula [3]. Nagpur mandarin fruit yield and quality were evaluated with procedures described by Ranganna [11]. The initial soil and leaf samples were collected from the different treatments during November, 2009, using procedures suggested by Srivastava et al. [27]. Finally leaf samples were digested in diacid mixture of H2SO4: HClO4 in 2.5:1 ratio. The leaf N was determined using alkaline permangate steam distillation method, P by vanadomolybdophosphoric acid method and K by flame photo metrically. The data on fruit yield and quality in different K fertigation treatments for 3 years were analyzed by analysis of variance method [6].

13.3 RESULTS AND DISCUSSION

13.3.1 TREE PERFORMANCE OF NAGPUR MANDARIN WITH K FERTIGATION

The growth of mandarin tree (tree height, stock girth, and canopy volume) was recorded during October of 2009–10, 2010–11 and 2011–12 (Table 1). The different Potassium (K) fertigation treatments using four different potash fertilizers affected the growth parameters of 14–16 years Nagpur mandarin during 2009–2012. According to the data in Table 1, among all the growth parameters, only canopy volume was significantly different among all K fertigation treatments. The tree height and stock girth were not significantly different among all K fertigation treatments. The highest mean tree height (5.53 m) and mean stock girth (77.25 cm) were recorded in mono-Potassium Phosphate fertigation [T4]. The lowest mean tree height (5.42 m) and mean stock girth (73.63 cm) were observed in K fertigation with Potassium Nitrate [T2] during 2009–2012. The significant canopy volume was observed ranging from 65.05 to 71.51m^3 in 2009–10, 77.51 to 84.88 m^3 in 2010–11 and 79.0 to 88.09 m^3 in 2011–12, respectively. The highest mean canopy volume (81.49 m^3) was recorded with K fertigation using mono-Potassium Phosphate [T4]. The lowest mean tree canopy volume (73.85 m^3) was observed in K fertigation with Potassium Nitrate [T2] during 2009–12. This may be due to fertigation Potassium fertilizers during the tree growth stages, frequent irrigation. The fertigation scheduling favored fruit growth development. The similar type of observations have been observed in the earlier studies on fertigation scheduling in Nagpur mandarin [14] and in acid lime [21] under the Central Indian conditions.

TABLE 1 The tree growth and canopy volume of Nagpur mandarin during 2009–2012.

Treatments	Tree height, m				Stock girth, cm				Canopy volume, m^3			
	2009–10	2010-11	2011-12	Mean	2009-10	2010-11	2011-12	Mean	2009-10	2010-11	2011-12	Mean
T1	5.31	5.54	5.65	5.50	74.13	75.75	75.90	75.26	67.31	78.57	80.25	75.46
T2	5.19	5.50	5.57	5.42	70.92	74.59	75.38	73.63	65.05	77.51	79.00	73.85
T3	5.26	5.53	5.70	5.50	73.32	78.25	80.04	77.20	70.38	81.37	84.86	78.87

TABLE 1 *(Continued)*

Treatments	Tree height, m				Stock girth, cm				Canopy volume, m³			
	2009–10	2010-11	2011-12	Mean	2009-10	2010-11	2011-12	Mean	2009-10	2010-11	2011-12	Mean
T4	5.26	5.64	5.68	5.53	75.07	77.83	78.86	77.25	71.51	84.88	88.09	81.49
LSD (P=0.05)	NS	NS	NS	—	NS	NS	NS	—	NS	0.28	0.67	—

T1 – Fertigation with Potassium Chloride; T2 – Fertigation with Potassium Nitrate;

T3 – Fertigation with Potassium Sulfate; and T4 – Fertigation with mono Potassium Phosphate.

13.3.2 LEAF NUTRIENT STATUS WITH K FERTIGATION USING FOUR POTASH FERTILIZERS

The periodic nutrient status of leaf was monitored to evaluate the effects of differential K fertigation with four different Potassium fertilizers treatments on leaf status and nutrient up-take. In all fertigation treatments, the initial and final leaf samples were analyzed for macronutrients (N, P and K) and micronutrients (Fe, Mn, Zn and Cu) during 2009–2012 (Table 2).

Before the initiation of K fertigation treatments, the leaf nutrient status was: N (1.86 to 2.08%), P (0.078 to 0.084%), K (0.97 to 1.18%), Fe (117.4 to 168.5 ppm), Mn (33.0 to 58.7 ppm), Cu (8.8 to 19.3 ppm), and Zn (16.6 to 28.2 ppm).

In the final leaf nutrient analysis, the K fertigation with mono Potassium Phosphate recorded the highest concentration of macronutrients (N, P and K) and micronutrients (Fe, Mn, Cu, and Zn) compared to rest of the other fertigation treatments (Table 2). The P and Cu values were significantly different among all the treatments, whereas N, K, Fe, Mn and Zn values were not significantly different. The fertigation with mono-Potassium Phosphate recorded the highest concentration of macronutrients (2.23% N, 0.095% P and 1.16% K) compared to rest of the fertigation treatments. Leaf N (2.14%), P (0.92%) and K (1.1%) contents were observed significantly higher with Potassium Sulfate fertigation than N (2.04%), P (0.087%) and K (1.08%) contents with Potassium Nitrate fertigation. The lowest leaf nutrient composition N (1.98%), P (0.09%) and K (1.01%) was observed with Potassium Chloride fertigation during 2009–2012. Similarly the final leaf analysis for micronutrients (Fe, Mn, Cu and Zn) was done during March 2012. The Fe, Mn and Zn elements were significantly different due to K fertigation scheduling, however, the Copper (Cu) element was not significant. The leaf analysis revealed that the K fertigation treatment with mono Potassium Phosphate recorded the highest concentration of micronutrients (127.1 ppm Fe, 60.1 ppm Mn, 10.9 ppm Cu and 26.2 ppm Zn) compared to rest of the fertigation treatments. Leaf Fe (120.8 ppm), Mn (53.1 ppm), Cu (10.8 ppm) and Zn (24.9 ppm) contents were observed significantly higher with Sulfate of potash fertigation than with Potassium Nitrate fertigation (Fe, 114.8 ppm), Mn (49.8 ppm), Cu (9.6 ppm) and Zn (23.1 ppm). The lowest leaf micronutrients nutrient composition with Fe (112.4 ppm), Mn (43.6

ppm), Cu (9.1 ppm) and Zn (19.2 ppm) content was observed with Potassium Chloride K fertigation during 2009–2012.

TABLE 2 The leaf nutrient status in different K fertigation treatments.

Treatments	Macronutrients (%)			Micronutrients (ppm)			
	N	P	K	Fe	Mn	Cu	Zn
Initial leaf nutrient status (2009–2010)							
T1	1.97	0.080	0.97	122.0	58.7	08.8	18.2
T2	1.86	0.078	0.88	168.5	37.5	10.2	18.4
T3	2.08	0.079	1.18	117.4	38.4	19.3	28.2
T4	2.08	0.084	1.01	142.5	33.0	09.3	16.6
LSD (P=0.05)	NS	NS	NS	NS	NS	NS	NS
Final leaf nutrient status (2011–2012)							
T1	1.98	0.090	1.01	112.4	43.6	09.1	19.2
T2	2.04	0.087	1.08	114.8	49.8	09.6	23.1
T3	2.14	0.092	1.10	120.8	53.1	10.8	24.9
T4	2.23	0.095	1.16	127.1	60.1	10.9	26.2
LSD (P=0.05)	0.08	NS	0.07	1.04	4.6	NS	1.47

Note: The treatments are defined in Table 1.

13.3.3. EFFECTS OF POTASSIUM FERTIGATION ON FRUIT PERFORMANCE OF NAGPUR MANDARIN

The Nagpur mandarin fruits were harvested during first fortnight of November in 2009, 2010 and 2011. Table 3 indicates the results for mean values of number of fruits per tree, yield per ha, TSS, juice content, and acidity during 2009–2012. The Potassium (K) fertigation with four different potash fertilizers affected significantly the yield and fruit quality of the Nagpur mandarin. The number of fruits per tree, fruit yield, fruit weight, total soluble solids (TSS), juice percentage, acidity, and TSS/acidity ratio were significantly different during 2010–2011 and 2011–2012. Yield and fruit quality were significantly affected by K fertigation treatments (Table 3). The average number of fruits per tree varied from 590 to 697 in all K fertigation treatments. The highest number of fruits per tree (697 fruits/tree) was in K fertigation with mono-Potassium Phosphate followed by K fertigation with of potash Nitrate (668 fruits/tree) and Potassium Sulfate (625 fruits/tree). The lowest number of fruits per tree was with K fertigation using Potassium Chloride (590 fruits/tree), may be due to single K element and not with nitrogen (N) or phosphorus (P) during the fruit development phases. The mean Nagpur mandarin yield varied from 24.32 to 31.13 tons/ha in all the K fertigation treatments.

The highest fruit yield per hectare was with mono Potassium Phosphate fertilizer (31.13 tons/ha) followed by K fertigation with Potassium Nitrate (29.40 tons/ha) and with Potassium Sulfate (26.77 tons/ha). The lowest fruit yield was with Potassium Chloride (murate of potash, 24.32 tons/ha), may be due to single K source and not with N and K elements in critical fruit growth development stages during 2009–2012 (Table 3). This clearly indicates that Potassium (K) fertigation with mono-Potassium Phosphate (MKP) is essential for production of good quality mandarin fruits. The micro irrigation in combination with K fertigation resulted in good quality fruits and gave higher yield.

TABLE 3 The performance of fruits of Nagpur mandarin during 2009–2012.

Treatments	No. of fruits	Yield	Average weight of fruit	TSS	Juice	Acidity	TSS/acidity Ratio
	No.	tons/ha	g	°Brix	%	%	Ratio
T1	590	24.32	154.96	10.07	37.16	0.85	11.8
T2	668	29.40	155.28	10.44	38.05	0.84	12.4
T3	625	26.77	155.33	10.48	37.55	0.80	13.1
T4	697	31.13	156.24	10.49	38.76	0.77	13.6
LSD (P=0.05)	31	1.72	0.03	2.81	0.52	NS	——

Note: The treatments are defined in Table 1.

13.3.4 EFFECTS OF POTASSIUM FERTIGATION ON FRUIT QUALITY OF NAGPUR MANDARIN

The mean fruit weight (156.24 g), TSS (10.49 °Brix), juice percent (38.76%) were highest and acidity (0.77) was lowest with mono Potassium Phosphate followed by K fertigation with Potassium Sulfate. The mean fruit weight (155.33 g), TSS (10.48 °Brix), juice percent (37.55%) and acidity (0.8) were observed in K fertigation with Potassium Sulfate. The lowest mean fruit weight (154.96 g), TSS (10.07 °Brix), juice percent (37.16%) and highest acidity (0.85) were observed in K fertigation with Potassium Chloride. The TSS to acidity ratio is an indicator of sweetness of the fruit. High TSS to acidity ratio implies that the fruits have more TSS (total soluble solids) and less acidity. The highest TSS to acidity ratio (13.6) was found in K fertigation with mono-Potassium Phosphate followed K fertigation with Potassium Sulfate with TSS to acidity ratio of 13.1. The TSS to acidity ratio was 12.4 with the K fertigation with Potassium Nitrate. The lowest TSS to acidity (11.8) was observed the K fertigation with Potassium Chloride Table 3. The similar results for fruit yield and quality have been observed in Nagpur mandarin [13, 28] and acid lime [15].

13.4 CONCLUSIONS

The quality and yield of Nagpur mandarin can be increased with Potassium fertigation. Potassium fertigation was successful with Potassium Chloride (KCL), Potassium Nitrate (KNO_3), Potassium Sulfate (K_2SO_4) and mono-Potassium Phosphate (KH_2PO_4) fertilizers in 14–16 years bearing Nagpur mandarin in Central India. The leaf nutrient uptake was high in fertigation with Mono Potassium Phosphate (150 g K_2O/tree) at 15 days interval from February to June. The mandarin yield was highest (31.13 tones/ha) with Mono Potassium Phosphate (150 g K_2O/tree) followed by Potassium Nitrate (150 g K_2O/tree) at 15 days interval (29.4 t/ha). The fruit quality was also affected with different sources of potash fertilizers. Highest fruit TSS (10.48–^{0}Brix) and fruit weight (156.24 g) were observed with mono Potassium Phosphate at 15 days interval.

The highest TSS to acidity ratio was observed in Mono Potassium Phosphate (13.6) followed by Potassium Sulfate (13.1). Thus the use of different potash (K) fertilizers through micro irrigation and fertigation technique is a sustainable solution for increasing the citrus production and fruit quality.

13.5 SUMMARY

At experimental farm of National Research Centre for Citrus (NRCC) – ICAR-Nagpur – India, the field experiment was set up on a 0.25 ha with 6 × 6 m tree spacing to evaluate effects of four different Potassium fertilizers on: Nutrient uptake due to K fertigation; growth and productivity of 12–14 years old bearing Nagpur mandarin during 2009 to 2012. The treatments consisted of fertigation with:

T1 Potassium Chloride (KCL) (150 g K_2O/tree) at 15 days interval;

T2 Potassium Nitrate (KNO_3) (150 g K_2O/tree) at 15 days interval;

T3 Potassium Sulfate of Potash (K_2SO_4) (150 g K_2O/tree) at 15 days interval; and

T4 Mono Potassium Phosphate (KH_2PO_4) (150 g K_2O/tree) at 15 days interval.

The recommended fertigation dose was 500:150:150 (N:P:K) and was given at 15 days interval. The fruit yield and quality were measured at harvest. Results showed the highest response of the fruit yield (31.13 t/ha) with treatment mono Potassium Phosphate. The total soluble solid was highest (10.49 ^{0}Brix) in K fertigation with mono Potassium Phosphate. Highest juice content (38.76%) and low acidity (0.77%) were found in K fertigation with mono Potassium Phosphate. The highest TSS to acidity ratio was observed in Mono Potassium Phosphate (13.6).

KEYWORDS

- acidity
- Brix
- citriculture
- citrus
- citrus grove
- evaporation
- fertigation
- fertilizer
- fruit quality
- fruit yield
- India
- irrigation scheduling
- juice content
- leaching
- lemon
- micro irrigation
- microjet irrigation
- mono Potassium Phosphate
- murate of potash
- Nagpur mandarin
- National Research Centre for Citrus – India
- NPK
- P_2O_5
- Potassium Chloride
- Potassium fertigation
- Potassium fertilizers
- Potassium Nitrate
- Potassium Sulfate
- root zone
- soil fixation
- total soluble solids, TSS
- TSS
- TSS/acidity ratio
- urea
- urea of Phosphate
- Valencia orange

REFERENCES

1. Beridze, T. R. (1990). The effect of organic fertilizers on lemon tree productivity. Sub tropicheskie Kul'tury, 3:83–86.
2. Bielorai, H., Deshberg, E., and Brum, M. (1984). The effect of fertigation and partial wetting of the root zone on production of shamouti orange. Proc. Int. Soc. of Citriculture, 1:118–120.
3. Castle, W. (1983). Growth, yield and cold hardiness of seven year old "Bearss" lemon on 20 seven root-stocks. Proc. Florida State Hort. Soc., 96:23–25.
4. Ferguson, J. J., Davies, F. S., and Bulger, J. M. (1990). Fertigation and growth of young 'Sunburst' tangerine trees. Proc. of Florida State Hort. Sci., 103:8–9.
5. Fouche, P. S., and Bester, D. H. (1987). The influence of water soluble fertilizer on nutrition and productivity of Navel orange trees under microjet irrigation. Citrus and Sub-tropical fruit. J., 62:8–12.
6. Gomez, K. A., and Gomez, A. A. (1984). *Statistical Procedures for Agriculture Research*. John Wiley & Sons, 664–665 pages.
7. Koo, R.C.J. (1981). Results of Citrus fertigation studies. Proc. of Florida State Horti. Sci., 93:33–36.
8. Koo, R.C. J., and Smjstrala, A. G. (1984). Effect of trickle irrigation and fertigation on fruit production and fruit quality of Valencia orange. Proc. Florida State Hort. Sci., 97:8–10.
9. Louse, F. (1990). Nitrogen fertigation of citrus summery of citrus research. Citrus Research Centre and Agricultural Station. University of California, Riverside, 20–22 pages.
10. Haynes, R. J. (1985). Principles of fertilizer use for trickle irrigated crops. Fert. Res., 6:235–255.
11. Ranganna, S. (1986). *Handbook of analysis and quality control for fruit and vegetable products*. Tata McGraw Hil Pub. Co. Ltd., New Delhi, 881–882 pages.
12. Shirgure, P. S., Lallan, R., Marathe, R. A., and Yadav, R. P. (1999). Effect of Nitrogen fertigation on vegetative growth and leaf nitrogen content of acid lime. Indian J. Soil Conserv., 27(1):45–49
13. Shirgure, P. S., Srivastava, A. K., and Singh, S. (2001a). Effect of pan evaporation based irrigation scheduling on yield and quality of drip irrigated Nagpur mandarin. Indian J. Agri. Sci., 71 (4), 264–266.
14. Shirgure, P. S., Srivastava, A. K., and Singh, S. (2001b). Growth, yield and quality of Nagpur mandarin (*Citrus reticulate* Blanco) in relation to irrigation and fertigation. Indian J. Agri. Sci. 71(8):547–550.
15. Shirgure, P. S., Srivastava, A. K., and Singh, S. (2001c). Effect of nitrogen fertigation and band placement fertilizer application on soil -leaf nutrient build-up and incremental growth of acid lime. J. Soil and Water Cons., 45 (3&4):176–181.
16. Shirgure, P. S., Srivastava, A. K., and Singh, S. (2001d). Effect of drip, microjets and basin irrigation method on growth, soil and leaf nutrient change in acid lime. Indian J. Soil Cons., 29 (3):229–234.
17. Shirgure, P. S., Srivastava, A. K., and Singh, S. (2003a). Evaluating micro irrigation systems in Nagpur mandarin under sub humid tropical climate. Trop. Agri. (Trinidad), 80(2):91–96.
18. Shirgure, P. S., Srivastava, A. K., and Singh, S. (2003b). Irrigation scheduling and fertigation in acid lime (*Citrus aurantifolia* Swingle). Indian J. of Agric. Sci., 73(7):363–367.
19. Shirgure, P. S., Srivastava, A. K., Singh, S., and Pimpale, A. R. (2004a). Micro irrigationscheduling growth, yield and quality of acid lime (*Citrus aurantifolia* Swingle). Indian J. of Agric. Sci., 74 (2):92–94.
20. Shirgure, P. S., Srivastava, A. K., and Singh, S. (2004b). Growth, yield and quality of acid lime under pan evaporation based micro irrigationscheduling. Indian J. of Soil Cons., 32 (1):32–35.
21. Shirgure, P. S., Srivastava, A. K., and Singh, S. (2004c). Integrated water and nutrient management in acid lime. Indian J. of Soil Cons., 32 (2):148–151.

22. Shirgure, P. S., and Srivastava, A. K. (2012). The effect of four under tree microjet irrigation (180–300°) systems on fruit yield and quality of Nagpur mandarin in Central India. Sci. J. Agri., 1(7):177–186.
23. Shirgure, P. S. (2012a). Effect of pulse irrigation scheduling with hybrid station controller on fruit yield and quality of Nagpur mandarin (*Citrus reticulate* Blanco). Sci. J. Crop Sci., 1(5):76–82.
24. Shirgure, P. S., 2012b. Micro-irrigation systems, automation and fertigation in Citrus. Sci. J. Rev., 1(5): 156–169.
25. Shirgure, P. S. (2013). Yield and fruit quality of Nagpur mandarin (*Citrus reticulata* Blanco) as influenced by evaporation based micro irrigationschedules. Sci. J. Crop Sci., 2(2):28–35.
26. Smith, M. W., Kenworthy, A. L., and Bedford, C. L. (1979). The response of fruit trees to injections of Nitrogen through a trickle irrigation system. J. Amer. Soc. Hort. Sci., 104: 311–313.
27. Srviastava, A. K., Ram, L., Huchche, A. D., Kohli, R. R., and Dass, H. C. (1994). Standardization of leaf sampling technique in Nagpur mandarin under subhumid tropical climate. Indian J. Hort. Sci. 51(1):32–36.
28. Srivastava, A. K., Shirgure, P. S., and Singh, S. (2003). Differential fertigation response of Nagpur mandarin (*Citrus reticulata* Balanco) on an alkaline Inceptisol under subhumid tropical climate. Trop. Agr. (Trinidad), 80(2):97–104.
29. Syvertsen, J. P., and Smith, M. L. (1996). Nitrogen uptake efficiency and Leaching losses from Lysimeter grown citrus trees fertilized at three nitrogen rates. J. Amer. Soc. Hort. Sci., 121(1):57–62.

CHAPTER 14

SENSOR BASED IRRIGATION SCHEDULING IN BLUEBERRIES

B. KEITH BELLINGHAM

CONTENTS

The author is thankful to Stevens Water Monitoring Systems, Inc. for the support. Printed with permission from: http://www.stevenswater.com/articles/irrigationscheduling.aspx.

14.1 INTRODUCTION

In the western United States, irrigation accounts for about 80% of the water consumed [8]. Concerns about changes in land use due to growing populations, climate change, and the protection of aquatic habitats are driving a need to conserve water. Optimization of irrigation will not only benefit the environment, but also benefit local economies. Over irrigation may lead to dangerous increases in the total maximum daily loads (TMDL) of temperature, nitrates, and salinity in natural waters [6]. Nitrate fertilizers leached out of the soils get transported to natural waters causing eutrophication and other aquatic impairments. Run off from over irrigation may affect water quality parameters such as pH, total suspended solids (TSS), and dissolved oxygen [18]. Other negative impacts associated with over irrigation include wastes of water and energy, and reduced crop yields. The negative impacts associated with under irrigation are more intuitive. Under irrigation may reduce crop yields, which will reduce profit margins. This chapter discusses a soil water balance model incorporated into a data acquisition system that is a power tool for scheduling and optimizing irrigation. A case study for blueberries is presented.

Advancements in computer microprocessors, memory and software development tools has improved data acquisition methods and made data acquisition system integration more reliable and more cost effective. The soil water balance model incorporates inputs of soil moisture, water application and evapotranspiration (ET). The soil moisture data acquisition system retrieves the input parameters via telemetry and populates software that accommodates the soil water balance model. The soil data acquisition software integrated with a soil water balance model is commercially available from Stevens Water Monitoring Systems, Inc.

14.2 SOIL MOISTURE BUDGET

To begin our discussion about soil moisture budgets, we first describe the components and the hydrological conditions of soil. In general, inorganic soil is composed of mixes of sands, silts and clays. Sands, silts and clays differ not only by particle size distribution, but also in the atomic arrangement and charge distribution at the molecular level [9]. Soil geomorphology is the process by which sands and silts chemically and physically transform into clays as the soil ages [2]. The soil textural class is determined by the gravimetric percentage of sand silt and clay. Figure 1 shows the soil texture classifications based on gravimetric percentage.

Sands, silts, clays and organics represent the solid particle composition of soil while air and water fill the pore spaces between the solid particles. When soil is completely saturated with water, the porosity will be equal to the volumetric soil moisture content [16]. The amount of organics in soil will affect the bulk density and the porosity. Some organic soils may have porosities of over 90%, but in general, most inorganic agricultural loams will have a porosity of near 50%. The pores can be nearly microscopic (micropores) or visible with the naked eye (macropores) [3].

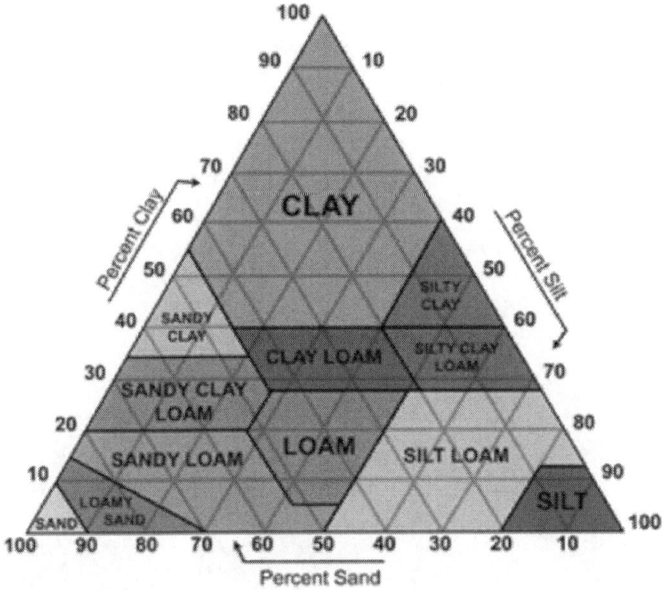

FIGURE 1 Soil textural classes based on the percentage of sand, silt, and clay.

The hydrologic properties of soil play an important role in a crop's ability to transpire water with their root systems. Knowledge of volumetric soil moisture content (θ , $m^3\ m^{-3}$) is an important input into the soil water balance model. Permanent wilting point (θ_{PW}) is the soil moisture level at which plants can no longer adsorb water from the soil. Plant transpiration and direct evaporation will decrease the moisture level in soil to a point below θ_{PW} and, in some cases, down to near dryness.

Field capacity (θ_{FC}) is defined as the threshold point at which the soil pore water will be influenced by gravity. Above field capacity, the gravitational force will overcome the capillary forces suspending the moisture in the pores of the soil allowing for down movement of water in the soil column. Below θ_{FC}, there will be a net upward movement of water driven by ET. Field capacity and permanent wilting point are heavily influenced by soil textural classes, particularly clay content [10].

Figure 2 shows an example of soil moisture at saturation, field capacity and permanent wilting point, for a typical soil. Clays interact with water in ways uniquely different from sand, silt and organics. Clays will have a physical and chemical affinity for water due to the negative charge distribution and the planner molecular lattice. The positive portion of the water molecule will be oriented toward the negatively charged clay lattice and the oxygen's lone electron pair will be pointed outwards [7]. The available water capacity (θ_{AC}) of soil is the water that is available to a plant. It represents the range of soil moisture values that lie above permanent wilting point and below the field capacity.

$$\theta_{PW} < \theta ac < \theta_{FC} \tag{1}$$

Table 1 shows the typical values for permanent wilting point and field capacity for common soil textural classes [10]. Plants are able to uptake water from soil if the soil moisture is above permanent wilting point. As the soil moisture approaches permanent wilting point, the plant will become increasingly stressed as the soil pore water becomes depleted. The point below field capacity where plants become stressed is called the maximum allowable depletion (MAD).

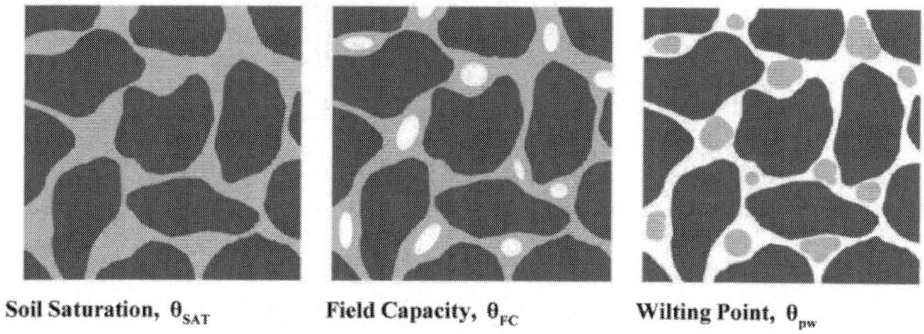

FIGURE 2 Soil moisture: Saturation, field capacity and permanent wilting point.

TABLE 1 Field capacity and permanent wilting point for common soil textural classes.

Soil Texture	Field capacity	Permanent wilting point
Clay	0.36	0.21
Loam	0.26	0.12
Loamy Sand	0.14	0.06
Sand	0.12	0.04
Sandy Clay Loam	0.33	0.175
Sandy Loam	0.23	0.1
Silt	0.32	0.165
Silt Loam	0.3	0.15
Silty Clay Loam	0.34	0.19
SiltyClay	0.36	0.21

The MAD value is expressed as a percent of the available water capacity. Table 2 shows typical MAD values for a few selected crops. Figure 3 shows the soil field capacities and the permanent wilting points for common soil textural classes. The topmost curve in Fig. 3 is the available water capacity showing 25%, 50% and 75% MADs. As shown in Fig. 3, the field capacity and the permanent wilting point will increase with the percentage of clay. With specific knowledge of field capacity, soil textural class and the maximum allowable depletion, a soil moisture target can be determined for irrigation optimization [4]. The soil moisture target is the range of soil moistures that lie above the MAD but below the field capacity. Below the MAD value the crop will still have the ability to receive water from the soil, however, the crop will become stressed after a period of time. If the crop becomes stressed due to the lack of water, the plant will have a reduced yield and become more susceptible to pathogens. If the soil moisture gets above field capacity, water will be transported downward by gravity potentially wasting water and leaching nutrients. Upper soil moisture target for the soils in the root zone will be the field capacity. The lower soil moisture target is determined by the MAD, θ_{FC}, and θ_{PW}:

$$\text{Lower Soil Moisture Target} = \theta_{FC} - (\theta_{FC} - \theta_{PW}) \times \text{MAD} \qquad (2)$$

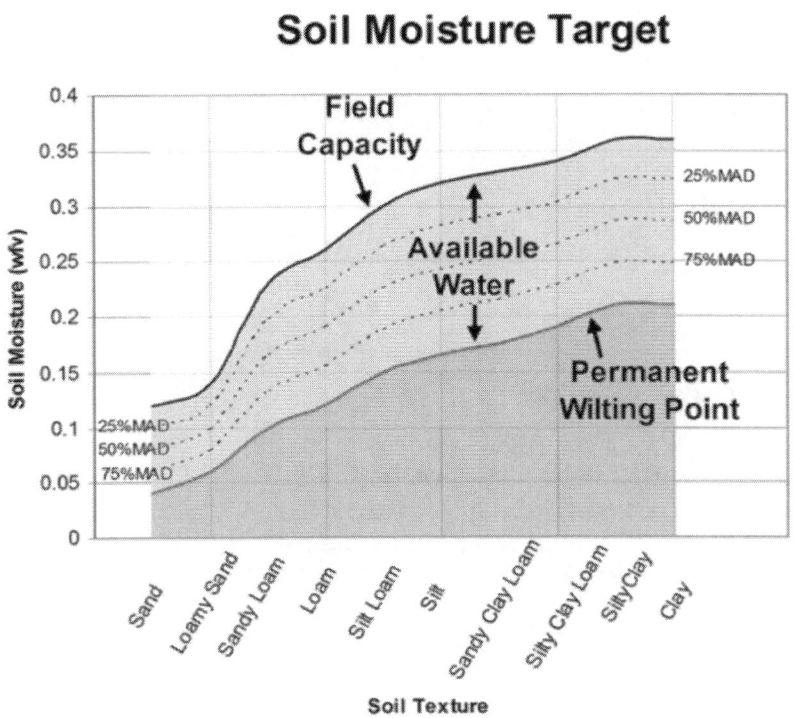

FIGURE 3 The relationship between soil textural classes and the hydrological thresholds θ_{PW}, θ_{AC}, and θ_{FC}. The 25%, 50% and 75% MAD levels are displayed in the available water capacity region.

For example, green beans with a MAD of 50% have a root zone depth of 18 inches. If the green beans are growing in a silt loam, the field capacity will be 0.3 water fraction by volume (wfv) and the permanent wilting point will be 0.15 wfv. Using Eq. (2), the lower soil moisture target will be 0.23 wfv. In this example, the soil moisture target for the green beans will lie between 0.23 wfv and 0.3 wfv from 5 inches to 18 inches deep adjacent to the root ball. It is important to note that the values in Table 1 are typical values and can vary slightly with bulk density of soil, mineralogy and organic content. Similarly, the MAD values in Table 2 are typical values and may vary by species, age of crop, region and soil chemistry.

TABLE 2 Maximum allowable depletion and effective root zone depth for selected crops [12].

Crop	Maximum allowable depletion (MAD)	Effective root Depth, inches
Blue berries Berries	50%	18
Orchard Apples	75%	36

TABLE 3 Typical values for sprinkler efficiencies for various sprinkler systems [12].

Irrigation system	Sprinkler efficiency (Ef)	Sprinkler efficiency (sprinkler spacing over 40×40 feet)
Solid set	0.70	0.63
Hand move or side roll	0.80	0.74
Pivot or linear move	0.90	0.81
Offset managed hand move	0.90	0.81

14.3 WATER APPLICATION

While soil moisture data provides information about the root zone, the measured application of water can be used concurrently with the soil moisture values to provide a more complete suite of tools for the irrigator. The measured application of water (D) is the amount of water applied to the crops with sprinklers, plus the amount of natural precipitation measured in inches/day. It is the total depth of water received by the crop.

14.3.1 SPRINKLER EFFICIENCY

In order to effectively use the application of water in a water budget model, a high sprinkler efficiency (Ef) is required. Sprinkler efficiency (Ef) is the measure of uniformity of water application. Ponding of irrigation water, and uneven application of water over the field is the result of poor sprinkler efficiency. Soil moisture data and rain gauge data are less meaningful if the monitoring site receives more or less water than the rest of the irrigation regime. Sprinkler efficiency is determined by placing catch cans or a set of containers of uniform size in the field. The catch cans can be placed in grid or uniformly distributed among the crops. After running the sprinklers for a length of time, the amount of water in the catch cans is measured. The sprinkler efficiency is expressed as a fraction and an *Ef* value of 1 is perfect uniformity. There are a number of methods for calculating *Ef*. The most common method for determining *Ef* involves averaging the lower 25% of the measured catchment of catch cans divided by the mean. An *Ef* value greater than 0.8 is preferred. Table 3 shows typical *Ef* values for several different types of sprinkler systems.

14.3.2 EVAPOTRANSPIRATION

An important factor for quantifying the water budget is the evapotranspiration rate (*ET*). Evapotranspiration is the water that is transpired out of the soil by the plant plus the amount of water lost to evaporation [1]. *ET* represents the rate of water consumed by the plant and lost by direct evaporation. The factors that affect the *ET* rate include wind, temperature, relative humidity, and solar radiation. The units for *ET* are inches/day. Based on the Penman Monteith model for *ET* estimations, *ET* is not measured directly for an individual crop, but rather it is determined from a standard reference grass and then adjusted for different crops and plants with a crop coefficient [1]. The evapotranspiration for a reference grass is referred to as the potential evapotranspiration (ET°). Potential evapotranspiration values will vary regionally and seasonally and are available in the literature. If literature values for ET° are not available or if the irrigator wishes to have a real time *ET* measurements, *ET* data acquisition systems are commercially available. ET data acquisition systems consist of weather sensors, telemetry and software that can retrieve the weather sensor inputs and perform the Penman Monteith model calculations. While an *ET* data acquisition system could potentially provide accurate real time ET° values, these systems are very expensive and do not necessarily represent microclimates. Because ET° is the *ET* for a standard reference grass, a crop coefficient (*Kc*) is necessary to determine the *ET* for the crop of interest. With information about sprinkler efficiency, crop coefficient and potential evapotranspiration, the water consumption (*ET*") for a specific crop (in inches per day) are calculated from the Eq. (3). Typically, *Kc* values will range from 0.75 to 1.25 depending on species of the plant, the growth stage of the plant, and vary regionally. In practice, ET° and *Kc* values can be obtained from a local government crop extension or a local crop advisor.

$$ET'' = [ET^\circ \, ' \, Kc] \, /\mathrm{Ef} \quad (3)$$

14.3.3 APPLIED WATER SCHEDULING

In general, the water application (D) in inches/day should be roughly equal to the system water loss (ET'') due to ET and sprinkler uniformity. The water loss calculated by Eq. (3) can be compared to the applied water measured with a rain gauge to set an irrigation target.

$$D \approx ET'' \tag{4}$$

It is difficult to keep $D \approx ET''$ on an hourly or daily basis due to factors such as pivot lap speed and soil infiltration rates. Eq. (4) should define a water application target on a weekly basis. In general, depending on the crop and the irrigation system, crops should be irrigated 3 to 7 times a week and net weekly sum of the daily D values should be roughly equal to the net weekly sum of the daily ET'' values. Figure 5 demonstrates a weekly water application target. In Fig. 4, there are three irrigation events and an ET'' rate of 0.26 inches per day. Based on an ET'' rate of 0.26 inches per day and the E_f, by the end of the week, 1.80 inches of water was consumed and approximately 1.80 inches would need to be applied. The application rate in Fig. 4 is 0.3 inches per hour for 2 h.

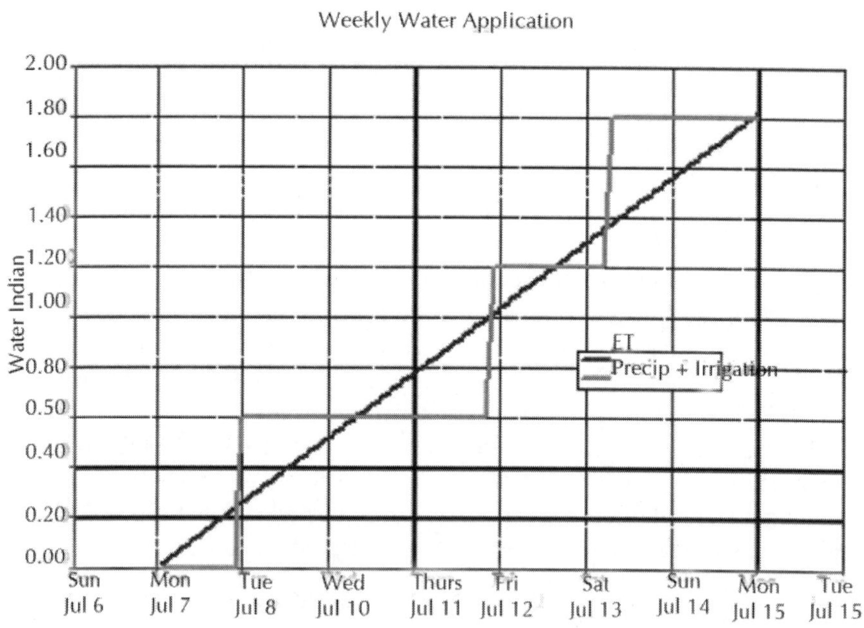

FIGURE 4 There are three irrigation events, and an ET'' rate of 0.26 inches per day. $D \approx ET''$ after the 3 irrigation event at the end of the week during the July, 2008.\

TABLE 4 Typical Infiltration rates based on soil texture.

Soil Texture	Typical infiltration rate inches/hour
Clay	0.05 to 0.25
Clay Loam	0.25 to 0.5
Loam	0.5 to 1
Sand	1.5 or more
Sandy Loam	1 to 1.5

To minimize the water loss due to direct evaporation, the irrigation events take place between sunset and sunrise. It is important to irrigate at a rate that is less than the infiltration rate of the soil. Runoff and ponding may occur if the rate of application exceeds infiltration rate of the soil. Table 4 provides infiltration rates of soils based on soil textural class [2]. The infiltration of water into soil will vary with texture, but it will also depend on soil moisture, vegetation, bulk density and soil geomorphology among other factors. Soil infiltration rates can be determined from tests and area soil surveys data.

14.4 DATA ACQUISITION

Data acquisition systems are the most effective tool for identifying and reaching soil moisture and water application targets for irrigation optimization. A data acquisition system with the water budgeting method was constructed and is commercially available from Stevens Water Monitoring Systems, Inc. The Stevens Agricultural Monitoring (SAM) Package integrates the input from sensors, displays the data from the remote field locations and integrates the water balance method described in the previous section. The SAM package includes rain gauges, the Stevens Hydra Probe Soil Sensor, a Stevens DL3000 data logger, telemetry and the software program. Described below is the engineering that collects field data (soil moisture and precipitation) and the software program that acquires the data from the data loggers through the telemetry. The data is either exported to the internet or is imported into the SAM software where it can be used to make informed decisions about irrigation scheduling.

14.4.1 SOIL MOISTURE DATA COLLECTION

The soil moisture is collected using the Stevens Hydra Probe. The Hydra Probe is the soil sensor used in the USDA's Soil Climate Analysis Network (SCAN) and NOAA's Climate Reference Network (CRN). The Hydra Probe uses electromagnetic waves to measure both the real and imaginary dielectric permittivity [5]. The real component of the dielectric permittivity represents the energy storage based on the high rotational dipole moment of water compared to that of dry soil [14]. The measured real dielectric permittivity ($\varepsilon\ r$) is used to accurately calculate the soil moisture in water fraction by volume () in most soils [11] with the calibration equation:

$$\theta = A\sqrt{\varepsilon_r} + B \tag{5}$$

where, A is 0.109 and B is equal to −0.179. The Hydra Probe is digital and Eq. (5) is written into the firmware of the probe. The digital communication between the Hydra Probe and the data logger is the standard communication format Serial Data Interface at 1200 Baud (SDI-12). The advantages of SDI-12 include connecting many sensors on a single serial addressable bus and cable lengths up to 1000 feet from the sensor to the data logger. Multiple digital sensors are "daisy chained" together and the longer cable lengths provide flexibility in the architecture of the system in the field. Up to 4 or more SDI-12 soil moisture profiles can be installed up to 1000 feet away from the data logger reducing the cost by using common data loggers and telemetry.

14.4.2 RAIN DATA COLLECTION

The precipitation and the irrigation from sprinklers are measured together with a tipping bucket rain gauge. A tipping bucket is a 6 to 10 inch in diameter cylinder with a screen at the top facing end and a drain out the bottom. Inside of the bucket is a dual sided tray that is located under a funnel. The tray will tip over and drain after receiving 0.01 inches of rain. After tipping, the other half of the tray will fill with water, tip and drain after receiving another 0.01 inches of water. Every time the tipping bucket's tray tips (0.01 inch of rain), an electrical pulse is sent to the DL3000 data logger. The data logger counts the tips and calculates the depth of rainfall over time. It is important that the tipping bucket remain level and is placed in a location that will receive a representative application of water from the sprinklers.

If an irrigation method is used that does not include the use of sprinklers such as furrow or drip irrigation, the method described in Fig. 4 and Eq. (4) will not be as applicable. In this case, one or no rain gauge would be used in the data acquisition package.

14.4.3 DATA LOGGER AND FIELD STATION

The Stevens Data Logic 3000 (DL3000) data collection platform resides inside a weather proof fiber glass enclosure located in the field. The cable from each SDI-12 Hydra Probe enters the enclosure by running through bulkhead bushings located on the bottom of the enclosure. The Hydra Probe power, ground and SDI-12 communication wires are "daisy chained" together with a multiplex inside the enclosure. A single SDI-12 communication wire runs from the multiplexer to the DL3000's SDI-12 communication port. The DL3000 will log data on a set time interval typically every 30 min, and will hold up to 2 Gigabytes of data. The wire from the tipping bucket also runs into the enclosure through a bulkhead and is wired into the DL3000's pulse port. The data logger has a wireless RS232 communication radio attached. A coaxial cable runs from the radio out of the enclosure through the bulkhead to an Omni directional antenna. Also contained in the field enclosure is a 9 Amp/hour 12 volt DC battery, and charge regulator for the solar panel power supply. Figure 5 describes a field station with a subsurface soil moisture monitoring profile.

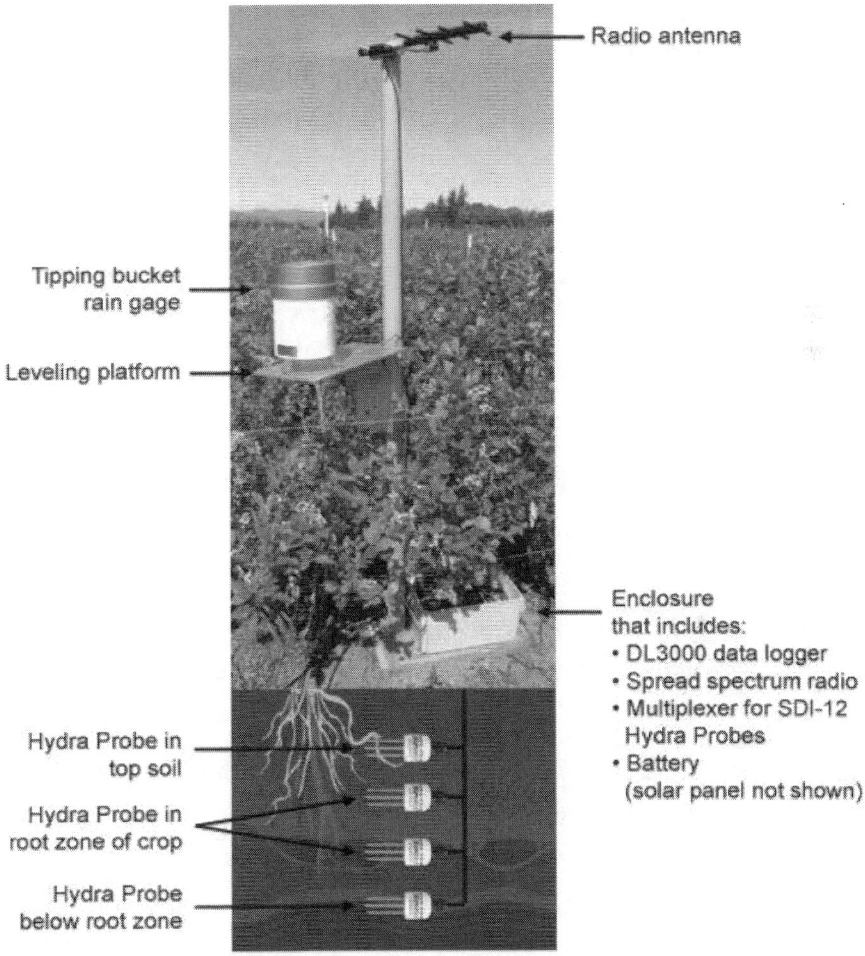

FIGURE 5 Typical soil moisture profile station, which includes four Hydra Probe Soil Sensors, Stevens DL3000 data logger, radio, antenna and accessories.

14.4.4 WIRELESS TELEMETRY

After the data from the sensors is received by the data logger, the data is transmitted from the field to the base station computer via radio. The frequency and type of radio would depend on the distance from the field to the base station computer. The radio communication between the field and the base station is usually line of sight. Large obstacles such as buildings, mountains and trees will impede the radio signal and prevent the signal from reaching its destination. If there is a large obstacle in the way, a repeater station could be installed, however, repeater stations will increase the overall cost of the system. Radio communication always takes place between two or more radios. The radio at the base station is called the server or master radio and the radios in the field are call client or slave radios. The master radio is connected to the base station computer and a directional Omni antenna. Each radio has a Media Access Control (MAC) address written into the radio's firmware, identifying it. When the master radio needs communication with a specific radio, the master radio will address the radio with the MAC address. Radios will only respond their specific MAC address from the master radio. In a network of radios, the master radio will communicate with each slave radio one by one and retrieve the sensor data from each logger individually.

Distance from the field site to the base station is the main factor determining the most appropriate radio and frequency. In most agriculture applications, 900 MHz Spread Spectrum radio with a 5 miles line of sight range is the most common. While satellite communication is common in the water resources industry, it is less common at the farm level due to licensing and hardware costs. Table 5 lists the different kinds of telemetry solutions, the ranges and the frequencies.

TABLE 5 Summary of telemetry options and ranges.

Radio	Range	Frequency
Blue Tooth	100 m	2,400 to 2,483.5 MHz
Cellular Modem	Cell Coverage	824.01 to 848.97 MHz
Geosynchronous Satellite	1/3 the of Earth	401.7010 to 402.0985 MHz
Low Earth Orbiting Satellite	Global Coverage	148 to 150.05 MHz
Spread Spectrum	5 miles	902 to 928 MHz
UHF	30 miles	300 to 1,000 MHz
VHF	30 miles	30 to 300 MHz
Wi-Fi	100 m	2.4 GHz
Wi-Max	30 miles	2.3 to 3.5 GHz

14.4.5 SOIL PROFILE

Soil moisture probes at different depths in the soil column are referred to as a soil profile. Depending on the root zone depth, the typical soil profile consists of four soil sensors. One probe in the top soil (2 to 4 inches) two probes in the root zone (6 to 30 inches) and one probe below the root zone (36 inches). The Hydra Probe in the topsoil will experience the greatest moisture fluctuation because it will be the most influenced by ET and downward flow. The topsoil may reach saturation or reach a soil moisture value over the field capacity thus conducting water downward into the root zone of the crop. The lower soil moisture target for the two Hydra Probes in the root zone however, are calculated from the MAD, θ_{FC} and θ_{PW} in Eq. (2) and the upper soil moisture target in the root zone will be the soil's field capacity. The soil sensor below the root zone should stay below field capacity. If the soil moisture below the root zone reaches values above field capacity, there will be downward conductance of water.

The soil profile should be placed in a location that will most represent the irrigated area. Soil moisture can be highly variable spatially [17]. The factors that affect soil moisture variability are slope, vegetation type, bulk density, soil type, microclimate, and other variables. An irrigation regime represents an area that is homogenous enough that the soil moisture variability will be low and the soil moisture data will represent the entire irrigation regime. There should be at least one soil profile for every irrigation regime. Irrigation regimes are determined by crop type, crop age, soil type, slope, and irrigation method. If the irrigation regimes are less than 1000 feet apart, it may reduce cost to tie multiple soil profiles into one data logger. By tying multiple profiles into a single data logger, the irrigator can save on the number of solar panels, batteries, radios, data loggers and other necessary accessories.

14.4.6 DATA ACQUISITION SOFTWARE

The central user interface of the data acquisition package is the software. The Stevens Agricultural Monitoring (SAM) Software is commercial available and can be subsidized by some energy and water conservation grants. The SAM software runs on a computer that is connected to the master radio. A master radio is not necessary if the system has a field cellular modem or satellite transceiver. The SAM Software acquires the sensor data in the field from a polling sequence. The polling sequence runs at a user specified time interval, which is usually every 15 or 30 min. Communication begins with a serial command from the software to the data logger to take a current a current reading from all of the sensors. The SAM sends the command to the master with instructions to use a specific slave radio. The data logger becomes active after receiving the command and takes a current reading from all of the sensors that are connected to it. Next the data logger sends a comma-delimitated string of sensor data back to the SAM software through the slave and master radio. The SAM software parses the data and populates the tables and graphical displays in the software.

The irrigator can then view the real time data and make decisions about when to irrigate based on the soil moisture targets and the rate of water consumption by the crop from the ET. Other features in the software include battery voltages for power management. In the SAM Software, a display of MAD, θ_{FC}, θ_{PW} and the lower soil moisture limit based on the calculations from Eq. (2) are superimposed unto the real time soil moisture data. The superimposed real time soil moisture onto the soil moisture targets are displayed on a screen similar to Fig. 6.

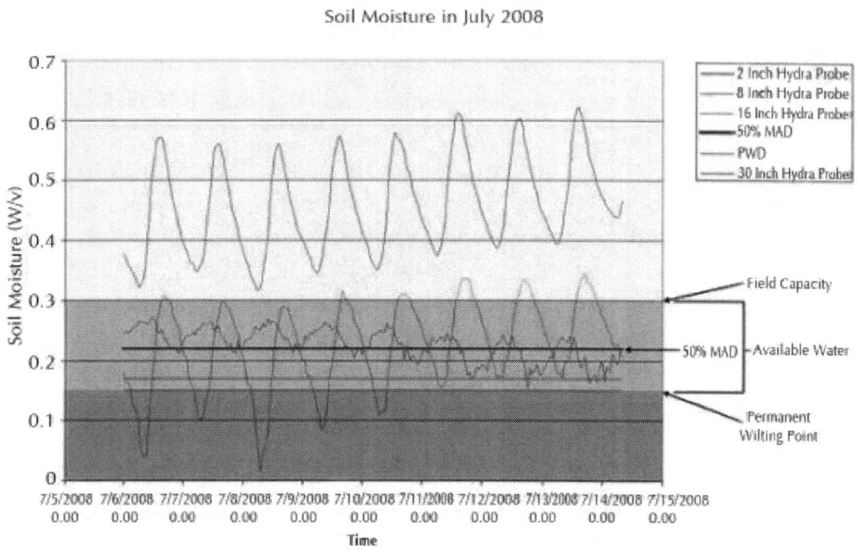

FIGURE 6 Soil moisture measurements in a profile 2, 8, 16 and 30 inches in depth. Daily irrigation events with subsequent decrease in soil moisture from a high ET rate.

At the beginning of the irrigation season, the irrigator can manually input the weekly ET values or the values from Eq. (3) into the SAM setup page. A real time display similar to Fig. 6 is displayed. With real time displays of the real time data superimposed onto the targets in a graphical representation will allow the irrigator to easily interpret the data.

14.4.7 SAM DATA ACQUISITION POLLING SEQUENCE FOR STATION 1

The flow chart below describes the process by which the SAM (Stevens Agriculture Monitoring) software communicates with the field stations. Figure 5 shows a diagram of a field station. The SAM Software will poll data from each station in consecutive order starting with the first field station. After retrieving the data from one field station the software will move on to the next field station.

1. The Polling Sequence initiates on a fixed time interval.
2. The Acquisition command "Take Current Readings Data Logger 1" along with a command to the master radio to communicate with radio 1 with its MAC address.

   These two commands are sent by the software out the serial port of the computer.

3. With an RS232 or USB connection to the computer, the Master Radio receives the “Take Current Readings Data Logger 1” message and transmits this message to slave radio 1 as commanded by the SAM software.
4. Slave radio 1 receives the “Take Current Readings Data Logger 1” and passes the message to the data logger via a RS232 cable.
5. Data Logger 1 receives the command “Take Current Readings Data Logger 1” from the slave radio and one by one collects the current data readings from each sensor that is connected to it.
6. Data Logger 1 sends a comma delimited data string back to the SAM software through the radios and serial ports.
7. The SAM software receives the data string, parses the data, and populates the graphical displays and tables in the software viewable by the user.
8. After the SAM software receives the data from data logger 1, it repeats steps 1 through 7 for data logger 2 and slave radio 2.

14.5 BLUEBERRY FARM IN WASHINGTON COUNTY, OREGON: CASE STUDY

A SAM Soil Moisture data acquisition package complete with telemetry and software was installed on a 200-acre blueberry farm in Washington County, Oregon. The soil unit is Woodburn Silt Loam with less than 3% slope and the soil taxonomic description is Typic Plinthoxeralf. There are two irrigation regimes based on the age of the crop. Two stations, one in each irrigation regime, were installed with four Hydra Probe soil sensors, a tipping bucket rain gauge, and an air temperature sensor. Soils data for this location and most locations in the United States are provide for free by the US Department of Agriculture’s Web Soil Survey Program, http://websoilsurvey.nrcs.usda.gov/app/WebSoilSurvey.aspx.

Figure 7 shows the annual precipitation and ET rate for blueberries in Washington County, Oregon [12]. The ET exceeds precipitation from April to October and this generally defines the irrigation season.

Each station is located 1 mile away from the computer with the master radio; therefore, this network uses spread spectrum radios. The stations each have one soil profile consisting of 4 Hydra Probes at various depths (2″, 8″ 16″ and 30″). The SDI-12 Hydra Probe Soil Sensors are wired into a multiplexer which is connected to the Stevens Data Logger. Each station is power with a solar panel and the enclosure houses the battery, multiplexer, charge regulator and radio. The radio antennas are mounted to the same mast as the tipping bucket. Figure 9 illustrates one of the field stations with the soil profile.

Using Tables 1 and 2, the permanent wilting point is 0.15 the field capacity is 0.3 and the MAD is 50%. The lower soil moisture target as calculated from Eq. (2) is 0.22.

Figure 6 shows the soil moisture for a warm week in July 2008. The topmost region of the chart represents soil moisture levels over field capacity, the middle region shows the range of soil moistures available to the crop (available water capacity) and the bottom region is below permanent wilting point. The two-inch deep soil moisture values fluctuate the most for downward conductivity and ET and stays above field capacity. This is typical because if the top 2 inches of the soil stayed below field capacity then the root zone would not receive the water. The 8-inch soil moisture values fluctuate widely due to ET and there is a 4 h lag time between the 2- and 8-inch soil moisture probes from the downward movement time of the wetting front. During extremely hot days, it is not uncommon to have the soil moisture values briefly drop below permanent wilting point between irrigation cycles. The 16-inch soil moisture mirrors the 8-inch values with a 4 h latency from the soil moisture values above it and the rise and fall of soil moisture values with the irrigation events. The 30 inch deep soil moisture probe below the root zone is remaining constant about 0.10 wfv indicating that water is not peculating downward to the water table.

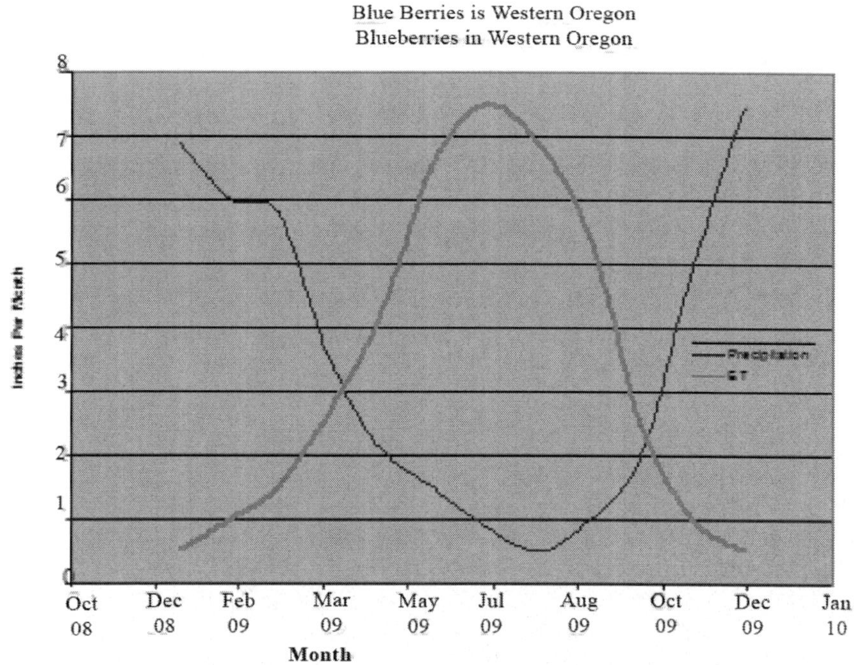

FIGURE 7 Typical values for monthly ET and precipitation for blue berries in western Oregon.

The solid set sprinklers rotator (with an efficiency of 0.90) apply water daily. For the month of July ET (=ET° ′ Kc) is 0.25 inches per day. Using Eq. (3), the daily water consumption will be 0.28 inches. A weekly display similar to Fig. 6 is displayed in the software, which will allow the irrigator to meet the soil moisture and water application targets.

14.6 CONCLUSIONS

As the demand for water increases, along with the need to protect aquatic habitats, water conservation practices for irrigation need to be effective and affordable. Precision irrigation will optimize irrigation by minimizing the waste of water, and energy, while maximizing crop yields. The most effective method for determining the water demands of crops is the based on the real time monitoring of soil moisture, and direct water application used in conjunction with the information about soil hydrological properties and evapotranspiration. The Stevens Agriculture Monitoring data acquisition system wirelessly acquires rain and soil data from the field and integrates the data into water management tools. The water management tools use information about evapotranspiration, soil and the crop to set specific irrigation targets. These irrigation targets will help the irrigator optimize the amount of water used on a weekly basis. Optimization of irrigation water will increase crop yields while conserving water resources.

14.7 SUMMARY

The water requirements of crops are dependent on ET, soil chemistry, and the MAD. Direct measurements of root zone soil moisture, water application along with published ET values and soil textures, can be used in a soil water balance model that can significantly optimize irrigation efficiency. Over the past five years, advancements in computer microprocessors, memory, and software development tools has improved data acquisition methods and made data acquisition system integration more reliable and more cost effective. This chapter presents an irrigation scheduling method based on a volumetric soil moisture balance model and data acquisition. An example of sensor-based irrigation scheduling in blueberries is discussed.

KEYWORDS

- **antenna**
- **available water capacity**
- **Baud rate**
- **blue berry**
- **Bluetooth**
- **capillary forces**
- **catch can**
- **cellular modem**
- **clay**
- **cloud based server**
- **computer software**

- **crop**
- **crop coefficient**
- **data acquisition**
- **data collection platform**
- **data logger**
- **data polling**
- **dielectric permittivity**
- **dipole moment**
- **enclosure**
- **eutrophication**
- **evapotranspiration (ET)**
- **field capacity (FC)**
- **geosynchronous satellite communication**
- **Grants**
- **green beans**
- **hydra probe soil sensor**
- **imaginary dielectric permittivity**
- **internet**
- **irrigation**
- **irrigation optimization**
- **irrigation regime**
- **irrigation scheduling**
- **irrigation system**
- **loam**
- **MAC address**
- **master radio**
- **Maximum Allowable Depletion (MAD)**
- **microclimate**
- **mineralogy**
- **natural waters**
- **omni directional antenna**
- **pathogens**
- **Penman-Monteith ET method**
- **Permanent Wilting Point (PWP)**
- **ponding**

- **porosity**
- **radio frequency**
- **rain gage**
- **real dielectric permittivity**
- **root zone**
- **RS232 communication**
- **salinity**
- **sand**
- **SDI-12 communication**
- **silt**
- **slave radio**
- **soil bulk density**
- **soil chemistry**
- **soil climate analyzes network**
- **soil geomorphology**
- **soil infiltration**
- **soil macro pores**
- **soil micro pores**
- **soil moisture**
- **soil moisture budget**
- **soil moisture sensor**
- **soil particle size**
- **soil saturation**
- **soil sensor**
- **soil survey**
- **soil textural class**
- **solar panels**
- **spatial variability**
- **spread spectrum radio**
- **sprinkler**
- **sprinkler efficiency**
- **Stevens Water Monitoring Systems**
- **telemetry**
- **tipping bucket**
- **total maximum daily load**

- **total suspended solids**
- **unsaturated soil**
- **volumetric soil moisture content**
- **water molecule**
- **web soil survey**
- **wetting front**
- **wireless telemetry**

REFERENCES

1. Allen, R. G., Pereira, L. S., Raes, D., Smith, M. (1998). *Crop Evapotranspiration Guidelines for Computing Crop Water Requirements*. FAO Irrigation and Drainage Paper 56. Food and Agriculture Organization of the United Nations.
2. Birkeland, P. W. (1999). *Soils and Geomorphology*. Third Edition. Oxford University Press.
3. Brady, N. C. (1974). *The Nature and Properties of Soils*. Eighth Edition. Macmillan Publishing Co., Inc.
4. Brouwer, C. (1988). *Irrigation Water Management: Irrigation Methods*. Training Manual Number Food and Agriculture Organization of the United Nations-Land and Water Development Division.
5. Campbell, J. E. Dielectric properties and influence of conductivity in soils at one to 50 Megahertz. *Soil Sci. Soc. Am. J.* (1990). *54,* 332–341.
6. Chapman, D. (1994). *Water Quality Assessments.* World Health Organization.
7. Grim, E. G. (1968). *Clay Mineralogy*. 2nd Edition. McGraw-Hill Co.
8. Hutson, S. S. (2000). *Estimated Use of Water in the United States in (2000).* USGS Circular (1268). US Geological Survey.
9. McBride, M. B. (1994). *Environmental Chemistry of Soils*. Oxford University Press.
10. Rowell, D. L. (1994). *Soil Science Methods and Applications*. John Wiley & Son Inc.
11. Seyfried, M. S., Grant, L. E., Du, E., Humes, K., (2005). Dielectric loss and Calibration of the Hydra Probe Soil Water Sensor. Vadose Zone Journal *4,* 1070–1079.
12. Smesrud, J. M., Hess, Selker, J. (1997). *Western Oregon Irrigation Guide*. Oregon State University.
13. Stevens Water Monitoring Systems, Inc.; www.stevenswater.com.
14. Topp, G. C., Davis, J. L., Annan, A. P. (1980). Electromagnetic Determination of Soil Water Content: Measurement in Coaxial Transmission Line. Water Resources Research *16,* 574–582.
15. US Department of Agriculture, NRCS, *Cooperative Web Soil Survey* http://websoilsurvey.nrcs.usda.gov/app/WebSoilSurvey.aspx
16. Warrick, A. W. (2003). *Soil Water Dynamics*. Oxford University Press.
17. Western, A. W., Zhou, S., Grayson, R. B., McMahon, T. A., Bloschl, G., Wilson, D. J. Spatial Correlation of Soil Moisture in Small Catchments and Its Relationship to Dominant Spatial Hydrological Processes. *J. Hydrology* (2004). *286,* 113–134.
18. Winter, T. C. (2002). *Ground Water and Surface water A Single Resource*. USGS Circular 1139 by US Geological Survey.

BIBLIOGRAPHY

This list is in addition to the references that are included at the end of each chapter in this volume. The mission of this collection is to help the user for in-depth study of a particular topic.

Caution: This list may not include all bibliographical references, all crops, all situations, and all topics related to drip/trickle or micro irrigation (surface or subsurface irrigation), throughout the world.

1. Aali, K. A., Liaghat, A., Dehghanisanij, H. (2009). The Effect of acidification and magnetic field on emitter clogging under saline water application. J. Agric. Sci., 1(1): 132–141.
2. Abbate, E., Dardanelli, J. L., Cantarero, M. G., Maturano, M. (2004). Climatic and water availability effects on water-use efficiency in wheat P. Suer. Crop Science Journal, 44: 474–483.
3. Abegaonkar, M. P., Karekar, R. N., Ayer, R. C. (1999). A microwave micro-strip ring resonator as a moisture sensor for biomaterials: Application to wheat grains. Measuring Science and Technology, 10: 195–200.
4. Abou Khaled, A., Sarraf, S., Vink, N. (1969). Evapotranspiration in the central Beqaa, with reference to the irrigation of potatoes and onions. Magon, IRAL, 26: 1–28.
5. Abou khaled, A. A., Alfaro, A., Smith, M. (1982). Lysimeters. Irrigation and Drainage Paper N° 39, FAO (Food and Agriculture Organization of the United Nations), Rome.
6. Abraham, N., Subramannian, S. (2000). Irrigation automation based on soil electrical conductivity and leaf temperature. Agricultural Water Management, 45(2): 145–157.
7. Abramowitz, M., and Stegun, I. A. (1964). *Handbook of mathematical functions*. Appl. Math. Ser. 55. U. S. Gov. Print. Office, Washington, DC.
8. Abramson, L. W., Thomas, S., S. Sharma and Boyce, G. M. (2001). Slope stability and stabilization methods. John Wiley. 138–157, 248, 263.
9. Abreu, V. M., Luís, S. P. (2002). Sprinkler irrigation systems design using ISADim. Paper number 02–2254, St. Joseph, MI: ASAE Annual Meeting.
10. Abrisqueta, J. M., A. Ruiz and Franco, J. A. (2001). Water balance of apricot trees (Prunus armeniaca L. cv. Belida) under drip irrigation. Agricultural Water Management Journal, 50(3): 211–227.
11. Abrol, I. P., Dixit, S., P. (1972). Studies of the drip method of irrigation. Exp. Agric. 8: 172–175.
12. Abtew, W., Obeysekera, J. (1995). Lysimeter study of evapotranspiration of cattails and comparison of three estimation methods. Trans. of ASAE, 38(1): 121–129.
13. Adams, F. (1984). Soil acidity and liming. Soil Science Society of America: Madison, WI Adamsen, F. J. (1989). Irrigation method and water quality effect on peanut yield and grade. Agronomy Journal, 81(4): 589–593.
14. Adamsen, F. J. (1992). Irrigation method and water quality effects on corn yield in the Mid-Atlantic Coastal Plain. Agronomy Journal, 84(5): 837–843.
15. Adamsen, F. J., Hunsaker, D. J., Perea, H. (2005). Border strip fertigation: Effect of injection strategies on the distribution of bromide. Trans. of ASAE, 48(2): 529–540.
16. Adin, A., Alon, G.1986. Mechanisms and process parameters of filter screens. J. Irrig. Drain. Eng.
17. Adin A. (1987). Clogging in irrigation systems reusing pond effluents and its prevention. Water.
18. Agassi, M., Tarchitzky, J., Keren, R., Chen, Y., Goldstein, D., Fizik, E. (2003). Effects of prolonged irrigation with treated municipal effluent on runoff rate. Journal of Environmental Quality, 32(3): 1053–1057.

19. Agehara, S., Warncke, D. D. (2005). Soilmoisture and temperature effects on nitrogen release fromorganic nitrogen sources.Soil Sci. Soc. of America J., 69: 1844–1855.
20. Agroeconomic analyses of drip Irrigation for sugarbeet production. Agron. J., 93(3): 517–523.
21. Ahmad, M and Ahmad, R. (1993). Advanced irrigation application techniques and their use in Balochistan. USAID, ISMR Project Report, P&I Directorate (South), WAPDA Report#389.
22. Ahmad, Yasin, S. M., Asghar, M. N., Ahmad, M. M. (2004). Root zone salinity management using fractional skimming wells with pressurised irrigation. In: Volume- V *Handbook on Pressurised Irrigation.* A report prepared by IWMI, MONA & WRR I for NDP, WAPDA.
23. Ahuja, L. R., Ma, L., Howell, T., A. (2002). *Agricultural system models in field research and technology transfer.* Boca Raton, Fl: John Wiley.
24. Ajayi, A. E., Olufayo, A. A. (2004). Evaluation of two temperature stress indices to estimate grain sorghum yield and evapotranspiration.Soil Sc. Soc. of America Journal, 68: 1282–1287.
25. Akbari, M., and Dehghanisanij H. (2008). The role of research on improvement and development of micro irrigation methods in Iran. Proc. of the Workshop on Pressurized Irrigation and Sustainable Development in Iran. 21 Feb. Karaj. Iran. (In Farsi).
26. Alam, M., Rogers, D. H. (1997). Tensiometer use in scheduling iirrigation. Kansas State University Agricultural Experiment Station and Coop. Extension Service. Manhattan, Kansas. 321–332.
27. Alam, M., Rogers, D. H. (2005). Field Performance of Subsurface Drip Irrigation (SDI) in Kansas. Proc. Irrigation Association International Irrigation Technical Conference, IA 05–1209. November 6–8, (2005). Phoenix, AZ. pages 1–5.
28. Alam, M., Rogers, D. H., Lamm, F. R., and Trooien, T. P. (2003). Filtration: A basic component for SDI to avoid clogging hazards. In proceedings of the Central Plains Irrigation Conference, Colby, KS, Feb. 4–5, (2003). CPIA, 760 N.Thompson, Colby, KS. pages 212–219.
29. Alam, M., Trooien, T. P., Rogers, D. H., Dumler, T. J. (2002). An efficient irrigation technology for alfalfa growers. J. Extension 40(3): 1–9.
30. Alam, M., Trooien, T. P., Dumler, T. J., Rogers, D. H. (2002). Using subsurface drip irrigation for alfalfa. J. American Water Resources Assoc. 38(6): 1715–1721.
31. Alam, M., Trooien, T. P., Lamm, F. R., Rogers, D. H. (2002). Filtration and maintenance considerations for SDI systems. In Proc. Central Plains Irrigation Short Course, Lamar, CO, Feb. 5–6, (2002). CPIA, 760 N.Thompson, Colby, KS. pages 110–121.
32. Alam, M., Trooien, T. P., Lamm, F. R., and Rogers, D. H. (2002). Filtration and maintenance considerations for subsurface drip irrigation (SDI) systems. KSU Cooperative Ext. Irrigation Mgmt. Series, MF-2361.
33. Alam, Mahmudul, Chamhuri Siwar, Murad, Wahid Molla, Rafiqul Islam & Mohd Ekhwan bin Toriman, (2010). Socioeconomic Profile Of Farmer In Malaysia: Study On Integrated Agricultural Development Area In North-West Selangor. Agricultural Economics and Rural Development. Vol. 7(2): 249–265.
34. Alam, Mahmudul, Mohd Ekhwan Toriman, Chamhuri Siwar, Rafiqul Islam Molla and Basri Talib, (2011). The Impact of Agricultural Supports for Climate Change Adaptation: A Farm Level Assessment. American Journal of Environmental Sciences. 7 (2): 178–182.
35. Albregts, E. E., Clark, G. A., Stanley, C. D., Zazueta, F. S., Smajstrla, A. G. (1991). Preplant fertilization of fruiting microirrigated strawberry. HortScience. 26(9): 1176–1177.
36. Albright, W. H., Gee, G. W., Wilson, G. V., Fayer, M. J. (2002). Alternative cover assessment project Phase I Report. DRI-41183. Desert Research Institute. Reno -NV. 533–556.
37. Alfalfa production with subsurface drip irrigation (SDI). KSU-NWREC spring field day, June 5, (2008).
38. Alizadeh, A. (2008). *Water, Soil and Plant Relations.* Ferdowsi University of Mashhad, 309 pages.
39. Alizadeh, A., Keshavarz A. (2005). Status of agricultural water use in Iran. Water Conservation, Reuse, and Recycling: Proceeding of an Iranian-American Workshop. Tehran. Iran. 94–105.
40. Al-Jamal, M. S., Sammis, T. W., Ball, S., Smeal, D. (2000). Computing the crop water production function for onion. Agricultural Water Management 46: 29–41.

41. Allen, I. H., Jones, P., Jones, J. W. (1985). Rising atmospheric CO_2 and evapotranspiration. Proceedings of the National Conference onAdvances in Evapotranspiration, December 16–17, Chicago, IL. St. Joseph, MI: ASAE.
42. Allen, L. H., Jr., Pan, D., Boote, K. J., Pickering, N. B., Jones, J. W. (2003). Carbon dioxide and temperature effects on evapotranspiration and water use efficiency of soybean. Soil Sci. Soc. of America J., 67: 1071–1081.
43. Allen, R. G., Pereira, L. S., Raes, D., Smith, M. (1998). *Crop Evapotranspiration Guidelines for Computing Crop Water Requirements*. FAO Irrigation and Drainage Paper 56, Food and Agriculture Organization of the United Nations, Rome.
44. Allen, R. G. (1995). Evaluation of procedures for estimating mean monthly solar radiation from air temperature. Report prepared for FAO, Water Resources Development and Management Service, FAO, Rome.
45. Allen, R. G., Brockway, C. E. (1983). Estimating consumptive use on a statewide basis. Irrigation and Drainage Specialty Conference at Jackson, WY. New York, NY: ASCE, 79–89.
46. Allen, R. G., Gichuki, F. N. (1989). Effects of Projected CO_2: Induced climatic changes on irrigation water requirements in the Great Plains States (Texas, Oklahoma, Kansas, and Nebraska). In The Potential Effects of Global Climate Change on the United States: Appendix C – Agriculture. Vol. 1. EPA-230-05-89-053, eds. Smith, J. B., Tirpak, D. A., 6: 1–42. Washington, DC: U. S. Environmental Protection Agency, Office of Policy, Planning and Evaluation.
47. Allen, R. G., Pruitt, W. O. (1986). Rational use of the FAO Blaney – Criddle formula. Journal Irrigation and Drainage Division of ASCE, 112(2): 139–155.
48. Allen, R. G., Pruitt, W. O. (1991). FAO-24 reference evapotranspiration factors. Journal of Irrigation and Drainage Engineering, ASCE, 117(5): 758–773.
49. Allen, R. G., Wright, J. L. (1977). Translating wind measurements from weather stations to agricultural crops. Journal of Hydrologic Engineering, ASCE, 2(1): 26–35.
50. Allen, R. G. (1986). A Penman for all seasons. Journal Irrigation and Drainage Division of ASCE, 112(4): 348–368.
51. Allen, R. G. (1992). Evaluation of a temperature difference method for computing grass reference evapotranspiration. Report submitted to UN-FAO Water Resources Development and Management.
52. Allen, R. G. (1995). Evaluation of procedures for estimating grass reference evapotranspiration using air temperature data only. Report prepared for FAO, Water Resources Development and Management Service, FAO, Rome.
53. Allen, R. G. (1996). Assessing integrity of weather data for use in reference evapotranspiration estimation. Journal ofIrrigation and Drainage EngineeringDivision, ASCE, 122(2): 97–106.
54. Allen, R. G. (1997). Self-calibrating method for estimating solar radiation from air temperature. Journal of Hydrologic Engineering, ASCE, 2(2): 56–67.
55. Allen, R. G. (2005). *Crop evapotranspiration: Guidelines for computing crop water requirements*. Food & Agriculture Organization of the United States. Publication 23765. 1–328.
56. Allen, R. G., Brockway, C. E., Wright, J. L. (1983). Weather station siting and consumptive use estimates. Journal of Water Resources Engineering and Management Division, ASCE, 109(2): 134–146.
57. Allen, R. G., J. Prueger and Hill, R. W. (1992). Evapotranspiration from isolated stands of hydrophytes: Cattail and Bulrush. Trans. of ASAE, 35(4): 1191–1198.
58. Allen, R. G., Smith, M., A. Perrier and Pereira, L. S. (1994). An update for the definition of reference evapotranspiration. ICID Bulletin, 43(2): 1–34.
59. Allen, R. G., Smith, M., Pereira, L. S., Pruitt, W. O. (1997). Proposed revision to the FAO procedure for estimating crop water requirements. In Proceedings 2nd International Symposium on Irrigation of Horticultural Crops, ed. K. S. Chartzoulakes. ISHS. Acta Horticulturae, I: 17–33.
60. Allen, R. G., Jensen, M. E., Wright, J. L., Burman, R. D. (1989). Operational estimates of reference evapotranspiration. Agronomy Journal, 81: 650–662.
61. Allen, R. G., Hill, R. W., Vemulapali, S. (1994). Evapotranspiration parameters for variably-sized wetlands. Paper presented at the 1994 Summer Meeting of ASAE No. 942132, 24.

62. Allen, R. G., Pruitt, W. O., Jensen, M. E. (1991). Environmental requirements for lysimeters. Proceedings of the ASCE International Symposium on Lysimetry, eds. Allen, R. G., Howell, T. A., Pruitt, W. O., Walter, L. A., Jensen, M. E., Lysimeters for Evapotranspiration and Environmental Measurements, Honolulu, HA. New York, NY: ASCE. 170–181.
63. Allen, R. G., Pruitt, W. O., Businger, J. A., Fritschen, L. J., Jensen, M. E., Quinn, F. H. (1996). Evaporation and Transpiration. Chapter 4: 125–252, In: *ASCE handbook of hydrology*. NY.
64. Allen, R. G., Pereira, L. S., Raes, D., and Smith, M. (1998). *Crop evapotranspiration–Guidelines for computing crop water requirements*. FAO Irrigation and Drainage Paper 56 (FAO: Rome).
65. Allen, R. G., Walter, I. A., Elliot, R. L., Howell, T. A., Itenfisu, D., Jensen, M. E., Snyder, R. L., eds, (2005). *The ASCE Standardized Reference Evapotranspiration Equation*. The American Society of Civil Engineers, Reston: Virginia Publication 23136. 1–216.
66. Alon, B. G., Lazorovitch, N., Shani, U. (2004). Subsurface drip irrigation in gravel-filled cavities. Vadose Zone Journal, 55: 2345–2654.
67. Alves, I. L. (1995). Modeling crop evapotranspiration: Canopy and aerodynamic resistances. PhD Dissertation, ISA, University of Tecnical Lisboa.
68. Amente, G., Baker, J. M., Reece, C. F. (2000). Estimation of soil solution electrical conductivity from bulk soil electrical conductivity in sandy soils. Soil Sc. Soc. of America J., 64: 1931–1939.
69. American Conference of Governmental Industrial Hygienists (ACGIH), (1999). TLVs and BEIs. Threshold limit values for chemical substances and physical agents. Biological Exposure Indices. Cincinnati, OH.
70. American Water Works Association, (1999). *Water quality & treatment: A handbook of community water supplies*. 5th ed. AWWA.
71. Amoozegar-Fard, A., Warrick, A. W., and Lomen, D. O. (1984). Design nomographs for trickle irrigation systems. J. Irrig. Drain. Eng., Am. Soc. Civ. Eng. 110: 107–120.
72. Anbumozhi, V., Matsumoto, K., Yamaji, E. (2001). Sustaining agriculture through modernization of irrigation tanks: An Opportunity and a Challenge for Tamilnadu, India. Agricultural Engineering International: the CIGR Journal of Scientific Research and Development, III(10): 2–12.
73. Aneela-Sardar, M. A., Akhtar, M. E. (2003). Effect of potash on N, P and K content of young mature leaves and nitrogen utilization efficiency in selected cotton varieties. *Pakistan J. Biol. Sci.* 6(9): 793–796.
74. Angelakis, A. N., M. H. F. Marecos do Monte, Bontoux, L., Asano, T. (1999). The status of wastewater reuse practice in the Mediterranean basin: Need for guidelines. Water Res. 33(10): 2201–2217.
75. Annandale, J. G., Stockle, C. O. (1994). Fluctuation of crop evapotranspiration coefficients with weather: a sensitivity analysis. Irrigation Science, 15(1): 1–7.
76. Anonymous, (1981). Questions and answers about tensiometers. Division of Agricultural Sciences, University of California.
77. Anonymous, (1982). Irrinews. ISSN 304–3606. Bet Degam-Israel. 25(1982): 12–13.
78. Anonymous, (1998). Farm and Ranch Irrigation Survey. In: Vol. 3. 1997 Census of Agriculture. USDA, National Agricicultur e Statistics Service, Washington, DC.
79. Anonymous, (2000). 1999 Annual irrigation survey. Irrigation Journal, 50(1): 16–31.
80. Anonymous, (2000). *Water reuse for irrigation*. CRC Press.
81. Anonymous, (2002). Installation of Water Systems. In Water reuse for irrigation. Eds. Valentina Lazarova and Akica Bahri Price, CRC Press, Florida – USA. 87–97.
82. Anonymous, (2003). Paddy and Water Environment. 1(4): 157–214.
83. Anonymous, (2001). National Committee on Plasticulture Applications in Horticulture, Progress Report – 2001, Department of Agricultural and Co-operation, Ministry of Agriculture, GOI, New Delhi, 98p.
84. Anonymous, (2004). National Committee on Plasticulture Applications in Horticulture, Review meeting od North Indian PFDC, October, 30. Department of Agricultural and Co-operation, Ministry of Agriculture, GOI, New Delhi.

85. Anonymous. Applying nutrients and other chemicals to trickle irrigated crops. Univ. of California Div. of Agric. Sci. Bull. 1893 (1981).
86. Anonymous. Computerized irrigation control system handout. Motorola Inc. Agro-Control Department, Fresn, CA, 93710.
87. Anonymous. Instruction manual – Irrigation Controler Model AG-7, Rain Bird, CA.
88. Anonymous. Instruction manual for installation and operation of "Free Flow" media filters. Water Management Products Division. P. O. Box 352 Corona, CA, 91720.
89. Anonymous. Irrigation water management using the neutron probe. Buchanan Circle Corp., Pacheco, CA, 94553.
90. Antunez, A., Martínez, J. P., Alfaro, C., Alé, M. (2011). Impact of surface and subsurface drip irrigation on yield and quality of "Honey Dew" Melon. Acta Horticulturae, (ISHS), 889, pages 417–422.
91. Aragués, R., Playán, E., Ortiz, R., Royo, A. (1999). A new drip-injection irrigation system for crop salt tolerance evaluation. Soil Sci. Soc. of America J., 63: 1397–1404.
92. Arbat, G., Lamm, F. R., A. A. Abou Kheira, (2010). Subsurface drip irrigation emitter spacing effects on soil water redistribution, corn yield and water productivity. Applied Engr. in Agric., 26(3): 391–399.
93. ASABE Standards, (2001). S526.2: Soil and water terminology. St. Joseph, MI: ASABE.
94. ASABE. (2000). Proceedings of the 4th Decennial National Irrigation Symposium. Evans, R. G., Benham, B. L., T. P. Trooieneds. St. Joseph, MI: ASABE.
95. Asadi, M. E. (2002). Impacts of fertigation via sprinkler irrigation of nitrate leaching and corn yield in an acid sulfate soil in Thailand. Doctoral Dissertation, Dissertation No: WM-00–02. Asian Institute of Technology, Bangkok, Thailand.
96. Asadi, R. (2011). Effect of drip irrigation system on water use efficiency and cotton yield in Kerman province. First National Conference on Drought Management and Water Shortages in Agriculture.
97. Asadi, Rasoul, Nader Kouhi, and Najme Yazdanpanah, (2012). Applicability of micro irrigation system on cotton yield and water use efficiency. Journal of Food, Agriculture & Environment, 10(1): 302–306.
98. ASAE EP405.1. 2010 ASAE Engineering Practice EP405.1, APR1988, Design and Installation of Micro irrigation Systems. ASAE, St. Joseph, Michigan. pages 1140–1144.
99. ASAE Standards, (1996). Field evaluation of micro irrigation systems. ASAE EP405.1 and EP458. St. Joseph, MI: American Society of Agricultural Engineers, 1–7.
100. ASAE Standards, (1999). S526.1: Soil and water terminology, 46th ed., St. Joseph, MI: ASAE.
101. ASAE Standards, (1999). EP-458: Field evaluation of micro irrigation systems. St. Joseph, MI: ASAE.
102. ASAE Standards. (2003a). EP405.1. Design and installation of micro irrigation systems. St. Joseph, Mich: ASAE.
103. ASAE Standards. (2003b). EP458. Field evaluation of micro irrigation systems. St. Joseph, MI.
104. ASAE, (2001). ASAE Standard S526.2, JAN01, In: *Soil and Water Terminology*, ASAE, St.Joseph, Michigan USA.
105. ASAE, (2008). Design and Installation of Micro irrigation Systems. ASAE EP405.1 April 1988 (R2008). ASABE, St Joseph, MI. 5 pages.
106. Asano, T., Levine, A. D. (1996). Wastewater reclamation, recycling, and reuse: Past, present, and future. Water Sci. Tech., 33(10–11): 1–14.
107. Ascough, G. W., Kiker, G. A. (2002). The effect of irrigation uniformity on irrigation water requirements. Journal of the Water South Africa, 28(2): 235–242.
108. Asghar, M. N., Yasin, M., Alam, M. M., Qureshi, A. S. (2004). Root zone salinity management using fractional skimming wells with pressurised irrigation: Volume- I Main Report. A report prepared by IWMI, MONA & WRR I for NDP, WAPDA.
109. Ashley, R. O., Neibling, W. H., King, B. A. (1998). Irrigation scheduling using water-use tables. Univ. of Idaho CIS 1039. 11 pages.

110. Assouline S., Möller, M., Cohen, S., M. A. Ben-Hur Grava, Narkis, K., Silber, A. (2006). Soil-plant system response to pulsed drip irrigation and salinity: Bell pepper case study. Soil Sci. Soc. of America J., 70: 1924–1927.
111. Assouline S., Cohen, S., Meerbach, D., Harodi, T., Rosner, M. (2002). Microdripirrigation of field crops: Effect on yield, water uptake and drainage in sweet corn soil. Soil Sci. Soc. of America J., 66: 228–235.
112. Assouline, S. (2002). The effects of microdrip and conventional drip irrigation on water distribution and uptake. Soil Sci. Soc. of America J., 66: 1630–1636.
113. Assouline, S., Möller, M., Cohen, S., Ben-Hur, M., Grava, A., Narkis, K., Silver, A. (2006). Soil-plant system response to pulsed drip irrigation and salinity: bell pepper case study. Soil Sci. Soc. of America J., 70(1): 1556–1568.
114. Assouline, S., Cohen, S., Meerbach, D., Harodi, T., Rosner, M. (2002). Microdrip irrigation of field crops: Effect on yield, water uptake and drainage in sweet corn. Soil Sci. Soc. of America J., 66(1): 228–235.
115. Avars, J., Bucks, D., Lamn, F., Nakayama, F. (2002). Micro irrigation installation. In Micro irigation for crop production. 22–43.
116. Avila, C. G., Trujillo, F. L., Estrada, C. A., T., Gaxiola, S. J., A., Juarez, I. (2003). Water consumption, water relations, and yield in alfalfa with subsurface drip irrigation. Agricultura Tecnica en Mexico 29(2): 113–123.
117. Aw, D., Diemer, G. (2005). Making a large irrigation scheme work: A case study from Mali. World Bank Publications, Infrastructure and Water Management. 7–9.
118. Axmann, H., Sebastianelli, A., Arrillaga, J. (1990). Sample preparation techniques of biological material for isotope analysis. In: Use of nuclear techniques in studies of soil-plant relationships (Edited by G. Hardarson), IAEA-Vienna,41–54 pages.
119. Ayars, J. E., Bucks, D. A., Lamm, F. R., and Nakayama, F. S. (2007). Introduction. Chapter 1 in *Micro irrigation for Crop Production – Design, Operation and Management*. Lamm, F. R., Ayars, J. E., and F. S. Nakayama (Eds.), Elsevier Publications. pages 1–26.
120. Ayars, J. E., Phene, C. J., Schoneman, R. A., Meso, B., Dale, F., Penland, J. (1995). Impact of bed location on the operation of subsurface drip irrigation systems. In Proc. Fifth International Micro irrigation Congress, ed. Lamm, F. R., pages 141–146. St. Joseph, MI: ASAE.
121. Ayars, J. E., Phene, C. J., Hutmacher, R. B., Davis, K. R., Schoneman, R. A., Vail, S. S., Mead, R. M. (1999). Subsurface drip irrigation of row crops: A review of 15 years of research at the water management research lab. Agric. Water Mgt. 42(1): 1–27.
122. Ayars, J. E., Phene, C. J., Hutmacher, R. B., Davis, K. R., Schoneman, R. A., Vail. S. S., Mead, R. M. (1999). USA: Subsurface drip irrigation of row crops: a review of 15 years of research at the Water Management Research Laboratory, University of California. Agricultural Water Management 42: 1–27, (1999).
123. Ayars, J. E., Phene, C. J., Schoneman, R. A., Meso, B., Dale, F., Penland, J. (1995). Impact of bed location on the operation of subsurface drip irrigation systems. In Proceedings 5th International Micro irrigation Congress, ed. Lamm, F. R., 141–146. St. Joseph, MI: ASAE.
124. Ayars, J. E., Phene, C. J., Hutmacher, R. B., Davis, K. R., Schoneman, R. A., Vail, S. S., Mead, R. M. (1999). Subsurface drip irrigation of row crops: A review of 15 years of research at the Water Management Research Laboratory. Agricultural Water Management, 42(2): 1–27.
125. Ayars, J. E., Schoneman, R. A., Dale, F., Meso, B., Shouse, P. (2001). Managing subsurface drip irrigation in the presence of shallow ground water. Agricultural Water Management Journal, 47(3): 243–264.
126. Ayars, J. E., Schoneman, R. A., Dale, F., Meso, B., Shouse, P. (2001). Managing subsurface drip irrigation in the presence of shallow ground water. Agricultural Water Management, 47(3): 243–264.
127. Ayers, R. S., Westcot, D. W. (1985). Water quality for agriculture. Irrigation and Drainage Paper 29, Rev. 1. Food and Agriculture Organization of the United Nations, Rome. 17.
128. Ayodele, E. A., Ayorinde, A., Olufayo, A., J. (2004). Evaluation of two temperature stress indices to estimate grain sorghum yield and evapotranspiration. Agronomy Journal, 96: 1282–1287.

129. Bacarello, V., Ferro, V., Provenzano, G., Pumo, D. (1999). Evaluating pressure losses in drip-irrigation lines. Journal of Irrigation and Drainage Engineering, 123(1): 1–7.
130. Baird, K. J. (2006). Development of a new methodology for estimating groundwater evapotranspiration. ProQuest / UMI. 1–194.
131. Bakhtiari, S. (2004). Cartographical Map of Provinces of Iran. Geography and Cartography Center, Tehran, 165 p.
132. Banta, E. R. (2000). MODFLOW (Electronic Resource): Modular Ground Water Model – Documentation of Packages for Stimulating Evotranspiration with a Segment Function (ETS 1) and Drains with Return Flow (DRT1). The U. S. Geological Survey.
133. Barber, S. A., Katupitiya, A., and Hickey, M. (2001). Effects of long-term subsurface drip irrigation on soil structure. Proceedings of the 10th Australian Agronomy Conference. Hobart, Tasmania.
134. Bar-Ram. Instructions for the Maintenance of BAR-RAM Drip Irrigation Systems. Israel.
135. Barrow, N. J. (1974). The slow reactions between soil and anions: Part I – Effects of time, temperature and water content of a soil on the decrease in effectiveness of phosphate for plant growth. Soil Science, 118: 380–386.
136. Barrow, N. J., and Shaw, T. C. (1975). The slow reactions between soil and anions: Part III – Effects of time and temperature on the decrease in isotopically exchangeable phosphate. Soil Science, 119: 190–197.
137. Barreras, J. T. (2001). Evaluation of retrievable drip tape irrigation systems. Irrigation Technology Center, Fresno – CATRC Paper No, (2001).
138. Barrett, R. A., and Parsons, S. A. (1998). The influence of magnetic fields on calcium carbonate precipitation. Journal of Water Research, 32 (3): 609–612.
139. Barth, H. K. (1995). Resource conservation and preservation through a new subsurface irrigation system. *Proceedings 5th International Micro irrigation Congress*, ed. Lamm, F. R., 168–174. St. Joseph, MI: ASAE.
140. Bar-Yosef, B. (1999) Advances in fertigation. Advances in Agronomy, 65: 2–77.
141. Bar-Yosef, B. Vegetable production and fruit yield and quality under combined trickle irrigation and fertigation. FAO/ECE Symposium on "Methods and concepts for the use of organic and chemical fertilizers". Geneva. 14–18 January, (1991).
142. Bar-Yosef, B., and Sagiv, B. (1982). Response of tomatoes to N and water applied via a trickle irrigation system. I. Nitrogen. Agronomy Journal, 74:633–636.
143. Bar-Yosef, B., Sagiv, B., Markovitch, T. (1989). Sweet corn response to surface and subsurface trickle phosphorus fertigation. Agronomy Journal 81(3): 443–447.
144. Bar-Yosef, B., Phene, C. J., Hutmacher, R. B. (1991). Plants response to subsurface trickle fertigation. BARD Project No. 1–1116–86 Final Report. Bet Dagan, Israel: BARD.
145. Basal, H., Dagdelen, N., Unay, A., Yilmaz, E. (2009). Effects of deficit drip irrigation ratios on cotton (Gossypium hirsutum) yield and fiber quality. Agronomy and Crop Science 159: 19–29.
146. Bascom, N. D., Lamm, F. R., and Manges, H. L. (1991). Designing for maximum length drip tape laterals. Presented at the 1991 Mid-Central ASAE Meeting. Paper no. MC91–101, ASAE, St. Joseph, MI. 14 pages.
147. Bassett, D. M., Anderson, W. D., Werkhoven, C. H., Dry matter production and nutrient uptake in irrigated cotton (Gossypium hirsutum). Agron. J. 62 (1997) 299–303.
148. Bastiaanssen, W. G.M. (1995). Regionalization of surface flux densities and moisture indicators in composite terrain. PhD Thesis, Wageningen Agricultural University, Wageningen, Page 273.
149. Batchelor, C. H. (1984). The accuracy of evapotranspiration functions estimated with the FAO modified Penman equation. Irrigation Science, 5(4): 223–234.
150. Batchelor, C. H., Lovell, C. J., M. Murata and McGrath, S. P. (1994). Improving water use effectiveness by subsurface irrigation. Aspects of Applied Biology, 38: 269–278.
151. Battam, M., Boughton, D., Hulme, P., Sutton, B. (2001). Drip irrigated cotton: Observing wetting patterns. Irrigation Journal, 51(4): 13–16.
152. Bauder, J. W., Bauer, A., Ramirez, J. M., Cassel, D. K. (1978). Alfalfa water use and production on dryland and irrigated sandy loam. Agron. J. 70(1): 95–99.

153. Bauder, J. W., King, L. D., Westesen, G. L. (1982). Scheduling irrigation with evaporation pans. Coop. Ext. Ser., Montana State Univ., Bozeman, Bulletin. 1262, Form Circular. 1211, 1–23.
154. Baumhardt, R. L., Lascano, R. J., Evett, S. R. (2000). Soil material, temperature and salinity effects on calibration of multisensor capacitance probes. Soil Sci. Soc. of America J., 64: 1940–1946.
155. Beard, J. R. (1985). An assessment of water use by turfgrass. In *Turfgrass Water Conservation*, eds. Gibeault, V. A., S. T. Cockerham. Publ. 21405, University of California Division of Agricultural and Natural Resources, Berkley, CA. Pages. 45–60.
156. Bebr, R. (1954). Agricultural hydraulics. Barcelona, S. A., Madrid: Salvat Publishing, Chapter 8: 212–220.
157. Belmans, C., Wesseling, J. G., Feddes, R. A. (1983). Simulation model of the water balance of a cropped soil: SWATRE Hydrology, I., 63: 271–286.
158. Ben Gal, A., Lazorovitch, N., Shani, U. (2004). Subsurface drip irrigation in gravel-filled cavities. Vadose Zone Journal, (3): 1407–1413.
159. Ben-Asher, J., Phene, C. J. (1993). Analysis of surface and subsurface drip irrigation using a numerical model. In *Subsurface Drip Irrigation-Theory, Practices and Application*, 185–202pages CATI Pub. No. 92–1001. Fresno, CA: California State University.
160. Ben-Gal, A., Dudley L. M. (2003). Phosphorus Availability under Continuous Point Source Irrigation. Soil Sci. Soc. Am. J. Vol. 67.
161. Ben-Gal, A., Lazorovitch, N., Shani, U. (2004). Drip irrigation in gravel-filled cavities.Vadose Zone Journal, 3(4): 1407–1413.
162. Ben-Gal, Alon and Naftali Lazarovitch. Beyond burying the lateral: Current issues in and future opportunities for subsurface drip irrigation. Netafim.
163. Benjamin, J. G., Porter, L. K., Ahuja, H. R., Butters, G. (1998). Nitrogen movements with furrow irrigation method and fertilizer band placement. Soil Sci. Soc. of America J., 62(1): 1103–1108.
164. Bergmann, L., Weismantel, G. E. (2000). *Filtration Technology Handbook.* North Carolina, NC: John Wiley.
165. Bernstein, L., Francois, L. E. (1973). Comparisons of drip, furrow and Sprinkler irrigation. Soil Sci. 115 73–86.
166. Bevan, K. (1979). A sensitivity analysis of the Penman-Monteith actual evapotranspiration estimates. Journal of Hydrology, 44: 169–190.
167. Beyazgul, M., Kayan, Y., Engelsman, F. (2000). Estimation methods for crop water requirements in the Gediz Basin of Western Turkey. Journal of Hydrology, 229: 19–26.
168. Beyer, C., and Holland, R.,1996. Guidelines to successful processing tomato production 1995/96. Agriculture Victoria: Melbourne – AU.
169. Bhattarai, S. P., Huber, S., and Midmore, D. J. (2004). Aerated subsurface irrigation water gives growth and yield benefits to zucchini, vegetable soybean and cotton in heavy clay soil. Ann. Appl. Biol., 144: 285–298.
170. Bhutta, M. N., and Azhar A. H. (2005). A proposal for sprinkler irrigation model farm at Mirani Dam Project (MDP), Turbat. IWASR I Internal Report No. 2005/01.
171. Bidwell, R. G.S. (1979). Plant physiology. New York: McMillan. 247–250.
172. Bidwell, O. W., Banbury, E. E., Barker, W. L., Muilenburg, G. E. (1980). The Colby Branch Experiment Station and agriculture in northwest Kansas with special mention of soils. KAES Bulletin 635. Manhattan, Kans.: Kansas Agricultural Experiment Station.
173. Biggs, J. S., Vallis, I., Kokot, S., and Keating, B. A. (1996). Effect of depth of urea application on loss of nitrogen by volatilization from acid soils. In: *Sugarcane – Research towards efficient and sustainable production* (Eds. Wilson, J. R., D. M. Hogarth Campbell, J. A., A. J. Garside). CSIRO, Division of Tropical Crops and Pastures: Brisbane – AU.
174. Binnie, C., Martin, K., George, S. (2002). Basic water treatment. London: IPC media.
175. Biran, I., Bravdo, B., Bushkin-Harav, I., Rawitz, E. (1981). Water consumption and growth rate of 11 turfgrasses as affected by mowing height, irrigation frequency and soil moisture. Agronomy Journal, 73: 85–90.

176. Bitteli, M., M. Flury and Campbell, G. S. (2003). A thermoelectric analyzer to measure the freezing and moisture characteristic of porous media. Water Resources Research, 39: 1041.
177. Black, A. S. (1992). Soil acidification in urine-and urea affected soil. Australian Journal of Soil Research, 30: 989–999.
178. Blad, B. L., Rosenberg, N. J. (1974). Lysimetric calibration of the Bowen-ratio energy balance method for evapotranspiration estimation in the Central Great plains. Journal Applied Meteorology, 13(2): 227–236.
179. Blanco, F. F., Folegatti, M. V., Casarini, E. (2003). Effects of nitrogen and potassium on dry matter production of tomato irrigated with saline water. Paper number 032235, ASAE Annual Meeting.
180. Blanco-Canqui, H., Lal, R., Post, W. M., Owens, L. B. (2006). Changes in long-term no-till corn growth and yield under different rates of stover mulch. Agronomy Journal, 98(1): 1128–1136.
181. Blaney, H. F., Criddle, W. D. (1950). Determining water requirements in irrigated areas from climatological and irrigation data. USDA Soil Conservation Service SCS-TP96. 44.
182. Blass, S. (1964). Sub-surface irrigation. Hassadeh 45: 1 (in Hebrew).
183. Blass, S. (1973). Water in strife and action (Hebrew). Israel: Massada limited.
184. Bogle, C. R., Hartz, T. K., Nunez, C. (1989). Comparison of subsurface trickle and furrow irrigation on plastic-mulched and bare soil for tomato production. Journal of Amercian Society for Horticultural Sciences, 114(1): 40–43.
185. Boland, A. M., Mitchell, P. D., Jerie, P. H., Goodwin, I. (1993). The effect of regulated deficit irrigation on tree water use and growth of peach. J. Hort. Sci. 68(2): 261–274.
186. Bolvin, H., Chambarel, A., Chanzy, A. (2004). Three-dimensional numerical modeling of a capacitance probe: Application to measurement interpretation. Soil Sci. Soc. of America J., 68: 440–446.
187. Bolvin, H., Chambarel, A., Chanzy, A. (2004). Three-dimensional numerical modeling of a capacitance probe: Application to measurement interpretation. Soil Sci. Soc. of America J., 68: 440–446.
188. Boman, B. J. (1995). Effects of orifice size on microsprinkler clogging rates. Applied Engr. in Agric. 11(6): 839–843.
189. Boman, B. J., Parsons, L. R. (1999). Micro sprinkler experiences in Florida citrus. Applied Engr. in Agric. 15(5): 465–475.
190. Boman, B. J., Bullock, R. C., Parson, M. L. (1995). Effectiveness of teflon and sevin sprays in controlling caterpillar damage to microsprinkler supply tubing. Applied Engr. in Agric. 11(4): 523–525.
191. Boman, B. J., Bullock, R. C., Parsons, M. L. (1995). Ant damage to micro sprinkler pulsator assemblies. Applied Engr. in Agric. 11(6): 835–837.
192. Boman, B. J., Zekri, M., Stover, E. (2005). Managing salinity in citrus. HortTechnol. 15(1).
193. Bomo, A. M., Husby, A., Stevik, T. K., Hanssen, J. F. (2003). Removal of fish pathogenic bacteria in biological sand filters. Water Resources, 37: 2618–2626.
194. Bonachela, S., Orgaz, F., Villalobos, F. J., Fereres, E. (2001). Soil evaporation from drip-irrigated olive orchards. Irrigation Science, 20(2): 65–71.
195. Bonan, G. B.,2002. *Ecological climatology: Concepts and applications*.Cambridge University Press. 141–146.
196. Bonet, L., Ferrer, P., Castel, J. R., Intrigliolo, D. S. (2010). Soil capacitance sensors and stem dendrometers. Useful tools for irrigation scheduling of commercial orchards?. Spanish Journal of Agricultural Research 8(S2): S52-S65.
197. Books, S. (2000). *Mulch it! A practical guide to using mulch in the garden and landscape*.North Adams, MA: Storey Publishing. 1–128.
198. Boote, K. J., Jones, J. W., Batchelor, W. D., Nafziger, E. D., Myers, O. (2003). Genetic coefficients in the CROPGRO–Soybean model: Links to field performance and genomics. Agro. J., 95: 32–51.

199. Bordovsky, J. P., Lyle, W. M. (1998). Cotton irrigation with LEPA and susurface drip systems on the southern high plains. In Proc. Beltwide Cotton Conf., San Diego, CA., Jan. 5–9, (1998). pages 409–412.
200. Bordovsky, J. P., Porter, D. (2003). Cotton response to pre-plant irrigation level and irrigation capacity using spray, LEPA and subsurface drip irrigation. Annual Meeting of the ASAE, Las Vegas, Nevada, July 27–30, (2003). ASAE Paper No. 032008. 11 pages.
201. Bordovsky, J. P., Porter, D. (2008). Effect of subsurface drip irrigation system uniformity on cotton production in the Texas High Plains. Appl. Engr. in Agric. 24(4): 465–472.
202. Camp, C. R., Lamm, F. R., Evans, R. G., and Phene, C. J. (2000). Subsurface drip irrigation: Past, present and future. Proc. of the 4th Decennial Symposium on Irrigation. 363–371 pages. ASAE.
203. Bordovsky, J. P., Porter, D. O. (2003). Comparison of Spray, LEPA, and Subsurface Drip Irrigated Cotton. Proc. ASAE 2003 Annual Int. Meetings. St. Joseph, MI: ASAE.
204. Bosch, D. J., Powell, N. L., Wright, F. S. (1992). An economic comparison of subsurface micro irrigation with center pivot sprinkler irrigation. Journal of Production Agriculture, 5(4): 431–437.
205. Bosen, J. F. (1958). An approximation formula to compute relative humidity from dry bulb and dew point temperatures. Monthly Weather Review, 86(12): 486.
206. Boss, M. G. (1985). Summary of ICID definitions of irrigation efficiency. Intl. Comm. Irr. And Drainage Bull. 34, pages 28–31.
207. Boswell, Michael J., James Hardie. Micro-Irrigation Design Manual. Toroagro, CA.
208. Bottcher, A. B., Miller, L. W. (1982). Automatic tensiometer scanner for rapid measurements. Trans. of ASAE, 25(5): 1338–1342.
209. Bouwer, E. J., Rijnaarts, H. H. M., Cunningham, A. B., Gerlach, R. (2000). Biofilms in porous media. In *Bio-films II: Process Analysis and Applications*, ed. Bryers, J. D., 123–158. New York: Wiley-Liss. 123–158.
210. Bracy, R. P., Edling, R. J., Moser, E. B. (1995). Drip-irrigation management and fertilization of bell pepper in a humid area. Proceedings 5th International Micro irrigation Congress, ed. Lamm, F. R., 181–186. St. Joseph, MI: ASAE.
211. Bralts, V. F. (1986). Field performance and evolution. In *Trickle irrigation for crop production. design, operation and management*. Nakayama, FS, Bucks, DA (eds). Elsevier, Netherlands, 3(4): 216–240.
212. Bralts, V. F., Edwards, D. M., Wu, I. P. (1987). Drip irrigation design and evaluation based on the statistical uniformity concept. In *Advances in Irrigation*, 4: 67–117.
213. Bralts, V. G., Keesme, C. D. (1982). Drip irrigation field Uniformity estimation. PaperNo. 82–2062. Summer meeting of American Society of Agricultural Engineers.
214. Brandle, (1996). Windbreak: An important component in a plasticulture system. Horticulture Technology, 6: 177–181.
215. Brandt R. A., W. B.William, (1988). Reliability in design. ASAE Distinguished Lecture No. 13, pages 1–27. Paper #913C0888. Winter Meeting of the American Society of Agricultural Engineers, December, Chicago-Illinois.
216. Brathwaite, O., Goyal, M. R. (1994). Water consumption by selected crops and climatology in Trinidad. AMA, Japan.
217. Braud, H. J. (1970). Subsurface irrigation in the Southeast. Proceedings of the National Irrigation Symposium, El-E9. St. Joseph, MI: ASAE.
218. Bravdo, B., Naor, A. (1996). Effect of water regime on productivity and quality of fruit and wine. Acta Horticulturae, 427, pages 15–26.
219. Breazeale, D., Neyfeld, J., Myer, G., Davidson, J. (2000). Breakeven analysis of alfalfa seed production using subsurface drip irrigation. Journal of Applied Irrigation Science, 35(1): 91–99.
220. Bremer, D. J. (2003). Evaluation of microlysimeters used in turfgrass evapotranspiration studies using the dual-probe heat-pulse technique.Soil Sci. Soc. of America J., 67: 1625–1632.
221. Bresbin Publications, Inc, (1962). The fantastic business of plastic plants and flowers. Modern Plastics, 39: 94–97, 205–206.

222. Bresler, E. (1977). Trickle – drip irrigation: principles and application to soil water management. Adv. Agron. 29: 343–393.
223. Bresler, E. (1978). Analysis of trickle irrigation with application to design problems. Irrig. Sci., 1: 3–17.
224. Broner, I. (2002). Micro-irrigation for orchard and row crops. University Cooperative Extension. Colorado, USA. 3–7.
225. Bronson, K. F., Onken, A. B., Booker, J. D., Lascano, R. J., Provin, T. L., Torbert, H. A. (2001). Irrigated cotton lint yields as affected by phosphorus fertilizer and landscape position. Soil Sci. Plant Anal. 32(11–12): 1959–1967.
226. Bronson, K. F., Onken, A. B., Keeling, J. W., Booker, J. D., Torbert, H. A. (2006). Nitrogen response in cotton as affected by tillage system and irrigation level. Soil Sci. Soc. of America J. 65: 1153–1163.
227. Bronson, K. F., Keeling, J. W., Booker, J. D., Chua, T. T., Wheeler, T. A., Boman, R. K., Lascano, R. J. (2003). Influence of landscape position, soil series, and phosphorus fertilizer on cotton lint yield. Agron. J. 95(4): 949–957.
228. Bronson, K. F., Booker, J. D., Bordovsky, J. P., Keeling, J. W., Wheeler, T. A., Boman, R. K., Parajulee, M. N., E. Segarra and Nichols, R. L. (2004). Site-specificirrigation and nitrogen management for cotton production in the southernhigh plains. Madison – WI: Agronomy Journal,98: 212–219.
229. Broughton, R. S. (1995). Economic, production and environmental impacts of sub-irrigation and controlled drainage. In *Sub-irrigation and Controlled Drainage* (Belchard HW, D'ltri FM eds). Lewish Publishers: Tokyo, 482 pages.
230. Brown, K. W., Thomas, J. C., Friedman, S. P., Meiri, A. (1996). Wetting patterns associated with directed subsurface irrigation. p. 806–811. Proc. Int. Conf. Evapotranspiration and Irrigation Scheduling, San Antonio, TX. 3–6 Nov, (1996). Am. Soc. Agric. Eng., St. Joseph, MI.
231. Brunt, D. (1952). *Physical and dynamical meteorology*, 2nd ed., Cambridge: Cambridge University Press. 428.
232. Brutsaert, W. H. (1982). Evaporation into the atmosphere. Dordrecht, Holland: R. Deidel Publishing Company.
233. Brutsaert, W. H. (2005). *Hydrology: An introduction*. Cambridge: Cambridge University Press. 253–275, 305, 362, 452–463.
234. Bryan, R. B., Hawke, R. M., Rockwell, D. L. (1999). Automated micro standpipe system for soil erosion research. Soil Sci. Soc. of America J., 63: 977–987.
235. Bryla, D. R., Trout, T. J., Ayars, J. E. (2003) Growth and production of young peach trees irrigated by furrow, microjet, surface drip, or subsurface drip systems. HortScience, 38(6): 1112–1116.
236. Bucks, D. A., Allen, S. G., Rorh, R. L., Gardner, B. R. (2005). Cotton under micro irrigation and level basin irrigation methods. American Society of Agronomy 87: 20–36.
237. Bucks, D. A., Allen, S. G., Rorh, R. L., Gardner, B. R. Short staple cotton under micro and level-basin irrigation methods. Irrig. Sci. 9 (1988) 161–176.
238. Bucks, D. A., Nakayama, F. S., Warrick, A. W. (1982). Principles, practices, and potentialities of trickle (drip) irrigation. In Advances in Irrigation, ed. Hillel, D., 219–299. New York, NY: Academic Press. 302.
239. Bucks. D. A., Nakayama, F. S. (1980). Injection of fertilizers and other chemicals for drip irrigation. In Proceedings of the agri-turf irrigation conference, Houston, Texas. Irrigation Association, Silver Spring, Md., USA, pp. 166–180.
240. Bucks, D. A., Davis, S. (1986). Historical Development. In *Trickle Irrigation for Crop Production*, eds. Nakayama, F. S., Bucks, D. A., pages 1–26. New York, N. Y.: Elsevier.
241. Bucks, D. A., Nakayama, F. S., Warrick, A. W. (1982). Principles, practices, and potentialities of trickle (drip) irrigation. In *Advances in Irrigation,* ed. Hillel, D., 219–299. New York, N. Y.: Academic Press. 302 pages.
242. Bucks, D. A. (1995). Historical developments in micro irrigation. In Proceedings of 5th International Micro irrigation Congress, ed. Lamm, F. R., 1–5. St. Joseph, MI: ASAE.

243. Bucks, D. A., Erie, L. J., French, O. F., Nakayama, F. S., Pew, W. D. (1981). Subsurface trickle irrigation management with multiple cropping. Trans. of ASAE, 24(6): 1482–1489.
244. Buclon, F. (1966). Comparisons of agricultural uses of plastics in France, Italy, Japan, Russia and the United States. Proceedings of National AgriciultureAgriculture Plastics Congress, 7: 21–33.
245. Buendía, B., Allende, A., Nicolás, E., Alarcón, J. J., Gil, M. I. (2008). Effect of regulated deficit irrigation and crop load on the antioxidant compounds of peaches. J Agric Food Chem., 56: 3601–3608.
246. Bui, W. (1990). Performance of "Turbo Model" drip irrigation tubes. In Proceedigs 3rd National Irrigation Symposum, 198–203. St. Joseph, MI: ASAE.
247. Bui, W., and Osgood, R. V. (1990). Subsurface irrigation trial for alfalfa in Hawaii. In Proceedigs 3rd National Irrigation Symposium, 658–660. St. Joseph, MI: ASAE.
248. Burman, R., Pochop, L. O. (1994). Evaporation, evapotranspiration and climatic data. Amsterdam: Elsevier Science B. V.
249. Burman, R. D., Wright, J. L., Jensen, M. E. (1975). Changes in climate and estimated evaporation across a large irrigated area in Idaho. Trans. of ASAE, 18(6): 1089–1091, 1093.
250. Burman, R. D., Jensen, M. E., Allen, R. G. (1987). Thermodynamic factors in evapotranspiration. Proceedings of the Irrigation and Drainage Engineering Specialty Conference, Portland, Ore., ASCE, 28–30.
251. Burman, R. D., Nixon, P. R., Wright, J. L., Pruitt, W. O. (1980). Water requirements. In *Design and operation of farm irrigation systems*, ed. Jensen, M. E., 189–232. St. Joseph, MI: American Society of Agricultural Engineers.
252. Burt, C. M., Clemmens A. J., Strelkoff, T. S., Solomon, K. H., Bliesner, R. D., Hardy, L. A., Howell, T. A., Eisenhauer, D. E. (1997). Irrigation performance measures: efficiency and uniformity. J. Irrig. Drain. Eng. 123, 423–442.
253. Burt, C., Styles, S. (2007). Chemical injection for water treatment. Chapter 11 pp 223–236. In: *Drip and micro irrigation design and management for trees, vines and field crops- Practice plus theory*, Burt, C., Styles, S., 3rd Ed., ITRC, Cal Poly, San Luis Obispo, CA. 396 pages.
254. Burt, C., Styles, S. (2007). Drip and micro irrigation design and management for trees, vines and field crops-practice plus theory. 3rd Ed., ITRC, Cal Poly, San Luis Obispo, CA. 396 pages.
255. Burt, C. M., Styles, S. W. (1999). *Drip and Micro Irrigation for Trees, Vines, and Row Crops: Design and Management* (with Special Sections on SDI). San Luis Obispo, Cal.: Cal Poly ITRC.
256. Burt, C. M., Styles, S. W. (2007). *Drip and Micro Irrigation Design and Management for Trees, Vines, and Field Crops: Practice plus Theory*. San Luis Obispo, Cal.: Cal Poly ITRC.
257. Burt, C. M., Isbell, B., Burt, L. (2003). Long-term salinity buildup on drip/micro irrigated trees in California. In Proc. Irrigation Assoc. Tech. Conf., 46–56. IA Paper No. IA03–0408. Falls Church, Va.: Irrigation Association.
258. Burt, C. (1999). *Drip and micro irrigation for trees, vines, and row crops*. California, CA: Irrigation Training and Research Center, 7–8.
259. Burt, C., O'Connor, K., Ruehr, T. (1998). Fertigation. Irrigation Training and Research Center, California Polytechnic State Univ. San Luis Obispo, CA. 83–148.
260. Burt, C. E. (1995). Is buried drip the future with permanent crops? Irrigation Business Technology 3(1): 20–22.
261. Burt, C. M., Styles, S. W. (2000). Updating drip irrigation knowledge.Irri. J., 50(2): 8–10.
262. Burt, C. M. (2000). Selection of irrigation methods for agriculture: Drip/micro irrigation. California, CA: Irrigation Training and Research Center (ITRC). 1–6.
263. Burt, C. M. (2005). Leaching of accumulated soil salinity under drip irrigation. Trans. of ASAE, 48(6): 2115–2121.
264. Burt, C. M., Clemmens, A., Bliesner, R., Hardy, L. (2000). Selection of irrigation methods for agriculture. American Society of Civil Engineers (ASCE), USA. Chapter 1: 1–26.
265. Burt, C. M., Styles, S., Walker, R. E., Parrish, J. (1999). Irrigation evaluation software. Irrigation Training and Research Center, Cal. Poly. San Luis Obispo, CA.

266. Burt, Charles M. *The Surface Irrigation Manual*, Waterman Industries, California, (1995).
267. Burton, D. J., Harned, F. H., Lesikar, B. J., Prochaska, J. F., Suchecki, R. J. (2001). On-site wastewater treatment. Proceedings of the Ninth National Symposium on Individual and Small Community Sewage.
268. Businger, J. A., Yaglom, A. M. (1971). Introduction to Obukhov's paper on turbulence in an atmosphere with a non-uniform temperature. Boundary-Layer Meteorology, 2: 3–6.
269. Businger, J. A. (1956). Some remarks on Penman's equations for the evapotranspiration. NetherlandsJournal of Agricultural Science, 4: 77.
270. Businger, J. A. (1988). A note on the Businger-Dyer profiles. Boundary-Layer Meteorology, 42: 145–151.
271. Caamal-Maldonado, J. A., Jiménez-Osornio, J. J., A. Torres-Barragán and Anaya, A. L. (2001). The use of allelopathic legume cover and mulch species for weed control in cropping system. Agronomy Journal, 93(1): 27–36.
272. Caballero, R., Bustos, A., Román, R. (2001). Soil salinity under traditional and improved irrigation schedules in Central Spain. Soil Sci. Soc. of America J., 65: 1210–1218.
273. Cai, X., Rosegrant M. (2003). world water productivity: current situation and future options. In *Water Productivity in Agriculture: Limits and Opportunities for Improvement.*
274. Caldwell, D. S., Spurgeon, W. E., Manges, H. L. (1994). Frequency of irrigation for subsurface drip-irrigated corn. Trans. of ASAE, 37(4): 1099–1103.
275. Caldwell, D. S., Spurgeon, W. E., and H. L. Manges. Frequency of Irr. for Drip-Irr. Corn. ASAE Paper No. 92–2574. Am. Soc. Agr. Engr., St. Joseph, MI.
276. Calif. Dept. Water Resour. Contract. No. B53812. Land, Air and Water Resources Pap. 10013-A, Univ. Calif., Davis. Pages III-36–III-59.
277. Camp, C. R. (1998) Subsurface drip irrigation: a review. Trans ASAE 41(5): 1353–1367.
278. Camp. CR, EJ Sadler, WJ Busscher (1989) Subsurface and alternate middle micro irrigation for the Southeastern Coastal Plain. Trans ASAE 32(2): 451–456.
279. Camp, C. R., Lamm, F. R. (2003). Irrigation systems, subsurface drip. In *Encyclopedia Water Science*, 560–564. New York, N. Y.: Marcel Dekker.
280. Camp, C. R., Lamm, F. R., Evans, R. G., Phene, C. J. (2000). Subsurface drip irrigation: Past, present and future. Proc. of the 4th Decennial Nat'l Irrigation Symp., Phoenix, Arizona, Nov. 14–16. pages 363–372.
281. Camp, C. R., Bauer, P. J., Hunt, P. G. (1997). Subsurface drip irrigation lateral spacing and management for cotton in the Southeastern Coastal Plain. Trans. ASAE 40(4): 993–999.
282. Camp, C. R., Bauer, P. J., Busscher, W. J. (1999). Evaluation of no-tillage crop production with subsurface drip irrigation on soils with compacted layers. Trans. ASAE 42(4): 911–917.
283. Camp, C. A., Lamm, F. R. (2003). Irrigation systems, subsurface drip. Encyclopedia of Water Science. pages 560–564.
284. Camp, C. R., Sadler, E. J., Yoder, R. E. (1996). Evapotranspiration and irrigation scheduling. Proceedings of the International Conference on Evapotranspiration and Irrigation Scheduling. St. Joseph, MI: American Society of Agricultural Engineers, ISBN 0–929355–82–2. Pages, (1166).
285. Camp, C. R., Sadler, E. J., Busscher, W. J. (1989). Subsurface and alternate-middle micro irrigation for the southeastern Coastal Plain. Trans. of ASAE, 32(2): 451–456.
286. Camp, C. R., Sadler, E. J., Busscher, W. J. (1997). A Comparison of uniformity measures for drip irrigation systems. Trans. of ASAE, 40(4): 1013–1020.
287. Camp, C. R., Garrett, J. T., Sadler, E. J., Busscher, W. J. (1993). Micro irrigation management for doublecropped vegetables in a humid area. Trans. of ASAE, 36(6): 1639–1644.
288. Camp, C. R., Hunt, P. G., and Bauer, P. J. (1992). Subsurface Micro irrigation and nitrogen management for Cotton in the Southeastern Coastal Plain. ASAE Paper No. 92–2572. ASAE., St. Joseph, MI 49085.
289. Camp, C. R., Bauer, P. J., Hunt, P. G. (1997). Subsurface drip irrigation lateral spacing and management for cotton in the southeastern coastal plain. Trans. of the ASAE, 40(4): 993–999.

290. Campbell, G. S. (1977). An introduction to environmental biophysics. N. Y. Springer Verlag, p. 159.
291. Cantona, Y., Solé-Benet, A., Domingo, F. (2004). Temporal and spatial patterns of soil moisture in semiarid badlands of SE Spain. Journal of Hydrology, 285: 199–214.
292. Capra A, Scicolone B. (2004). Emitter and filter tests for wastewater reuse by drip irrigation.
293. Capra, A., Scicolone, B. (1998). Water quality and distribution uniformity in drip/trickle irrigation systems. J Agric. Eng. Res., 70: 355–365.
294. Capra, A., and Tamburino, V. (1995). Evaluation and control of distribution uniformity in farm irrigation systems. Proceedings of 46th Inter national Executive Council Meeting, ICID, special technical session, Roma, Italy.
295. Caraballo, E., Goyal, M. R., C. Chao de Báez, (1986). Effects of water application rates and planting density on growth parameters of drip irrigated onions. J. Agric. U. P.R., 70(2): 135–42.
296. Carcione, J. M., Serian, G. (2000). An electromagnetic modeling tool for the detection of hydrocarbons in the subsoil. Geophysica Prospecting, 48: 231–256.
297. Cardenas-Laihacar, B., Dukes, M. D., Miller, G. L. (2005). Sensor-based control of irrigation in Bermuda grass. Gainsville, FL: Agricultural and Biological Engineering Deptartment, University of Florida. 23–55.
298. Carlile, B. L., A. Sanjines. Subsurface Trickle Irrigation System for On-Site Wastewater dispersal and Reuse. Proceedings of the Texas On-Site Waste Water Treatment and Res. Conf. Austin, TX. Aug. 9–11, (1992).
299. Carpena, R. M., Dukes, M. D., Li, Y. C., Klassen, W. (2005). Field comparison of tensiometer and granular matrix sensor automatic drip irrigation on tomato. Tropical Research and Education Center, IFAS, University of Florida, 15(3): 1–12.
300. Carrijo, O. A., Cuenca, R. H. (1992). Precision of evapotranspiration estimates using neutron probe. Journal oflrrigation and Drainage Engineering, ASCE, 118(6): 943–953.
301. Carsel, R. F., and Parrish, R. S. (1988). Developing joint probability distributions of soil water retention characteristics. Water Resour. Res., 24: 755–769.
302. Caruso, M., Tarantino, A. (2004). A shearbox for testing unsaturated soils at medium to high degrees of saturation. Géotechnique Technical Note, 54, No. 4, 281–284.
303. Casey, A., Francis, X. M., E. Derby Nathan, (2002). Improved design for an automated tension infiltrometer. Soil Sci. Soc. of America J., 66: 64–67.
304. Cason, K. (2002). Elevated drip irrigation system works. Countryside & Small Stock Journal, 87(3): 1–20.
305. Cason, K. (2003). Elevated drip irrigation system works. Countryside & Small Stock Journal, 87(5): 20.
306. Cassel, S. F., Sharmasarkar, S., Held, L. J., Miller, S. D., Vance, G. F., Zhang, R. (2001). Agroeconomic analysis of drip irrigation for sugar beet production. Agronomy Journal, 93(3): 517–523.
307. Castel, J. R., Buj, A. (1994). Growth and evapotranspiration of young, drip irrigated Clementine trees, Proceedings of International Congress of Citriculture. Seventh meeting of the International Society of Citriculture, 8–13 March 1992 Acireale, Italy 2: 651–656.
308. Castel, J. R., Buj. A. (1990). Response of Salustiana oranges to high frequency deficit irrigation. Irrg. Sci., 11: 121–127.
309. Cavanaugh, P. (1992). Raisin grower buries drip. Grape Grower, 24 (4): 6–8.
310. Cavero, J., Plant, R. E., Shennan, C., Friedman, D. B., Williams, J. R., Kiniry, J. R., and Benson, V. W. (1999). Modeling nitrogen cycling in tomato safflower and tomato – wheat rotations. Agricultural Systems, 60: 123–135.
311. CD – SDI website. Western Irrigation Operation and Maintenance Workshop, Feb 24, 2004, Garden City, Kansas.
312. Center pivot sprinkler and SDI economic comparisons. SDI meeting, Sublette, Kansas, March 3, (2005).

313. Cetin, O., Bilget, L. (2002). Effects of different irrigation methods on shedding and yield of cotton. Agricultural Water Management 54: 1–15.
314. Chamran, F., Gessler, P. E., Chadwick, O. A. (2002). Spatially explicit treatment of soil-water dynamics along a semiarid catena. Soil Sci. Soc. of America J., 66: 1571–1583.
315. Chandler, D. G., Seyfried, M., M. Murdock and McNamara, J. P. (2004). Field calibration of water content reflectometers. Soil Sci. Soc. of America J., 68: 1501–1507.
316. Chao, C., Goyal, M. R. (1990). Class A pan evaporation versus evapotranspiration at seven locations in Puerto Rico. J. Agric. U. P.R., 74(3): 343–346.
317. Charlesworth, P. B., Muirhead, W. A. (2003). Crop establishment using subsurface drip irrigation: A comparison of point and area sources. Irrig. Sci. 22(3–4): 171–176.
318. Charman, P. E., Murphy, B. W. (2005). *Soils – Their Properties and Management.* 2nd Edition, Oxford University Press, South Melbourne, Vic. – AU.
319. Chartzoulakis, K., Bertaki, M. (2001). Towards Sustainable Water Use on Mediterranean Islands: Addressing Conflicting Demands and Varying Hydrological, Social and Economic Conditions WORK PACKAGE 2, In: Investigation of irrigation methods – Recommendations. Deliverable D14 and D22. Project No EVK1 – CT – 2001–00092. Funded by the European Commission.
320. Chase, R. G. (1985). Phosphorus application through a subsurface trickle system. In Proceedings 3rd International Drip/Trickle Irrigation Congress, Vol. 1: 393–400. St. Joseph, MI: ASAE.
321. Chase, R. G. (1985). Subsurface trickle irrigation in a continuous cropping system. In Proceedings of the 3rd International Drip/Trickle Irrigation Congress, ol. 2: 909–914. St. Joseph, MI: ASAE.
322. Chaudhari, AB and Kothari RM, (2009). Soil conditioners as a pivotal biotech input for integrated farming and contingency income. In: *Molecular Biotechnology*, (Eds. Varma, A., N. Verma), I. K. International Publ. Pvt. Ltd., New Delhi.
323. Chaudhari, AB, Phirke NV, Patil MG, Talegaonkar SK and Kothari RM, (2008). Bio- fertilizers and soil conditioner for organic farming. In: Bio-fertilizers (Ed. P. C. Trivedi), Pointer publisher, Jaipur, pages 39–79.
324. Chertkov, V. Y., Ravina, I., Zadoenko, V. (2004). An approach for estimating the shrinkage geometry factor at moisture content. Soil Sci. Soc. of America J., 68: 1807–1817.
325. Chibowski, E., Hotysz, L., Szczes, A. (2003). Adhesion of in situ precipitated calcium carbonate in the presence and absence of magnetic field in quiescent conditions on different solid surfaces. Journal of Water Research. 37: 4685–4692.
326. Chidenga, E. E. (2003). Irrigation technology choices and operations and maintenance in smallholder systems in Zimbabw. Wageningen: Wageningen Academic Publishers. 1–205.
327. Chieng, S., Ghaemi, A. (2003). Uniformity in a micro irrigation with partially clogged emitters. Paper number 032097, 2003 ASAE Annual Meeting.
328. Chiew, F. H.S., Kamadalasa, N. N., Malano, H. M., McMahon, T. A. (1995). Penman-Monteith: FAO-24 reference crop evapotranspiration and class-A pan data in Australia. Agricultural Water Management, 28: 9–21.
329. Chih-Pin, L. (2003). Frequency domain versus travel time analyses of TDR waveforms for soil moisture measurements. Soil Sci. Soc. of America J., 67: 720–729.
330. Chincholkar, SB, Chaudhari BL, Talegaonkar SK and Kothari RM, (2000). Microbial iron chelators: A sustainable tool for the biocontrol of plant diseases. In: Biocontrol potential and its exploitation in sustainable agriculture. (Eds. Upadhyay, R. K., Mukerji, K. G., B. P. Chamola), Kluwer Acad / Plenum Publ., New York, Vol 1, 49–70.
331. Choudhury, B. J., Idso, S. B., Reginato, R. J. (1987). Analysis of an empirical model for soil heat flux under a growing wheat crop for estimating evaporation by an infrared temperature based energy balance equation. Agricultural and Forest Meteorology, 39: 283–297.
332. Christians, N. E. (2000). The mathematics of turfgrass maintenance. AnnArbor – MI: ., Ann Arbor Press. Page 1–176.
333. Christiansen, J. E. (1968). Pan evaporation and evapotranspiration from climatic data. Journal of Irrigation and Drainage Engineering, ASCE, 94: 243–265.

334. Christopherson, S. H., Anderson, J. L., Gustafson, D. M. (2001). Evaluation of recirculating sand filters in Minnesota. American Society of Agricultural Engineers, St. Joseph, MI. 207–214.
335. Claerhout, L. (1996). SDI research: Taking water conservation one drop at a time. Kansas State Agriculturalist, Spring, (1996). Page 19.
336. Clark, R. N. (1979). Furrow, sprinkler, and drip irrigation efficiencies in corn. 1979 summer meeting of the ASAE and the CSAE, University of Manitoba, Winnipeg, Canada. ASAE paper no. 79–2111, ASAE, St Joseph, MI, USA.
337. Clark, G. A. (1990). Measurement of soil water potential. HortScience. 25(12): 1548–1551.
338. Clark, G. A. (1992). Drip irrigation management and scheduling for vegetable production. HortTechnology, 2(1): 32–37.
339. Clark, G. A. (2000). Subsurface Drip Irrigation. In: *Soybean Management & The Land: A Best Management Practices Handbook for Growers.* American Soybean Association, Washington Representation, Washington, DC. pages 52–53.
340. Clark, G. A. (2003). Irrigation design steps and elements. In: *Encyclopedia of Water Science.* Stewart, B. A., Howell, T. A., ed., Marcel Dekker, Inc., New York. pages 454–458.
341. Clark, G. A., Lamm, F. R., and Rogers, D. H. (2005). Sensitivity of thin- walled drip tape emitter discharge to water temperature. Appl. Engr in Agric. 21(5): 855–863.
342. Clark, G. A., Stanley, C. D., Smajstrla, A. G., Zazueta, F. S. (1999). Micro irrigation design considerations for sandy soil vegetable production systems. International Water and Irrigation, 19(4): 12–17.
343. Clark, G. A., Stanley, C. D., Maynard, D. N. (1993). Surface vs. subsurface drip irrigation of tomatoes on a sandy soil. Proc. Florida State Hort. Soc. 106: 210–212.
344. Clark, G. A., Stanley, C. D., Maynard, D. N. (2000). Municipal solid waste compost (MSWC) as a soil amendment in irrigated vegetable production. Trans. ASAE 43(4): 847–853.
345. Clark, G. A., Rogers, D. H., Dogan, E., Krueger, R. (2004). The IrriGage: A non-evaporating in-field precipitation gage. Applied Engineering in Agriculture. 20 (4): 463–466.
346. Clark, G. A., Maynard, D. N., Stanley, C. D. (1996). Drip-irrigation management for watermelon in a humid region. Applied Engr. in Agric. 12(3): 335–340.
347. Clark, G. A., Albregts, E. E., Stanley, C. D., Smajstrla, A. G., Zazueta, F. S. (1996). Water requirements and crop coefficients of drip-irrigated strawberry plants. Trans. ASAE 39(3): 905–913.
348. Clark, G. A., Lamm, F. R., Rogers, D. H. (2005). Sensitivity of thin-walled drip tape emitter discharge to temperature. Applied Engineering in Agriculture. 21(5): 855–863.
349. Clark, G. A., Srinivas, K., Rogers, D. H., Stratton, R., Martin, V. L. (2003). Measured and simulated uniformity of low drift nozzle sprinklers. Trans. of ASAE. 46(2): 321: 330.
350. Clark, G. A., Smajstrla, A. G. (1982). Water distributions in soils as influenced by irrigation depths and intensities. Soil Crop Sci. Soc. Fla. Proc. Vol. 42. 157–165.
351. Clark, G. A., Smajstrla, A. G. (1996). Design considerations for vegetable crop drip irrigation systems. HortTechnology, 6(3): 155–159.
352. Clark, G. A., Smajstrla, A. G. (1996). Injecting chemicals into drip irrigation systems. HortTechnology, 6(3): 160–164.
353. Clark, G. A., Smajstrla, A. G. (1992). Treating Irrigation Systems with Chlorine. Circular, (1039). Fla. Coop. Ext. Ser., Univ. of Florida.
354. Clark, G. A., Stanley, C. D. (1992). Subirrigation by Micro irrigation. Applied Engineering in Agriculture. 8(5): 647–652.
355. Clark, G. A., Reddell, D. L. (1991). Construction details and microclimate modifications of a permanently rain sheltered lysimeter system. Trans. of ASAE. 34(2): 429–435.
356. Clark, G. A., Maynard, D. N. (1992). Vegetable production on various bed widths using drip irrigation. Applied Engineering in Agriculture. 8(1): 28–32.
357. Clark, G. A., Smajstrla, A. G., Zazueta, F. S., Haman, D. Z. (1994). Injecting chemicals into micro irrigation systems. Proc. Amer. Soc. Plasiculture Drip Irrigation Workshop. Lexington, KSep, Y. (1994). ASHS. 23–28 pages.

358. Clark, G. A., Smajstrla, A. G., and Haman, D. Z. (1989). Water Hammer in Irrigation Systems. Circular 828. Fla. Coop. Ext. Ser., Univ. of Florida.
359. Clark, G. A., Smajstrla, A. G., Haman, D. Z., and Zazueta, F. S. (1990). Injection of Chemicals into Irrigation Systems: Rates, Volumes, and Injection Periods. Bulletin 250. Fla. Coop. Ext. Ser., Univ. of Florida.
360. Clark, G. A., Stanley, C. D., Smajstrla, A. G., Zazueta, F. S. (1994). Micro irrigation Design Considerations For Vegetable Production. Proc. Amer. Soc. Plasiculture Drip Irrigation Workshop. Lexington, KSep, Y. (1994). ASHS. p. 6–11.
361. Clark, G. A., Stanley, C. D., Maynard, D. N. (2000). Municipal solid waste compost (MSWC) as a soil amendment in irrigated vegetable production. Trans. of ASAE. 43(4): 847–853.
362. Clark, G. A., Stanley, C. D., Maynard, D. N. (1993). Surface vs. subsurface drip irrigation of tomatoes on a sandy soil. Proceedings of the Florida State Horticultural Society, 106: 210–212.
363. Clark, G. A., Stanley, C. D., and Zazueta, F. S. (1993). Qualitative sensing of water movement from a point-source emitter on a sandy soil. Applied Engineering in Agriculture. 9(3): 299–303.
364. Clark, G. A., Stanley, C. D., Maynard, D. N., Hochmuth, G. J., Hanlon, E. A., Haman, D. Z. (1991). Water and fertilizer management of microirrigated fresh market tomatoes. Trans. of ASAE, 34(2): 429–435.
365. Clark, G. A., Maynard, D. N., Stanley, C. D. (1996). Drip-irrigation management for watermelon in a humid region. Applied Engineering in Agriculture, 12(3): 335–340.
366. Clark, G. A., Haman, D. Z., Prochaska, J., Yitayew, M. (2007). General System Design Principles. In: *Micro irrigation for Crop Production*. Lamm, F., Ayars, J., F. Nakayama ed., Elsevier, Amsterdam, The Netherlands. pages 161–220.
367. Clark, G. A., Maynard, D. N., Stanley, C. D., Hochmuth, G. J., Hanlon, E. A., and Haman, D. Z. (1990). Irrigation Scheduling and Management of Microirrigated Tomatoes. Circular 872. Fla. Coop. Ext. Ser., Univ. of Florida.
368. Clark, G. A., Dogan, E., Rogers, D. H., Martin, V. L. (2006). Evaluation of IrriGage Collectors to Measure Irrigation Depths from Low Pressure Sprinklers. Applied Engineering in Agriculture. 22(1): 63–72.
369. Clark, G. A., Albregts, E. E., Stanley, C. D., Smajstrla, A. G., Zazueta, F. S. (1996). Water requirements and crop coefficients of drip irrigated strawberry plants. Trans. of ASAE, 39(3): 905–913.
370. Clark, G. A., Lamm, F. R., and Rogers, D. H. (2004). Water temperature effects on the discharge rate of collapsible emitting hose. Proc. Irrigation Assn. Int'l. Irrigation Technical Conf., November 14–16, 2004, Tampa, FIA, L., Falls Church, VA. IA04–1043. pp 424–434.
371. Clark, G. A., Barnes, P. L., Lamm, F. R., McVey, S. (1997). Chemigation uniformity in micro irrigation materials. Presented at the 1997 Int'l. ASAE meeting. ASAE Paper no. 972176, St. Joseph, MI. 10 pages.
372. Clemmens, A. J. (2006). Canal automation. Resource Magazine, ASABE. 7–8.
373. Cline, J. F., Burton, F. G., Cataldo, D. A., Skiens, W. E., Gano, K. A. (1982). Long- term biobarriers to plant and animal intrusions of uranium tailings. DOE/UMT-0209, PNL-4340, UC-70. U. S. Dept. of Energy Rep. under contract DE-AC06–76RLO 1830. Sep, (1982). Pacific Northwest Nat'l. Lab., Richland, Washington. 60 pages.
374. Clothier, B. E., Green, S. R. (1994). Rootzone processes and the efficient use of irrigation water. Agric. Water Management 25: 1–12.
375. Clothier, B. E., S. R. Green.1997. Roots: The big movers of water and chemicals in soil. Soil Sci. 162: 534–543.
376. Clothier, B. E., Clawson, K. L., Pinter, P. J., Moran, M. S., Reginato, R. J., Jackson, R. D. (1986). Estimates of soil heat flux from net radiation during the growth of alfalfa. Agricultural and Forest Meteorology, 37: 319–329.
377. Clow, B. D. (1992). Sizing Irrigation Reservoirs for Treated Domestic Wastewater. Proc. of Urban and Agric.Water Reuse, Orlando, Fl, June 28 – July 1, (1992). Water Env. Fed. Alexandria, Virginia 22314, (1994). pp 73–88.

378. Coal, F. J., Sanchez, C. A. Effective distance of nutrient acquisition for sugarcane grown on everglades Histosols. J.Am.Soc.Sugarcane Technol., 10(1990) 39–44.
379. Coates, R. W., Delwiche, M. J., Brown, P. H., Shackel, K. A. (2004). Precision irrigation/ fertilization in orchards. ASAE Paper Number 042249. ASAE/CSAE Annual International Mtg., Ottawa, Ontario, Canada, 1–4 August, (2004).
380. Coelho, E. F., Or, D. (1999). Root distribution and water uptake patterns of corn under surface and subsurface drip irrigation. Plant Soil J., 206: 123–136.
381. Coelho, F. E., Or, D. (1996). A parametric model for two-dimensional water uptake intensity by corn roots under drip irrigation. Soil Sci. Soc. of America J. 60(4): 1039–1049.
382. Colaizzi, P. D., Schneider, A. D., Evett, S. R., Howell, T. A. (2004). Comparison of SDI, LEPA, and spray irrigation performance for grain sorghum. Trans. ASAE 47(5): 1477–1492.
383. Colaizzi, P. D., F. R Lamm, Howell, T. A., Evett, S. R. (2006). Crop production comparison under various irrigation systems. Proc. Central Plains Irrigation Conference, Colby, KS., Feb. 21–22, (2006). CPIA, 760 N.Thompson, Colby, KS. pages 189–207.
384. Colaizzi, P. D., Gowda, P. H., Marek, T. H., Porter, D. O. (2009). Irrigation in the Texas High Plains: A brief history and potential reductions in demand. J. Irrig and Drain. Eng., 58(3): 257-.
385. Colaizzi, P. D., Evett, S. R., Howell, T. A. (2005). Cotton production with SDI, LEPA, and spray irrigation in a thermally-limited climate, CD-ROM. Paper No, (1249). Irrigation Association Annual Meeting, 6–8 Nov, Phoenix, AZ. Falls Church, Va.: Irrigation Assoc.
386. Colaizzi, P. D., Evett, S. R., Howell, T. A., R. L Baumhardt, (2010). Crop production comparison with spray, LEPA, and subsurface drip irrigation in the Texas High Plains. In Proc. 5th Decennial Nat'l. Irrigation Conf. ASABE Paper No. IRR10–9704. 23 pages.
387. Colaizzi, P. D., F.R Lamm, Howell, T. A., and Evett, S. R. (2006). Crop production comparison under various irrigation systems. In: Proc. Central Plains Irrigation Conference, Colby, KS, Feb. 21–22, (2006). CPIA, 760 N.Thompson, Colby, KS. pages 189–207.
388. Colaizzi, P. D., Evett, S. R., Howell, T. A., and Baumhardt, R. L. (2009). Comparison of grain sorghum, soybean, and cotton production under spray, LEPA, and SDI. Proc. 21st Annual Central Plains Irrig. Conf., Colby Kansas, February 24–25, (2009). CPIA, 760 N.Thompson, Colby, KS. pages 122–139.
389. Colombo, A. (2006). Plant water accessibility function: A design and management tool for trickle irrigation. Agricultural Water Management, 82 (1/2): 45–62.
390. Communar, G., and Friedman, S. P. (2010). Relative water uptake rate as a criterion for trickle irrigation system design: I, Coupled source–sink steady water flow model. Soil Sci. Soc. Am. J. 74: 1493–1508.
391. Communar, G., and Friedman, S. P. (2010). Relative water uptake rate as a criterion for trickle irrigation system design: II, Surface trickle irrigation. Soil Sci. Soc. Am. J. 74: 1509–1517.
392. Communar, Gregory, and Shmulik Friedman, P., Relative water uptake rate as a criterion for trickle irrigation system design: III, Subsurface Trickle Irrigation. Soil Sci. Soc. Am. J. 74: 1518–1525.
393. Comparing the economics of center pivot and subsurface drip irrigation systems. Water and the Future of Kansas conference, March 2008, Topeka, Kansas.
394. Congress of Computers in Agriculture and Natural Resources. Iguacu Falls – Brazil, 13–15.
395. Constantini, E. A.C., Castelli, F., Raimondi, S., Lorenzoni, P. (2002). Asssessing soil moisture probe. Soil Sci. Soc. of America J., 66: 1889–1896.
396. Cook, F. J., P. Fitch and Thorburn, P. J. (2006). Modeling trickle irrigation: Comparison of analytical and numerical models for estimation of wetting front position. CSIRO. 1–8 pages.
397. Cornelis, W., Ronsyn, J., M. Van Meirvernne, Hartmann, R. (2001). Evaluation of pedotransfer functions for predicting the soil moisture retention curve. Soil Sci. Soc. of Am. J., 65: 638–648.
398. Cornell, J. T. (1989). The recycling of plastic in agriculture. Proceedings of National Agriciulture Plastics Congress, 21: 60–64.
399. Cornish, G., Bosworth B., Perry C., and Burkeet J. (2004). Water charging in irrigated. agriculture; An analysis of international experience. FAO Water Reports 28. Food and Agriculture Organization of the United Nations. Rome, (2004).

400. Cote, C. M., Bristow, K. L., Charlesworth, P. B., Cook, F. J., and Thorburn, P. J. (2003). Analysis of soil wetting and solute transport in subsurface trickle irrigation. Irrig. Sci. 22: 143–156.
401. Crabb, P (1997). Murray-Darling Basin resources. Murray-Darling Basin Commission: Canberra – AU.
402. Cranmer, A. M., Bordovsky, J. P., Mustian, J. T., and D.M Nesmith, (2009). Soil amendment and tillage evaluation to improve germination with subsurface drip irrigation. 2009 Proceedings of the Beltwide Cotton Conference, San Antonio, Texas. Jan 5–8, (2009). 411–421 pages.
403. Crescimanno, G., Garofalo, P. (2006). Management of irrigation with saline water in cracking clay soils. Soil Sci. Soc. of America J., 70(1): 1774–1787.
404. Crespo Ruiz, M., Goyal, M. R., C. Chao de Baez and Rivera, L. E. (1988). Nutrient uptake and growth characteristics of nitrogen fertigated sweet peppers under drip irrigation and plastic mulch. Journal of Agriculture of the University of Puerto Rico, 72(4): 575–584.
405. Crooke, W. M., and Simpson, W. E. (1971). Determination of ammonium on Kjeldahl digests of crops by an automated procedure. Journal of the Science of Food and Agriculture, 22: 9–10.
406. Cruse, R., Wiegand, C. L., Swason. W. A. (1982). The effect of rainfall and irrigation management on citrus juice quality in Texas. J. Am. Soc. Hortic. Sci., 107, 767–770.
407. Cuenca, R. H., Nicholson, M. T. (1982). Application of the Penman equation wind function. Journal of Irrigation and Drainage Engineering, ASCE, 108(1): 13–23.
408. Cullum, R. F., Moore, M. T., Cooper, C. M. (2006). Assimilation of agrichemicals and sediments in runoff within drainage ditches and constructed wetlands. Paper number 062004, ASAE Annual Meeting.
409. Cumming, R. W. (1991). Long-term effects of lime in extensive pasture areas of Australia. In: *Plant-soil interactions at low pH*. 453–464 pages. Kluwer Academic Publishers: Dordrecht, the Netherlands.
410. Current irrigation technologies and their role in efficient use of water. Water and Future of Kansas conference, Topeka, Kansas, March 16, (2006).
411. CWC (1991). Drip Irrigation System and Its R elevance in Better Water Use. Irrigation Research Management Improvement Organization, Central Water Commission (CWC), New Delhi, April, (1991).
412. Czemerda, K., McConnell, T. (2003). Drip irrigation: Extension service. West Virginia University.
413. Dagdelen, N., Basal, H., Yilmaz, E., Gurbuz, T., Akcay, S. (2009). Different drip irrigation regimes affect cotton yield, water use efficiency and fiber quality in Western Turkey. Agricultural Water Management 69: 111–120.
414. Dalrymple, D. G. (1973). A global review of greenhouse food production. USDA Report 89.
415. Darusman, Khan, A. H., Stone, L. R., and Lamm, F. R. (1997). Water flux below the root zone vs. drip-line spacing in drip irrigated corn. Soil Sci. Soc. Am. J., 61(6): 1755–1760.
416. Darusman, A. H.K., Stone, L. R., Spurgeon, W. E., and Lamm, F. R. (1997). Water flux below the root zone versus irrigation amount in drip-irrigated corn. Agronomy Journal 89: 375–379.
417. Darwish, T. M. (1995). Implementation and Perspective of Fertigation in Crop Production in Lebanon. Proceeding of the Short Advanced Course on Fertigation. FAO/RNE, CIHEAM/IAM-B and LEBANESE UNIVERSITY. Nov. 26– Dec.3, 1995: 207–220.
418. Das, B. S., Wraith, J. M., Inskeep, W. P. (1999). Nitrate concentrations in the root zone estimated using time domain reflectometry. Soil Sci. Soc. of America J., 63: 1561–1570.
419. Dasberg S., Or, D. (1999). *Drip irrigation*. Springer-Verlag Berlin Heidelberg, Germany.159p.
420. Dasberg, S. (2000). Drip irrigation. New York: Springer-Verlag.
421. Dasberg, S., Bar-Akiva, A., Spazisky, S., Cohen, A. (1988). Fertigation vs broadcasting in an orange grove. Fertilizer Research 15: 147–154.
422. David, C., Nielsen, K. (2004). Forage yield and quality under varying water availability. Agronomy Journal, 96: 204–213.
423. Davies, J. W. (1975). Mulching effects on plant climate and yield. Technical Note no. 136. Secretariat of the World Meteorological Organization, Geneva-Switzerland. 1–92.

424. Davis, K. R., Phene, C. J., McCormick, R. L., Hutmacher, R. B., Meek, D. W. (1985). Trickle frequency and installation depth effects on tomatoes. Proceedings 3rd International Drip/ Trickle Irrigation Congress, 2: 896–902. St. Joseph, MI: ASAE.
425. Davis, S, (1967). Subsurface irrigation-How soon a reality? Agricultural Eng., 48(11): 654–655.
426. Davis, S. (1974). History of drip irrigation. Agribusiness News 10(7): 1.
427. Davis, S., Nelson, S. D. (1970). Subsurface irrigation easily automated. Journal of the Irrigation and Drainage Division, ASCE 96(IR-1): 47–51.
428. Davis, S., Nelson, S. D. (1970). Subsurface irrigation today and tomorrow in California. In Proceedings of the National Irrigation Symposium, Hl-H8. St. Joseph, MI: ASAE.
429. Davis, S., Pugh, W. J. (1974). Drip irrigation: Surface and subsurface compared with sprinkler and furrow. In Proceedings of the 2nd International Drip Irrigation Congress, 109–114. Riverside, CA: University of California.
430. De Kreij, Van der Burg, A. M.M., Runia, W. T. (2003). Drip irrigation emitter clogging in Dutch greenhouses as affected by methane and organic acids. Agric. Water Management, 60: 73–85.
431. De Neve, S., Hoffman, G. (2000). Influence of soil compaction on C and N mineralization from soil organic matter and crop residues. Biology and Fertility of Soils, 30: 544–549.
432. De Wit, C. T. (1999). The efficient use of labor, land and energy in agriculture. Agricultural Systems, 4(October): 279–228.
433. Decoteau, D. R., Kasperbaur, M. J., Hunt, P. G. (1989). Mulch surface color affects yield of fresh-market tomatoes. Jounal of American Society for Horticultural Science, 114: 216–219.
434. Deficiency-enhanced uptake rate of phosphorus. Physiologia Plantarum 68(3): 483–490.
435. Dehghanisanij, H. (2008). Introducing of Nation Strategic Plan for sustainable development of pressurized irrigation systems in Iran. Proceeding of the Workshop on Pressurized Irrigation and Sustainable Development in Iran. 21 Feb. Karaj. Iran. (In Farsi).
436. Dehghanisanij, H., Akbari, M. (2006). Influences of irrigation management on efficiency of drip irrigation systems in Iran. Second Workshop on Micro irrigation. Karaj, Iran.
437. Dehghanisanij, H., Akbari, M. (2008). Importance of daily climate information on improvement of crop water productivity in arid and semiarid regions. 9th Int. Conf. on Dryland Development. Nov. 7–10. Alexandria, Egypt.
438. Dehghanisanij, H., Agassi, M., Anyoji, H., and Eneji, A. E. (2006). Improvement of saline water use under drip irrigation system. Agricultural Water Management Journal. 85(3): 233–242.
439. Dehghanisanij, H., Anyoji H., Riyahi, H., El-Hassan, W. A. (2007). Effect of emitter characteristics and irrigation schemes on emitter clogging under saline water use. Journal of Arid Land Studies. 16(4): 225–233.
440. Dehghanisanij, H., Oweis, T., Qureshi, A. S. (2006). Agricultural water use and management in arid and semi-arid areas: Current situation and measures for improvement. Annals of Arid Zone. 45: 355–378.
441. Dehghanisanij, H., Taherpour, M., and Yamamoto, T. (2002). Evaluation of factors affecting emitter clogging of micro irrigation in Southeast of Iran. Transaction of the Japanese Society of Irrigation, Drainage and Reclamation Engineering Journal, 217: 1–8.
442. Dehghanisanij, H., Yamamoto, T., Ould-Ahmad, B. A., Fujiyama, H., and Miyamoto, K. (2005). The effect of chlorine on emitter clogging induced by algae and protozoa and the performance of drip irrigation. Trans. ASAE, 48(2): 519–527.
443. Dehghanisanij, H., Yamamoto, T., Rasiah, V., Utsunomiya, J., and Inoue, M. (2004). Impact of biological clogging agents on filter and emitter discharge characteristics of micro irrigation system. Journal of Irrigation and Drainage, 53: 363–373.
444. Dehghanisanij, H., Mehdi, A. (2013). Micro irrigation in Iran- current statues and future needs. Irrigation Technology, Agricultural Engineering Research Institute, Karaj, Alborz, Iran. P. O.Box: 31585–845.
445. Demir, V., and Uz, E. (1994). Damla sulama sistemlerinde kullan lan filtreler [Filters used in drip Design and management considerations for SDI]. Agricultural Engineering Research Institute, Karaj, Alborz, Iran. P. O.Box: 31585–845.

446. Design, Installation, and Performance of Trickle Irrigation Systems, (2004). ASAE Engineering Standard No. ASAE-EP-405.
447. DeTar, W. R., Browne, G. T., Phene, C. J., Sanden, B. L. (1996). Real-time irrigation scheduling of potatoes with sprinkler and subsurface drip systems. In Proc. Intl. Conf. on Evapotranspiration and Irrigation Scheduling, 812–824. St. Joseph, MI: ASAE.
448. DeTar, W. R., Phene, C. J. (1991). Subsurface drip irrigation vs. furrow irrigation of cotton. Cal. Irr. Inst. 29th Annual Meeting Jan. 30–31, 1991, Sacramento, CA.
449. DcTar, W. R., Phene, C. J., Clark, D. A. (1994). Subsurface drip vs furrow irrigation: 4 years of continuous cotton on sandy soil. In Beltwide Cotton Conference, 542–545. Memphis, TN: National Cotton Council.
450. DeTar, W. R., Browne, G. T., Phene, C. J., Sanden, B. L. (1996). Real-time irrigation scheduling of potatoes with sprinkler and subsurface drip systems. In Proceedings International Conference on Evapotranspiration and Irrigation Scheduling, eds. Camp, C. R., Sadler, E. J., and Yoder, R. E., 812–824. St. Joseph, MI: ASAE.
451. Devasirvatham, Viola, (2009). *A review of Subsurface Drip Irrigation in Vegetable Production.* In: Irrigation Matters Series no 03/09. Cooperative Research Center for Irrigation Futures. IF Technologies Pty. Ltd., Australia.
452. Devitt, D. A., Miller, W. W. (1988). Subsurface drip irrigation of bermudagrass with saline water. Applied Agricultural Research, 3(3): 133–143.
453. Deybe, D., and Flichman, G. (1991). A regional agricultural model using a plant growth simulation program as activities generator. An application to a region in Argentina. Agricultural Systems, 37: 219–227.
454. Dhuyvetter, K. C., Lamm, F. R., Rogers, D. H. (1995). Subsurface drip irrigation (SDI) for field corn-An economic analysis. Proceedings of the 5th Internatonal Micro irrigation Congress, ed. Lamm, F. R., 395–401 pages. St. Joseph, MI: ASAE.
455. Dhuyvetter, K. C., Lamm, F. R., and Rogers, D. H. (1994). Subsurface drip irrigation for field corn: An economic analysis. KSU Cooperative Ext. Irrigation Management Series, L-909. 6 pages.
456. Dhuyvetter, K. C., Lamm, F. R., and Rogers, D. H.,1995. An economic comparison of subsurface drip irrigation (SDI) and center pivot irrigation for field corn. In Proc. Central Plains Irrigation Short Course, Garden City, KS, Feb. 7–8. KSU Ext. Agric. Engr., Manhattan, KS. pages 139–147.
457. Dhuyvetter, K. C., Lamm, F. R., and Rogers, D. H. (1995). Subsurface drip irrigation (SDI) for field corn – An economic analysis. In proceedings of the Fifth International Micro irrigation Congress, Orlando, FL, April 2–6, pages 395–401. ASAE, St. Joseph, MI.
458. Dingkuhn, M. (1994). Climatic determinants of irrigated rice performance in the Sahel. III. Characterizing environments by simulating crop phenology. Agricultural Systems, 48: 435–456.
459. Dirksen, C. (1978). Transient and steady flow from subsurface line sources at constant hydraulic head in anisotropic soil. Trans. of ASAE, 21(5): 913–919.
460. DNR Water Facts, W21: Micro Irrigation System.
461. Dogan, E., G. A. Clark., Rogers, D. H., Martin, V. L., Vanderlip, R. L. (2006). On-Farm scheduling studies and CERES-Maize simulation of irrigated corn. Applied Engineering in Agriculture, 22(4): 509–516.
462. Dolman, A. J., Stewart, J. B. (1987). Modelling forest transpiration from climatological data. In *Forest hydrology and watershed management*, eds. Swanson, R. H., Bernier, P. Y., Woodard, P. D., 167: 319–327. IAHS Publisher, Wallingford, Oxfordshire, OX10–8BB, UK.
463. Domingo, R., Ruiz-Sánchez, M. C., Sánchez-Blanco, N. J., Torrecillas, A. (1996). Water relations, growth and yield of Fino lemon trees under regulated deficit irrigation. Irrigation Science 16, pages115–123.
464. Donatelli, M., Stöckle, C. O., Nelson, R. L., Bellocchi, G. (2003). ET-CSDLL: A dynamic link library for the computation of reference and crop evapotranspiration. Soil Sci. Soc. of America J., 67: 1334–1336.

465. Doorenbos, J., Kassam, A. H. (1979). *Yield response to water*. FAO Irrigation and Drainage Paper No. 33, FAO, Rome, Italy. 193.
466. Doorenbos, J., Pruitt, W. O. (1975). *Guidelines for predicting crop water requirements*. Irrigation and Drainage Paper 24, Food and Agriculture Organization of the United Nations, Rome. Page 179.
467. Doorenbos, J., Pruitt, W. O. (1977). *Crop water requirements*. FAO Irrigation and Drainage Division of ASCE. Paper 24, Food and agriculture organization of the United Nations, Rome. 1–156.
468. Doss, B. D., Evans, L. E. (1980). Irrigation methods and in row chiseling for tomato production. J. American Asoc. Hort. Sci., 105: 611–614.
469. Dougherty, M., Fulton, J., Burmester, C., Curtis, L., Monks, D. (2007). Precision fertilization using sub-surface drip irrigation (SDI) for site-specific management of cotton. ASABE Paper No. 072194 at ASABE International Meeting, Minneapolis, MN., 17–20 June, (2007).
470. Douglas, L. A., Sochtig, H., and Flaig, W. (1978). Calorimetric determination of urea in soil extracts using an automated system. Soil Science Society of America Journal, 42: 291–292.
471. Downey, D., Graham, W. D., Clark, G., A. (1994). An inexpensive system to measure water level rise during the Bouwer and Rice slug test. Trans. of ASAE, 10(2): 247–253.
472. Dragoni, D., Lakso, A., N. (2009). An apple-specific ET model. *Acta Horticulturae*.
473. Dragoni, D., Lakso, A. N., Piccioni, R. M. (2005). Measuring transpiration in apple trees using heat pulse sap flow gauges calibrated with whole-canopy gas exchange chambers. Ag. Forest Meteorology, 130: 85–94.
474. Drip Irrigation Management, (1981). Division of Agricultural Sciences at University of California. Berkley, CA.
475. Droogers, P. (2000). Estimating actual evapotranspiration using a detailed agro-hydrological model. Journal of Hydrology, (Amsterdam), 229: 50–58.
476. Drotleff, L. (2006). The Father of drip. American Vegetable Grower, 54(5): 18.
477. Du, T., Kang, S., Zang, J., Li, F. (2008). Water use and yield responses of cotton to alternate partial root-zone drip irrigation in the arid area of China. Irrigation Science, 26: 147–159.
478. Duffie, J. A., Beckman, W. A. (1991). *Solar engineering of thermal processes*. 2nd ed., New York, NY: John Wiley and Sons. 994.
479. Dukes, M. D., Haman, D. Z., Lamm, F. R., Buchanan, J. R., Camp, C. R. (2005). Site selection for subsurface drip irrigation systems in the humid region. Proc. World Water and Environmental Resources Congress, Reston Va.: ASCE Environmental and Water Resources Institute.
480. Dukes, M. D., Scholberg, J. M. (2004). Soil moisture controlled subsurface drip irrigation on sandy soils. Appl. Engr. in Agric., 21(1): 89–101.
481. Dukes, M. D., Scholberg, J. M. (2004). SDI automatic irrigation, arachis hypogaea, zea mays, Paper number 042188, ASAE Annual Meeting.
482. Dukes, M. D., Scholberg, J. M. (2005). Soil moisture controlled subsurface drip irrigation on sandy soils. Applied Engineering in Agriculture, 21(1): 89–101.
483. Dukes, M. D., Haman, D. Z., Lamm, F., Buchanan, J. R., Camp, C. R. (2005). Site selection for Subsurface Drip Irrigation Systems in the Humid Region. Proc. ASCE-EWRI Water Congress, May 15–19, 2005.
484. Dukes, M. D., Scholberg, J. M., Hanselman, T. A. (2005). Quantification of nitrogen movement under drip irrigated vegetable production. Paper number 052240, ASAE Annual Meeting.
485. Dumler, T. J., D. M. O'Brien, Martin, K. L. (2011). Center-pivot-irrigated corn cost-return budget in Western Kansas. KSU Farm Management Guide, MF-585. Manhattan, Kansas. 4 pages.
486. Duncan, J. (1993). Buried drip gains more ground. Grape Grower, 25(4): 20–23.
487. Duppong, L. M., Delate, K., Liebman, M., Horton, R., Romero, F., Kraus, G., J. Petrich and Chowdbury, P. K. (2004). The effect of natural mulches on crop performance, weed suppression and biochemical constituents of catnip. Crop Science Journal, 44(3): 861–869.
488. Dyer, A. J. (1974). A review of flux-profile relationships. Boundary Layer Meterol., 7: 363–372.
489. Dyer, A. J., Hicks, B. B. (1970). Flux-gradient relationships in the constant flux layer. Quarterly Journal of the Royal Meteorological Society, 96: 715–721.

490. Ebel, R. C., Proebsting, E. L., Evans, R. G. (1995). Deficit irrigation to control vegetative growth in apple and monitoring fruit growth to schedule irrigation. HortScience, 30: 1229–1232.
491. Economics of SDI for Corn at Central Plains Irrigation Conference, February 19–20, 2008, Greeley, Colorado.
492. Economics of SDI for Corn. Victory Electrical Coop annual irrigation meeting, Dodge City, Kansas, February 8, 2008.
493. Edstrom, J., Schwankl, L. (1998). Weed suppression in almond orchards using subsurface drip irrigation. 35–36 pages. Las Cruces, N. M.: Western Society of Weed Science.
494. Edwards, D. M., Eastin, J. D., Olson, R. A., German, R. (1970). Subsurface irrigation today and tomorrow in the Midwest. Proceedings of the National Irrigation Symposium. St. Joseph, MI: ASAE.
495. Edwards, J. H., Bruce, R. R., Horton, B. D., Chesness, J. L., and Wehunt, E. J. (1982). Soil cation and water distribution as affected by N applied through a drip irrigation system. Journal of the American Society Horticultural Science, 107: 1142–1148.
496. Eigenberg, R. A., Nienaber, J. A., Woodbury, B. L., Ferguson, R. B. (2006). Soil conductivity as a measure of soil and crop status. Soil Sci. Soc. of America J., 70: 1600–1611.
497. El-Bably, A. Z. (2002). Effect of irrigation and nutrition of copper and molybdenum on Egyptian clover (trifolium alexandrnium 1.). Agronomy Journal, 94(1): 1066–1070.
498. El-Gindy, A. M., El-Araby, A. M. (1996). Vegetable crop response to surface and subsurface drip under calcareous soil. Proceedings International Conference on Evapotranspiration and Irrigation Scheduling, eds. Camp, C. R., Sadler, E. J., and Yoder, R. E., 1021–1028 pages. St. Joseph, MI: ASAE.
499. El-Hafedh, A. V. O. M., Daghari, H., Maalej, M. (2001). Analysis of several discharge rate-spacing-duration combinations in drip irrigation system. Agric. Water Mgt., 52(1): 33–52.
500. Eliades, G. (1994). Response of grapefruit to different amounts of water apllied by drippers and minisprinklers. *Acta Horticulturae*, 365: 129–146.
501. Elliott, R. L., Harp, S. L., Grosz, G. D., Kizer, M. A. (1988). Crop coefficients for peanut evapotranspiration. Agricultural Water Management, 15: 155–164.
502. El-Otmani, M., Ait-Oubahou, A., El-Hassainate, F., Kaanane, A., Lovatt, C. J. (2004). Effect of Gibberellic acid, urea and KNO_3 on yield and on composition and nutritional quality of clementine mandarin fruit juice. *Acta Horticulturae* (ISHS), 2: 149–157.
503. Emblenton, T. W., Reitz, H. J., Jones, W. W. (1973). Citrus fertilization. In: Citrus Industry. Reuther W. (Ed) 3: 122–181.
504. Enciso, J. M. (2004). Installing a subsurface drip irrigation system for row crops. Texas Cooperative Extension, Texas A&M University System, College Station, TX-USA.
505. Enciso, J. M., Colaizzi, P. D., Multer, W. L. (2005). Economic analysis of subsurface drip irrigation lateral spacing and installation depth for cotton. Trans. ASAE, 48(1): 197–204.
506. Enciso, J. M., Unruh, B. L., Colaizzi, P. D., Multer, W. L. (2003). Cotton response to subsurface drip irrigation frequency under deficit irrigation. Applied Eng. in Agriculture, 19(5): 555–558.
507. Enciso, J. M., Colaizzi, P. D., Multer, W. L., Stichler, C. R. (2007). Cotton response to phosphorus fertigation using subsurface drip irrigation. Applied Eng. in Agric. 23(3): 299–304.
508. Enciso, J. M., Multer, W. L., Lamm, F. R. (2009). Impact of management on the life expectancy of drip systems. Proc. Irrigation Association Technical Conference, San Antonio, Texas, December 2–5, (2009). IA, Falls Church, VA. 11 pages.
509. Enciso, J. M., Multer, W., Lamm, F. R. (2009). Evaluation of old subsurface drip irrigation systems in Texas. ASABE Paper No. 095725. St. Joseph, MI: ASABE.
510. Endale, D. M., and Fipps, G. (2001). Simulation-based irrigation scheduling as a water management tool in developing countries. Irrig. Drain., 50: 249–257.
511. English, M., and Raja, S. N. (1996). Perspectives on deficit irrigation. Agric. Water Manage., 32: 1–14.
512. Entry, J. A., Sojka, R., E. (2003). The efficacy of polyacrylamide to reduce nutrient movement from an irrigated field. Trans ASAE, 46(1): 75–83.

513. Environ, J. (2006). Analysis of irrigation systems using sustainability-related criteria. Plant Soil Journal, 30: 1150–1153.
514. Environmental Protection Agency, (1999). Alternative disinfectants and oxidants guidance manual. 30–31.
515. Environmental Protection Agency, (1999). Trouble shooting guide for small ground water systems with hypochlorination.
516. EP405, (1995). Design, installation and performance of trickle irrigation systems. St. Joseph, MI: Standards of American Society of Agricultural Engineering, 49085.
517. Ernst, T. (1998). SDI reduces water losses from evaporation, runoff. KSU Natural Resources and Environmental Management, Fall, (1998). 4.
518. Escoe, A. K. (2006). *Piping and pipelines assessment guide*. Maryland Heights, Mo: Elsevier Gulf Professional Publishing.
519. Eslami, A. (2011). Study on brackish water use for a pistachio orchards using subsurface drip irrigation system. Research Report. AERI. 110 pages.
520. Estimation of water consumption for country main crops. Technical publication No. 1006–1997 Soil and Water Research Institute, Agricultural Ministry of Iran, (1997).
521. European Communities, EUR 10869.
522. Everson, D. O., M. Faubion and Amos, D. E. (1978). Freezing temperatures and growing seasons in Idaho. University of Idaho Agricultural Experiment, 18.
523. Evett, S., Laurent, J. P. (2000). Soil water components based on capacitance probes in a sandy soil. Soil Sci. Soc. of America J., 64: 311–318.
524. Evett, S. R., Howell, T. A., Schneider, A. D., Upchurch, D. R., Wanjura, D. F. (1996). Canopy temperature based automatic irrigation control. In: Evapotranspiration and Irrigation Scheduling. Camp, C. R., Sadler, E. J., R. E. Yoder (eds). Proc. International Conf., San Antonio, TX., ASAE, St. Joseph, MI. pages 207–213.
525. Evett, S. R., Laurent, J. P., Cepuder, P., Hignett, C. (2002). Neutron scattering, capacitance and TDR soil water content measurements on four continents. 17th World Congress of Soil Science (WCSS), August 14–21, Bangkok-Thailand, pages 1010–1021.
526. Evett, S. R., Howell, T. A., Schneider, A. D., Wanjura, D. F., D. R. Upchurch. 2002 Automatic drip irrigation control regulates water use efficiency. International Water and Irrig., 22(2): 32–37.
527. Evett, S. R., Howell, T. A., Schneider, A. D. (1995). Energy and water balances for surface and subsurface drip irrigated corn. Proceedings of the 5th International Micro irrigation Congress, ed. Lamm, F. R., 135–140 pages. St. Joseph, MI: ASAE.
528. Evett, S. R., Howell, T. A., Schneider, A. D., Upchurch, D. R., Wanjura, D. F. (2000). Automatic drip irrigation of corn and soybean. Proceedings of the 4th Decennial National Irrigation Symposium, November 14–16. 401–408.
529. Experimental analysis of local pressure losses for micro irrigation laterals. Journal of Irrigation and Drainage Engineering, July 2004, 130(4): 318–324.
530. Fakher, Fatemeh Majidi, Farzad Paknejad, Mohammad Nabi Ilkaee, Mohammad Nasri and Alireza Pazoki, (2012). Simulation of wheat cultivar response to irrigation treatments using of CERES-Wheat Model. American Journal of Agricultural and Biological Sciences, 7(2): 135–142.
531. Fan, T., Stewart, B. A., Payne, W. A., Yong, W., Luo, J., Gao, Y. (2005). Long-term fertilizer and water availability effects on cereal yield and soil chemical properties in Northwest China.Soil Sci. Soc. of America J., 69: 842–855.
532. Fangmeier, D. D., Elliot, W. J., Workman, S. R., Huffman, R. L., Schwab, G. O. (2005). *Soil and Water Conservation Engineering.* Fifth Edition. Delmar Cengage Learning, 502 pages.
533. Fangmeier, D. D., D. J. Garrot Jr., Husman, S. H., Perez, J. (1989). Cotton water stress under trickle irrigation. Trans. of ASAE, 32(6): 1955–1959.
534. Fanning, J. L. (2001). A field and statistical modeling study to estimate irrigation water use at benchmark farms study sites in Southwestern Georgia. U. S. Dept. of the Interior, U. S. Geological Survey.
535. FAO (1993). Year book, volume 47.

536. Farahani, H. J., Shaner, D. L., Buchleiter, G. W., Bartlett, G. A. (2006). Evaluation of a low volume agro-chemical application system for center pivot irrigation. Applied Engineering in Agriculture, 22(4): 517–528.
537. Fares, A., Alva, A. K. (2000). Evaluating the capacitance probes for optimal irrigation of citrus through soil moisture monitoring in an entisol profile. Irrigation Science, 19: 57–64.
538. Fawaz, M (1992). Water resources in Lebanon. National Seminar on water resources in Lebanon. 27–28 November, 1992 Beirut.
539. Feddes, R. A. (1987). Crop factors in relation to Makkink reference crop evapotranspiration. Technical Bulletin No. 67, 33–45. Institute for Land and Water Management Research.
540. Feigin, A., Letey, J., Jarrell, W., M. (1982). Nitrogen utilization efficiency by drip irrigated celery receiving preplant or water applied N fertilizer. Agron. J., 74: 978–983.
541. Felsot, A. S., Cone, W., Yu, J., Ruppert, J. R. (1998). Distribution of imidacloprid in soil following subsurface drip chemigation. Bulletin of Environmental Contamination & Toxicology, 60(3): 363–370.
542. Felsot, A., Evans, R. G., Tallman, L. (2000). Soil distribution and plant uptake of imidacloprid under drip and furrow irrigation. Proceedings of the 4th Decennial National Irrigation Symposium, Phoenix, AEvans, Z. R., G., Benham, B. L., Trooien, T. P., eds. American Society of Agricultural Engineers, St. Joseph, MI. pages 416–427.
543. Feng, G. L., Meiri, A., Letey, J. (2003). Evaluation of a model for irrigation management under saline conditions: I. Effects on plant growth. Soil Sci. Soc. of America J., 67(1): 71–76.
544. Fereres, E. (1981). *Drip irrigation management*. Cooperative Extension, University of California, Berkeley, CA, Leaflet No. 21259.
545. Fereres, E., Soriano, M. A. (2007). Deficit irrigation for reducing agricultural water use. Journal of Experimental Botany, 58: 147–159.
546. Fernández, J. E., Díaz-Espejo, A., Infante, J. M., Durán, P., Palomo, M. J., Chamorro, V., Girón, I. F., Villagarcía, L. (2006). Water relations and gas exchange in olive trees under regulated deficit irrigation and partial rootzone drying. Plant and Soil, 284: 273–291.
547. Ferrell, J. E. (1990). Less water, more crops. World Magazine. The San Francisco Chronicle, March 11, (1990). 17–18.
548. Filter, emitter and location on clogging when using effluents. Agric.Water Mgt., 96: 67–69.
549. Fischbach, P. E., Thompson, T. L., Stetson, L. E. (1978). Electric controls for automatic surface irrigation systems with reuse system. Trans. of ASAE, St. Joseph – MI.
550. Flanagan, D. C., Norton, L. D., Shainberg, I. (1997). Effect of water chemistry and soil amendments on a silt loam soil, Part 1: Infiltration and runoff. Trans. ASAE, 40(6): 1549–1554.
551. Flerchinger, G. N., Pierson, F. B. (1991). Modeling plant canopy effects on variability of soil temperature and water. Agricultural and Forest Meteorology, 56: 227–246.
552. Follett, R. F., Porter, L. K., Halvorson, H. D. (1991). Border effects on nitrogen–15 fertilized winter wheat microplots grown in the greatplains. Agron. J., 83: 608–612.
553. Ford, H. W. (1979). A key for determining the use of sodium hypochlorite (liquid chlorine) to inhibit iron and slime clogging of low pressure irrigation systems in Florida. Lake Alfred AREC Research Report-CS79–3. Mimeograph. 5 pages.
554. Ford, H. W. (1979). Water quality tests for low volume irrigation. Lake Alfred AREC Research Report CS79–6. University of Florida.
555. Ford, H. W. (1979). The present status of research on slimes of sulfur in low pressure irrigation systems and filters. Fruit Crops Mimeo FC79–2. University of Florida.
556. Ford, H. W. (1979). The present status of research on iron deposits in low pressure irrigation systems. Fruit Crops Mimeo FC79–3. University of Florida.
557. Ford, H. W. (1979). The use of chlorine in low pressure systems where bacterial slimes are a problem. Fruit Crops Mimeo FC79–5. University of Florida.
558. Ford, H. W. (1979). The use of surface water for low pressure irrigation systems. Fruit Crops Mimeo FC79–1. University of Florida.

559. Franco, J. A., Abrisqueta, J. M., Hernansaez, A., Morena, F. (2000). Water balance in a young almond orchard under drip irrigation with water of low quality. Agricultural Water Management Journal, 43(1): 75–98.
560. Frank, A. B. (2003). Evapotranspiration from northern semiarid grasslands.Soil Sci. Soc. of America J., 67: 1504–1509.
561. Frank, A. B. (2003). Evapotranspiration from northern semiarid grasslands.Soil Sci. Soc. of America J., 67: 1504–1509.
562. Frére, M., Popov, G. F. (1979). Agrometeorological crop monitoring and forecasting. FAO Plant Production and Protection Paper No. 17. FAO, Rome, Italy. 38–43.
563. Frevert, D. K., Hill, R. W., Braaten, B. C. (1983). Estimation of FAO evapotranspiration coefficients. Journal of Irrigation and Drainage Engineering, ASCE, 109(2): 265–270.
564. Friel, R., Or, D. (1999). Frequency analysis of time-domain reflectometry (TDR) with application to dielectric spectroscopy of soil constituents. Geophysics, 64: 707–718.
565. Fritschen, L. J., Fritschen, C. L. (1991). Design and evaluation of net radiometers. 7th Symposium on Meteorology Observations and Instrumentation, New Orleans, LA., 5.
566. Fuehring, H. D., Ghurayyib, A., A. (1969). Fertilizers for irrigated potatoes in the Bekaa plain of Lebanon. F. A.S., A. U.B., Publication No. 36.
567. Fundenburg, E., Kovar, J., Smith, C., Elston, R. (1996). A comparison of three soil test P extracts on an alkaline Louisiana soil. Proc. Beltwide Cotton Conf., 2: 1428–1429. Memphis, Tenn.: Nat. Cotton Council.
568. Gajri, P. R., Arora, V. K., Chaudhary, M. R. (1994). Maize growth, response to deep tillage, straw mulching and farmyard manure in coarse textured soils of N. W. India. Soil Use Manag., 10: 15–20.
569. Gamayunov, N. I. (1983). Coagulation of suspension after magnetic treatment. Journal of Applied Chemistry. 56: 975–982.
570. García-Tejero, I. (2010). Deficit irrigation for sustainable Citrus cultivation in Guadalquivir river basin. Ph. D. Thesis. Universidad de Sevilla, Spain, 285 pages.
571. García-Tejero, I., Durán-Zuazo, V. H., Jiménez-Bocanegra, J. A., Muriel-Fernández, J. L. (2011). Improved water-use efficiency by déficit irrigation programmes: Implications for saving water in citrus orchards. *Scientia Horticulturae* 128: 274–282.
572. Gardenas, A., Hopmans, J. W., Hanson, B. R., Simunek, J. (2005). Two-dimensional modeling of nitrate leaching for various fertigation scenarios under micro irrigation. Agric. Water Manage. 74: 219–242.
573. Gardner, B. R., R. L. Roth. Applying nitrogen in irrigation water. 493–506. In. R. D. Hauck (ed.) *Nitrogen in crop production*. American Society of Agronomy. CSSA and SSSA, Madison, WI.
574. Gardner, W. H. (1979). How water moves in the soil. Crops and Soils, 32(2): 13–18.
575. Gardner, W. R. (1958). Some steady-state solutions of the unsaturated moisture flow equation with application to evaporation from a water table. Soil Sci., 85: 228–232.
576. Garratt, J. R., Hicks, B. B. (1973). Momentum, heat and water vapourvapor transfer to and from natural and artificial surfaces. Quarterly J. of the Royal Meteorological Soc., 99: 680–687.
577. Garratt, J. R. (1992). *The atmospheric boundary layer*. Cambridge: Cambridge University Press, 316.
578. Gash, J. H.C., Shuttleworth, W. J., Lloyd, C. R., André, J. C., Goutorbe, J. P., Gelpe, J. (1989). Micrometeorological measurements in Les Landes forest during HAPEX-MOBILHY. Agricultural and Forest Meteorology, 46: 131–147.
579. Gehl, R. J., Schmidt, J. P., Stone, L. R., Schlegel, A. J., Clark, G., A. (2005). In-situ measurements of nitrate leaching implicate poor nitrogen and irrigation management on sandy soils. J. Environ. Qual. 34: 2243–2254.
580. Gelly, M., Recasens, I., Mata, M., Arbones, A., Rufat, J., Girona, J., Marsal, J. (2003). Effects of water deficit during stage II of peach fruit development and postharvest on fruit quality and ethylene production. J. Hort. Sci. Biotechnology, 78: 324–330.
581. George, W., Pruitt, W. O., Dong, A. (1985). Evapotranspiration modeling. In*: California Irrigation Management Information System.* Final Report by Snyder, R., Henderson, D. W., Pruitt, W.

O., Dong, A., California. Department Water Resources. Contract. No. B53812. Land, Air and Water Resources.

582. Gerrish, P. J., Bralts, V. F., Shayya, W. H. (1996). An improved analysis of micro irrigation hydraulics using a virtual emitter system. Trans. ASAE, 39(4): 1403–1410.
583. Ghaemi, A. A., Tabarzad, A. (2013). Micro Irrigation Design Software (MIDS) to evaluate Technica and Economical Aspects of MIS. Water Eng. Dept., College of Agric., Shiraz University, Iran.
584. Ghali, G. S., Svehlik, Z. J. (1988). Soil-water dynamics and optimum operating regime in trickle-irrigated fields. Agricultural Water Management, 13: 127–143.
585. Gibson, W. (1974). Hydraulics, mechanics and economics of subsurface and drip irrigation of Hawaiian sugarcane, 2: 639–648. Proceedings of the 15th Congress International Society Sugar Cane Technology, Honolulu, Hawaii: Hawaiian Sugar Planters' Assoc.
586. Gideon, O., DeMalach, Y., Gillerman, L., and David, I. (1995). Pear response to saline water application under subsurface drip irrigation. Proc. Fifth International Microirrgation Congress, Orlando, FL, April 2–6, pages 97–103.
587. Gideon, O., DeMalach, Y., Gillerman, L., David, and I., Lurie, S. (2002). Effect of water salinity and irrigation technology on yield and quality of pears. Biosystems Eng., 81: 237–247.
588. Gilbert, R. G., Ford, H. W. (1986). Emitter clogging. Chapter 3.1 in *Trickle Irrigation for Crop Production.* Nakayama, F. S., Bucks, D. A., Eds. Elsevier Pubications, Amsterdam. 383 pages.
589. Gilbert, R. G., Nakayama, F. S., and Bucks, D. A. (1979). Trickle irrigation: prevention of clogging. Trans. of ASAE, 22(3): 514–519.
590. Gillespie, V. A., Phillips, A. L., I-Pai Wu, (1979). Drip irrigation design equations. Journal Irrigation and Drainage, ASCE, 105 (IR3): 247–257, Paper # 14819.
591. Gilley, J. R., Allred, E. R. (1974). Infiltration and root extraction from subsurface irrigation laterals. Trans. of the ASAE, 17(5): 927–933.
592. Gilley, J. R., Allred, E. R. (1974). Optimum lateral placement for subsurface irrigation systems. In Proceedings of the Second International Drip Irrigation Congress, 234–239 pages. Riverside, CA: University of California.
593. Gilley, J. R., Supalla, R. J. (1999). Economic analysis of energy saving practices in irrigation. Trans. of ASAE, 26: 1784–1792.
594. Ginestar, C., Castel, J. R. (1996). Responses of young Clementine citrus trees to water stress during Different phenological periods. Journal of Horticultural Science, 71: 551–559.
595. Gish, T. J., Dulaney, W. P., Kung, K. J. S., Daughtry, C. S. T., Doolittle, J. A., Miller, P. T. (2002). Evaluating use of ground-penetrating radar for identifying subsurface flow pathways. Soil Sci. Soc. of America J., 66: 1620–1629.
596. Giuseppina, C., Garofalo, P. (2006). August. Management of irrigation with saline water in cracking clay soils. Soil Sci. Soc. of America J., 70: 1774–1787.
597. Glenn, D. M. (1999). Physiological effects of incomplete root-zone wetting on plant growth and their implications for irrigation management. Hort. Science, 35: 1041–1043.
598. Glossary of Soil and Water Terms, (1967). Special Publication SP-04–67. St. Joseph – Michigan: American Society of Agricultural and Biological Engineers.
599. Glossary of Soil Science Terms, (1979). Soil Science Society of America. 677 South Segoe Road. Madison-Wisconsin 53711, USA.
600. Goldberg, D., Shmueli, M. (1970). Drip irrigation – A method used under arid and desert conditions of high water and soil salinity. Trans. of the ASAE, 13(1): 38–41.
601. Goldberg, D., Gornat, B., Rimon, D. (1976). *Drip Irrigation Principles, Design and Agricultural Practices*. Kfar Shmaryahu, Israel: Drip Irrigation Scientific Publications.
602. Goldhamer, D. A., Beede, R. H. (2004). Regulated deficit irrigation effects on yield, nut quality and water-use efficiency of mature pistachio trees. Journal of Horticultural Science and Biotechnology, 79: 538–545.
603. Goldhamer, D. A., Salinas, M. (2000). Evaluation of regulated deficit irrigation on mature orange trees grown under high evaporative demand. Proceedings of the International Society of Citriculture, IX Congress. Orlando, FL: ISC, pages 227–231.

604. Goldhamer, D. A., Viveros, M. (2000). Effects of preharvest irrigation cut off durations and postharvest water deprivation on almond tree performance. Irrigation Science, 19: 125–131.
605. Goldhamer, D. A., Viveros, M., Salinas, M. (2006). Regulated deficit irrigation in almonds: effects of variations in applied water and stress timing on yield and yield components. Irrigation Science, 24: 101–114.
606. Goldman, A. P. (2004). The compleat squash: A passionate grower's guide to pumpkins, squashes, and gourds. Artisan Journal Series by Peter Pauper, Powell Books. Page 18. October.
607. Golfinopoulos, S. K. (2002). Formation of organic by-products during chlorination of natural waters. Journal of Environmental Monitoring, 4(4): 910–916.
608. González-Altozano, P., Castel, J. R. (1999). Regulated deficit irrigation in Clementina de Nules' citrus trees. I. Yield and fruit quality effects. Journal of Horticultural Science and Biotechnology 74: 706–713.
609. González, E. A., Goyal, M. R. (1988). Estimation of irrigation requirement of sweet corn (Zea mays cv. Sursweet) in the south coast of Puerto Rico. J. Agric. U. P.R., 72(2): 277–284 (Spanish).
610. González, E. A., Goyal, M. R., C. Chao de Báez, (1988). Effects of three irrigation levels on growth and yield of sweet corn (Zea mays cv. Suresweet) under drip irrigation. J. Agric. U. P.R., 72(4): 565–574.
611. Goorahoo, D., Carstensen, G., Zoldoske, D. F., Norum, E., Mazzei, A. (2002). Using air in subsurface drip irrigation (SDI) to increase yields in bell peppers. Int. Water Irrig., 22: 39–42.
612. GOP, (2005). Mid Term Development Framework 2005–2010, Planning Commission, Govt. of Pakistan, May.
613. Gosse, G., Perrier, A., Itier, B. (1977). Etude de l'évapotranspiration réelle d'une culture de blé dans le bassin parisien. Annual Agronomy, 28(5): 521–541. (in French).
614. Goyal, D., Sharma, R. K., Ramamurthy, V., Kothari, R. M. (2001). An integrated biotech approach for the sustainable improvement in agro-forestry systems. In: *Innovative Approaches in Microbiology* (Eds. Maheshwari, D. K., R. C. Dubey), B. Singh Publ., Dehradun, pages 367–377.
615. Goyal, M. R. (1979). Stresses generated in soil crust by emerging dicot seedlings. Ph.Dissertation, D., The Ohio State University, Columbus, Ohio, Aug, (1979). 121.
616. Goyal M. R. (1979). Crusted soils affect seedling emergence. The Crops and Soils Magazine, Madison, WI, 31 (6): 27. April-May issue.
617. Goyal, M. R. (1980). Annual Report 1/PR/H326 (S-143)/1979–80 – Trickle Irrigation in Puerto Rico. Agricultural Experiment Station, UPR-RUM, Río Piedras. 159.
618. Goyal, M. R. (1980). Drip system uses much less water. San Juan Star, San Juan, PR, Jan. 27. Page B8.
619. Goyal, M. R. (1981). Annual Report 1/PR/H326 (S-143)/1979–80 – Trickle Irrigation in Puerto Rico. Agricultural Experiment Station, UPR-RUM, Río Piedras. 129.
620. Goyal, M. R. (1981). Bibliography – Drip/Trickle Irrigation. National Agricultural Plastics Association (NAPA) Tech. Bull. No. 3 and Special publication #1175 of Agricultural Experiment Station, UPR RUM, Mayagüez. Salisbury, Maryland. 79.
621. Goyal, M. R. (1981). Try new irrigation system for commercial vegetable production (Spanish). El Mundo, San Juan. 61 (349): B8.
622. Goyal, M. R. (1982). Annual Report 1/PR/H326 (S-143)/1979–80 – Trickle Irrigation in Puerto Rico. Agricultural Experiment Station, UPR-RUM, Río Piedras. 95.
623. Goyal, M. R. (1983). AMA Abstracts and Index (1971–80). AMA by FMIRC, Japan. 75.
624. Goyal, M. R. (1983). Annual Report 1/PR/H326 (S-143)/1979–80 – Trickle Irrigation in Puerto Rico. Agricultural Experiment Station, UPR-RUM, Río Piedras. 102.
625. Goyal, M. R. (1983). Labor input requirements for experimental production of summer peppers under drip irrigation. J. Agric. U. P.R., 67(1): 23–27.
626. Goyal, M. R. (1987). Title and abstracts, "Trickle irrigation in humid regions: Puerto Rico." Agriculture Experiment Station, Río Piedras. 1–20.

627. Goyal, M. R. (1985). Subject Index of Drip/Trickle Irrigation in Action Vol. I and II. Third International Drip/Trickle Congress by ASAE, Fresno, CA. Nov. 16–20. 473–475.
628. Goyal, M. R. (1987). Insitu recovery of drip irrigated cabbage roots. J. Agric. U. P.R., 71(4): 419–422.
629. Goyal, M. R. (1988). Potential evapotranspiration for south coast of Puerto Rico with the Hargreaves and Samani technique. J. Agric. U. P.R., 72(1): 57–64.
630. Goyal, M. R. (1988). Potential evapotranspiration for Vieques Island of Puerto Rico with the Hargraves-Samani model. Res. Note, J. Agric. U. P.R., 72(1): 177–178.
631. Goyal, M. R. (1989). Estimation of monthly water consumption by selected crops in the semiarid and humid regions of Puerto Rico. Monograph by AES-UPR-RUM, Río Piedras. 454. ISBN 0–9621805–0-5 and Library of Congress #88–51661. 1–454.
632. Goyal, M. R. (1989). Golden rules and drip irrigation future in Puerto Rico. Paper No. PRS89–04 at Spring Meeting of the Puerto Rico Section Inc. of ASAE at Hato Rey. May 26, pages 1–6.
633. Goyal, M. R. (1989). Irrigation Research and Extension Progress in Puerto Rico. ISBN O-9621805–1-3.
634. and Library of Congress #88–51939 by Agric. Exp. Sta. of UPR-RUM. Symposium proceedings on first congress on "Irrigation in Puerto Rico". March 8, (1989). 1–549.
635. Goyal, M. R. (1990). Trouble shooting in trickle irrigation. Third Irrigation Symposium by ASAE at Phoenix-AZ. October 28 -November 1, (1990). 464–468.
636. Goyal, M. R. (1990). Drip Irrigation Management Book (Spanish)-Review. Third Irrigation Symposium by ASAE at Phoenix-AZ, October 28-November 01, (1990). 651–653.
637. Goyal, M. R. (1990). Estimation of total water consumption of vegetable crops in the semiarid and humid regions of Puerto Rico. Proc. Int. Symposium on tropical Hydrology and 4th Caribbean Islands Water Resources Congress by Am. Water Res. Assoc., San Juan-PR. July 22–27. 207–215.
638. Goyal, M. R. (1990). Management of Drip Irrigation (Spanish: Manejo de Riego por Goteo). Servicio de Extensión Agrícola, RUM, Rio Piedras. IA 80. ISBN-0–9621805–2-3- Páginas 1–548.
639. Goyal, M. R. (1991). Irrigation Requirements Estimations in Puerto Rico, In: Achievements in Tropical and Subtropical Agricultural Research under PL 89–106 Special Research Grants, Caribbean Basin Administrative Group: Univ. of Florida. 21–22.
640. Goyal, M. R., et. al, (1981). Moisture and soybean seedling emergence. Trans. ASAE, 24 (6): 1432–35.
641. Goyal, M. R., et. al, (1982). Stresses generated in soil crust by emerging dicots. Trans. ASAE, 25(3): 556–562.
642. Goyal, Megh R. (2006). Water consumption by selected crops and climatology: Case study in Trinidad. Corporación Universitaria Santa Rosa de Cabal (UNISARC), Colombia. Boletín: Investigaciones de UNISARC, Volumen 4 No 1.
643. Goyal, Megh R. (2012). Management of Drip/Triclkle or Micro Irrigation. New Jersey, USA: Apple Academic Press Inc.
644. Goyal, Megh R. (2013). Evapotranspiration: Principles and Applications for Water Management. New Jersey, USA: Apple Academic Press Inc.
645. Goyal, Megh R. (2014). Management of Drip/Triclkle or Micro Irrigation. New Jersey, USA: Apple Academic Press Inc.
646. Goyal, Megh R. (2014). Book Series: Research Advances in Sustainable Micro Irrigation. Volume 1: Sustainable Micro Irrigation: Principles and Practices; Volume 2: Sustainable Practices in Surface and Subsurface Micro Irrigation; Volume 3: Sustainable Micro Irrigation Management for Trees and Vines. New Jersey, USA: Apple Academic Press Inc.
647. Goyal, M. R., Allison, W. F. (1983). Summer drip irrigation requirements for cucumber. J. Agric. U. P.R., 67(3): 328–334.
648. Goyal, M. R., Bellerive-Martelly, P. (1987). Agriculture in Haiti: With emphasis in irrigation. AMA Japan, 18(4): 54–57.

649. Goyal, M. R., Chao, C. (1989). Tomato and sweet pepper response to furrow-, microsprinkler- and drip irrigation. J. Agric. U. P.R., 73(3): 239–254.
650. Goyal, M. R., Chao, C. (1991). Estimation of potential evapotranspiration with the Hargreaves and Samani model at selected locations in Puerto Rico. J. Agric. U. P.R., 75(1): 107–109.
651. Goyal, M. R., Colberg, O., Acosta, A. (1985). Xylem irrigation: Principles, practices and problems. Paper #85–2620 at ASAE Winter Meeting, Chicago, IL., Dec. 17–20. 18.
652. Goyal, M. R., M. Crespo-Ruíz and Rivera, L. E. (1988). Root distribution of nitrogen fertigated sweet peppers under drip irrigation. J. Agric. U. P.R., 72 (1): 51–56.
653. Goyal, M. R., Drew, L. O., Nelson, G. L., Logan, T. J. (1977). Effects of environmental and seed characteristic factors on soybean seedling force. ASAE Paper #77–1011 at Annual Meeting of American Society of Agricultural Engineers (ASAE), Raleigh, NC, June 21. 32.
654. Goyal, M. R., Drew, L. O., Nelson, G. L., Logan, T. J. (1982). Critical time for soybean seedling emergence force. Trans. ASAE, 23 (4): 831–35.
655. Goyal, M. R., Drew, L. O., Nelson, G. L., Logan, T. J. (1982). Soybean seedling emergence force. Trans. ASAE, 23 (4): 836–39.
656. Goyal, M. R., González, E. A. (1986). Automation in drip irrigation. Agricultural Extension Service, UPR-RUM, Mayagüez. Serie XI, IA75. 1–57.
657. Goyal, M. R., González, E. A. (1988). Irrigation requirements of sorghum in the south and north coasts of Puerto Rico. J. Agric. U. P.R., 72 (4): 585–598 (Spanish).
658. Goyal, M. R., González, E. A. (1988). Seasonal consumptive use of peppers water by bell and cubanelle peppers in semiarid and humid coastal sites in Puerto Rico. J. Agric. U. P.R., 72(4).
659. Goyal, M. R., González, E. A. (1988). Water requirements for vegetable production in Puerto Rico. National Irrigation and Drainage Congress by Am. Soc. Civil Engrs. at Lincoln, NE, July 19–22, (1988). 385–392.
660. Goyal, M. R., González, E. A. (1989). Monthly consumptive use of rice in the semiarid and humid regions of Puerto Rico. J. Agric. U. P.R., 73(1): 31–44.
661. Goyal, M. R., González, E. A. (1989). Monthly consumptive use of sugarcane in four climatic regions of Puerto Rico. J. Agric. U. P.R., 73(1): 93–95.
662. Goyal, M. R., González, E. A. (1989). Monthly water consumptive use of papaya in climatic regions of Puerto Rico. J. Agric. U. P.R., 73(2): 175–176.
663. Goyal, M. R., González, E. A. (1989). Pan coefficients for Puerto Rico. J. Agric. U. P.R., 73(1): 89–92.
664. Goyal, M. R., González, E. A. (1989). Evapotranspiration (Spanish). Servicio de Extensión Agrícola, UPR-RUM, Rio Piedras. Serie XIV, IA72 1–45.
665. Goyal, M. R., González, E. A., C. Chao de Báez, (1987). Potential evapotranspiration estimation for.
666. Goyal, M. R., González, E. A., C. Chao de Báez, (1988). Temperature versus elevation relationships for Puerto Rico. J. Agric. U. P.R., 72(3): 449–467.
667. Goyal, M. R., E. A. González Fuentes y J. A. Santiago Rivera, (1989). Riego por goteo: Estudio de viabilidad. Servicio de Extensión Agrícola-Recinto Universitario de Mayagüez, Río Piedras, PR. IA78 Serie 17, páginas 1–18.
668. Goyal, M. R., González, E. A., Rivera, L. E., Chao, C. (1987). Sweet pepper response to drip-, microsprinkler-and furrow irrigation. ASAE Paper #87–2523 at 1987 Winter Meeting of American Society of Agricultural Engineers at Chicago, IL. December 15–18, (1987). 13.
669. Goyal, M. R., Guadalupe-Luna, R., E. Recio de Hernández and Rivera, L. E. (1984). Effects of plastic mulch types on sizing and crop performance of drip irrigated winter and summer peppers. J. Agric. U. P.R., 68 (3): 297–305.
670. Goyal, M. R., Guadalupe-Luna, R., E. Recio de Hernández, Rivera, L. E., Caraballo, E. (1985). Effects of water application rates and planting density on size arrangements of drip irrigated onions. J. Agric. U. P.R., 69(3): 383–389.
671. Goyal, M. R., Guadalupe-Luna, R., Hernández, E. R., C. Chao de Báez, (1989). Post harvest evaluation of nitrogen fertigated sweet peppers under drip irrigation and plastic mulch. J. Agric. U. P.R., 73(2): 109–114.

672. Goyal, M. R., Luna, R. G., E. Recio de Hernández and Rivera, L. E. (1984). Effects of plastic mulch types on sizing and crop performance of drip irrigation winter and summer peppers. Journal of Agriculture of the University of Puerto Rico, 68(3): 297–305.
673. Goyal, M. R., Nelson, G. L., Drew, L. O., Carpenter, T. G., Logan, T. J. (1979). Stresses generated in soil.
674. crusts by emerging dicot seedlings. Paper No. 79–1043 at Annual Meeting of ASAE, Manitoba, June 25–27. 22.
675. Goyal, M. R., Poventud-Suárez, A. (1987). Soil moisture potential distribution around an emitter. J. Agric. U. P.R., 71(4): 423–426.
676. Goyal, M. R., Ravalo, E. J., González, E. A. (1989). Normal monthly class A pan evaporation for agricultural experiment substations in Puerto Rico. J. Agric. U. P.R., 73(1): 85–78.
677. Goyal, M. R., Rivera, L. E. (1982). Crop response of trickle irrigated vegetables in Puerto Rico. Paper #82–2058 at Summer Meeting of ASAE held at Madison, WI. 29.
678. Goyal, M. R., Rivera, L. E. (1984). Drip irrigation: Chemigation (Spanish). Agricultural Extension Service. UPR-RUM, Mayagüez. Serie IV, IA61. 1–30.
679. Goyal, M. R., Rivera, L. E. (1985). Bibliography -Drip/Trickle Irrigation (Supplement). College of Engineers and Surveyors of Puerto Rico, Hato Rey (Publ. No. 84–13 of Agric. Exp. Sta., UPR-RUM). 60.
680. Goyal, M. R., Rivera, L. E. (1985). Drip irrigation: Service and Maintenance (Spanish). Agricultural Extension Service UPR-RUM, Mayagüez. Serie V, IA64. 1–12.
681. Goyal, M. R., Rivera, L. E. (1985). Evaluation of field uniformity for water application in drip irrigation (Spanish). Agricultural Extension Service UPR-RUM, Mayagüez. Serie VI, IA65. 1–18.
682. Goyal, M. R., Rivera, L. E. (1985). Trickle irrigation scheduling in vegetables. In: Drip/Trickle Irrigation in Action. Third Internacional Drip/Trickle Congreso by American Society of Agricultural Engineers at Fresno-CA. Vol. II. Paper No. M-7. 838–843.
683. Goyal, M. R., Rivera, L. E. (1986). Depths to locate tensiometers (Spanish). Agricultural Extension Service, UPR-RUM, Mayagüez. Serie IX, IA68. 1–2.
684. Goyal, M. R., Rivera, L. E. (1986). Design of drip irrigation systems (Spanish). Agricultural Extension Service UPR-RUM, Mayagüez. Serie VIII, IA70. 1–72.
685. Goyal, M. R., Rivera, L. E. (1988). Potassium fertigation in drip irrigated peppers in Puerto Rico. National Irrigation and Drainage Congress by Am. Soc. Civil Engrs. at Lincoln, NE, July 19–22, (1988). 87–90.
686. Goyal, M. R., Rivera, L. E., E. Caraballo and Santiago, C. L. (1985). Growth characteristics of trickle irrigated vegetables, In: Drip/Trickle Irrigation in Action. Third International Drip/Trickle Congress by ASAE, Fresno, CA. November 16–20. Vol. I, Paper #F-9. 249–254.
687. Goyal, M. R., Rivera, L. E., Martínez, M., Rojas, N. (1984). Principles of drip irrigation (Spanish). Agricultural Extension Service UPR-RUM, Mayagüez. Serie III, IA62. 1–52.
688. Agricultural Extension Service, UPR-RUM. Mayagüez. Serie XV, IA 76. 1–22.
689. Goyal, M. R., Rivera, L. E., Santiago, C. L. (1985). Nitrogen fertigation in drip irrigated peppers, tomatoes and eggplant, In: Drip/Trickle Congress by ASAE, Fresno, CA. Nov. 16–20. Vol. I, Paper #I-10. 388–392.
690. Goyal, M. R., Román, J., F. Gallardo and Luis Rivera, E. (1983). Chemigation via trickle irrigation system in vegetables and fruit orchards. Paper #83–1008 at Summer Meeting of ASAE held at Bozeman, MT, June. 12.
691. Goyal, M. R., Román, J., Montalvo, R., Gallardo, F. (1984). Pestigation via trickle irrigation systems. Paper No. 64 at Symposium on Advances in Pesticide Application Technology at 188th Meeting of American Chemical Society, Philadelphia, PA. August 29–31. 41.
692. Goyal, M. R., Román, J., Zapata, R. M., Gallardo, F. (1990). Pestigation in tickle irrigation. Third Irrigation Symposium by ASAE at Phoenix-AZA. Oct. 28-Nov 1, (1990). 464–468.
693. Goyal, M. R., J. Santaella-Pons and Rivera, L. E. (1982). Tensiometer: Use, Installation and Maintenance (Spanish). Agricultural Extension Service UPR-RUM, Mayagüez. Serie I, IA73. 1–18.

694. Goyal, M. R., Santiago, C. L., Chao, C. (1984). How plastic mulch types affect growth parameters of drip irrigated summer peppers. J. Agric. U. P.R., 8(4): 365–373.
695. Goyal, M. R., T. Persaud and Rivera, L. E. (1988). Labor input requirements for experimental production of drip irrigated vegetables. J. Agric. U. P.R., 72(1): 41–50.
696. Goyal, M. R., Snyder, V. A., Rivera, L. E. (1983). Solute movements patterns in drip irrigated tomatoes at Santa Isabel. J. Agric. U. P.R., 67 (4): 486–493.
697. Goyal, Megh R., Shih, S., F. (2007). Generation of missing climatic data in Puerto Rico. Corporación Universitaria Santa Rosa de Cabal (UNISARC), Colombia. Boletín: Investigaciones de UNISARC, 5(1): 1–11.
698. Goyal, M. R., Torres-Sepúlveda, A. (1982). Drip irrigation is a beneficial technique (Spanish). Agricultural Extension Service, UPR-RUM, Mayagüez. Serie II, IA60. 1–2.
699. Goyal, Megh R. (2012). Appendices. 317–332. In: *Management of Drip/Trickle or Micro Irrigation* edited by Megh R Goyal. New Jersey, USA: Apple Academic Press Inc.
700. Grabow, G. L., Harrison, K., Smith, W. B., Vories, E., Zhu, H., Khalilian, A. (2005). Design of subsurface drip irrigation systems in humid areas. Proc. World Water and Environmental Resources Congress. Reston Va.: ASCE Environmental and Water Resources Institute.
701. Grainger, G. (1995). Micro Irrigation System. Rural Water Advisory Services, W22, October.
702. Grant, D. R. (1975). Comparison of evaporation from barley with Penman estimates. Agricultural Meteorology, 15: 49–60.
703. Grattan, S. R., Schwankl, L. J., Lanini, W. T. (1990). Distribution of annual weeds in relation to irrigation method. Proceedings of the 3rd National Irrigation Symposium, 148–153. St. Joseph, MI: ASAE.
704. Grattan, S. R., Schwankl, L. J., and Lanini, W. T. (1988). Weed control by subsurface drip irrigation. CaliforniaAgricultural, 42(3): 22–24.
705. Grattan, S. R., Bowers, W., Dong, A., Snyder, R. L., Carroll, J. J., George, W. (1998). New crop coefficients estimate water use of vegetables, row crops. California Agriculture, 52(1): 16–21.
706. Grimes, D. W., Wiley, P. L., Sheesley, W. R. (1992). Alfalfa yield and plant water relations with variable irrigation. Crop Sci., 32(6): 1381–1387.
707. Grimes, D. W., Munk, D. S., Goldhamer, D. A. (1990). Drip irrigation emitter depth placement in a slowly permeable soil. In Proceedings of the 3rd National Irrigation Symposum, pages 248–254. St. Joseph, MI: ASAE.
708. Grismer, M. E. (2001). Regional alfalfa yield, ETc and water value in western states. J. Irrig. and Drain. Eng., 127(3): 131–139.
709. Grismer, M. E. (2002). Regional cotton lint yield, ETc and water value in Arizona and California. Agricultural Water Management, 54(3): 227–242.
710. Guitjens, J. C. (1993). Alfalfa irrigation during drought. Irrig. Drain. Eng. 119(6): 1092–1098.
711. Gunston, H., Batchelor, C. H. (1983). A comparison of the Priestley-Taylor and Penman methods for estimating reference crop evapotranspiration in tropical countries. Agricultural Water Management, 6: 65–77.
712. Gushiken, E. C. (1995). Irrigating with reclaimed water through permanent subsurface drip irrigation systems. In: *Micro irrigation for a changing world.* Proc 5th Intl. Micro irrigation Congress. Lamm, F., R. (Ed.), St. Joseph, Michigan. ASAE. pages 269–274.
713. Gushiken, E. C. (1992). Nonpotable subsurface drip irrigation system at Holoholokai beach Park. First Annual Pan Pacific Green Industry Conference, Jul. 30, (1992). Honolulu, HI.
714. Hagedorn, C. (2003). Subsurface drip irrigation for onsite/decentralized wastewater systems. HATCH Project VA-135691.
715. Hair, M., Coit, L., Boland, T. (2001). Michigan gardener's guide (Revised ed.), Cool Springs Press. 1–272.
716. Halevy, J., Kramer, O. (1986). Nitrogen fertilizer management of cotton grown under drip irrigation in grumusol. Irrig. Sci., 7: 63–72.
717. Hall, B. J. (1985). History of drip/trickle irrigation. Proc. Third Int'l. Drip/Trickle Cong., Nov. 18–21, 1985, Fresno, California. ASAE, St. Joseph, Michigan. 1–7.

718. Hall, B. J. (1963). Continous polyethylene tube covers for cucumbers. Proceedings of National Agriciulture Plastics Congress, 4: 112–132.
719. Hall, B. J. (1971). Comparisson of drip and furrow irrigation for market tomatoes. Proceedings of National Agriciulture Plastics Congress, 10: 19–27.
720. Hallowell, L. (1992). Drip irrigation in corn? A KSU study indicates it may be feasible. NC+ Down to Earth Ideas, 26(2): 6–7.
721. Haman, D. Z., Sorenson, R. B., Ross, D. S., Evans, R. O., Tacker, P. (2005). Critical management issues when using SDI in humid areas. Proc. World Water and Environmental Resources Congress, Reston Va.: ASCE Environmental and Water Resources Institute.
722. Haman, D. Z., Pritchard, R. T., Smajstrla, A. G., Zazueta, F. S., Lyrene, P., M. (1997). Evapotranspiration and crop coefficients for young blueberries in Florida. Applied Engr. in Agric., 13(2): 209–216.
723. Haman, D. Z., Smajstrla, A. G., and Zazueta, F. S. (1987). Media filters for trickle irrigation in Florida. Fact Sheet AE-57. Fla. Coop. Ext. Ser., Univ. of Florida.
724. Haman, D. Z., Smajstrla, A. G., and Zazueta, F. S. (1988). Screen filters in trickle irrigation systems. Fact Sheet AE-61. Fla. Coop. Ext. Ser., Univ. of Florida.
725. Haman, Doroto Z., and Forrest Izuno, T. (1989). Principles of micro-irrigation. AE-24, May. Fla. Coop. Ext. Ser., Univ. of Florida.
726. Hamdy, A. (1995). Fertilizers and their efficient use. *Advanced short course on fertigation*, Nov. 26– Dec. 3, (1995). FAS-UL. Beirut Lebanon, pages 83–138.
727. Hamdy, A. (1989). Fertigation prospects and problems. 10th Session of the FAO Regional Commission on Land and Water Use in the Near East. Amman, Jordan, 10–14 December.
728. Hamze, M., Serhal, A., El-Moubayed, L. (1991). Fertigation and chemigation: practices and prospects in Lebanon. Proceedings of the Expert Consultation on fertigation/Chemigation. 8–11 September, Cairo, pages 253–258.
729. Handley, D. (2000). Drip irrigation: The system of choice for California grape growers. Irrigation Journal, 50(2): 22–24.
730. Hanks, R. J. (1984). Prediction of crop yield and water consumption under saline conditions. In: *Soil salinity under irrigation: Processes and management*, eds. Shainberg, I., Shalhevet, J., Section 8.2: 272–283. Berlin, Springer-Verlag.
731. Hansen, R. C., Pasian, C. C. (1999). Using tensiometers for precision micro irrigation of container-grown roses. Applied Engr. in Agric., 15(5): 483–490.
732. Hanson, B. R., Fipps, G., Martin, E. C. (2000). Drip irrigation of row crops: What is the state of the art? Proc. 4th Decennial Natl. Irrigation Symp. (eds. Evans, R. G., Benham, B. L., B. L., T. P. Trooien), 391–400 pages. St. Joseph, MI: ASAE.
733. Hanson, B. (2006). Keeping the lines open. American Vegetable Grower, 54(5): 20.
734. Hanson, B., Putnam, D. (2000). Can alfalfa be produced with less water? Proc. 29th Natl. Alfalfa Symp. and 30th California Alfalfa Symp. Davis, CA: University of California, Department of Agronomy and Range Science Extension.
735. Hanson, B., Schwankl, L., Grattan, S. R., Prichard, T. (1997). Drip irrigation for row crops. Davis, CA: University of California Cooperative Extension.
736. Hanson, B., Bowers, W., Davidoff, B., Kasapligil, D., Carvajal, A., Bendixen, W. (1995). Field performance of micro irrigation systems. Proc. 5th Int'l. Micro irrigation Cong., F. R. Lamm (Ed.), Apr. 2–6, 1995, Orlando Florida. ASAE, St. Joseph, Michigan. pages 769–774.
737. Hanson, B. R., Bendixen, W. E. (1993). Salinity under drip irrigation of row crops. Proceedings of the International Exposition and Technical Conference, 196–202 pages. Arlington, VA: Irrigation Association.
738. Hanson, B. R., May, D. M., and Bendixen, W. E. (1997). Wetting patterns under surface and subsurface drip irrigation. ASAE Paper No. 97–2178. St. Joseph, MI: ASAE.
739. Hanson, E. G., Williams, B. C., Fangmeier, D. D., Wilke, O., C. (1970). Influence of subsurface irrigation on crop yields and water use. In Proc. National Irrigation Symposium, pages D1-D13. St. Joseph, MI: ASAE.

740. Hanson, E. G., Patterson, T. C. (1974). Vegetable production and water-use efficiency as influenced by drip, sprinkler, subsurface, and furrow irrigation methods. In Proceedings of the 2nd International Drip Irrigation Congress, 97–102 pages. Riverside, CA: University of California.
741. Hansonn, F., Martin, J. (2002). Drip irrigation of row crops: What is the state of the art? Proceedings of the 4th Decennial Irrigation Symposium, ASAE. 4–20.
742. Haque, BU, Saleem, M., and Bhutta, M. N. (2005). Evaluation and comparison of sprinkler, drip and bed and furrow irrigation. Mona Reclamation Experimental Project, Bhalwal, MR-EP Internal Report No. 2005/01.
743. Hargert, G. W., Frank, K. D., Rehm, G., W. (1978). Anhydrous ammonia and N-serve for irrigated corn. University of Nebraska-Lincoln Agronomy Dept. Soil Sci., Res. Report.
744. Hargreaves, G. H. (1974). Estimation of potential and crop evapotranspiration. Trans. of ASAE, 17: 701–704.
745. Hargreaves, G. H. (1975). Moisture availability and crop production. Trans. of ASAE, 18(5): 980–984.
746. Hargreaves, G. H. (1983). Discussion of application of Penman wind function by Cuenca, R. H., M. J.J. Nicholson. Journal of Irrigation and Drainage Engineering, ASCE, 109(2): 277–278.
747. Hargreaves, G. H., Samani, S. A. (1985). Reference crop evapotranspiration from temperature. Applied Engineering in Agriculture, ASAE, 1(2): 96–99.
748. Hargreaves, G. H., Samani, Z. A. (1986). World water for agriculture precipitation management. International Irrigation Center, Utah State University, Logan, UT, USA. Pages. 1–617.
749. Hargreaves, G. L., Hargreaves, G. H., Riley, J. P. (1985). Agricultural benefits for Senegal river basin. Journal of Irrigation and Drainage Engineering, ASCE, 111: 113–124.
750. Harmson, E. W., Caldero, J., Goyal, M., R. (2001). Consumptive water use estimates for pumpkin and onion at two locations in Puerto Rico. Proceedings of the Sixth Caribbean Islands Water Resources Congress. Editor: Walter F. Silva – Araya. University of Puerto Rico, Mayaguez, Puerto Rico.
751. Harmson, Eric W., Goyal, M. R., S. Torres – Justiniano, (2001). Comparison of potential evapotranspiration with the Hargreaves and Samani model and modified Penman – Monthlith model at selected locations in Puerto Rico. J. Agric. U. P.R. 86(1–2): 35–54.
752. Harmsen, E. W., Gonzaléz, A. (2005). Technical Note: A computer program for estimating crop evapotranspiration in Puerto Rico. J. Agric. Univ. P. R. 89(1–2): 107–113.
753. Harmsen, E. W., Mecikalski, J., Mercado, A., P. Tosado Cruz, (2010). Estimating evapotranspiration in the Caribbean Region using satellite remote sensing. Proceedings of the AWRA Summer Specialty Conference, Tropical Hydrology and Sustainable Water Resources in a Changing Climate. San Juan, Puerto Rico. August 30-September 1, (2010). 42–47.
754. Harmsen, E. W., Mecikalski, J., Cardona-Soto, M. J., A. Rojas González, Vásquez, R. (2009). Estimating daily evapotranspiration in Puerto Rico using satellite remote sensing. WSEAS Transactions on Environment and Development, 6(5): 456–465.
755. Harmsen, E. W. (2003). Fifty years of crop evapotranspiration studies in Puerto Rico. Journal of Soil and Water Conservation, 58(4): 214–223.
756. Harper-Lore, B., Wilson, M. (2000). *Roadside use of native plants*. Island Press. 1–665.
757. Hillel, D. (2003). *Introduction to environmental soil physics*. Academic Press. 1–494.
758. Harrison, L. P. (1963). Fundamentals concepts and definitions relating to humidity. In *Humidity and moisture*, ed. Wexler, A., Vol. 3. NY: Reinhold Publishing Co.
759. Hashemi, F., Habibian, M. T. (1979). Limitations of temperature based methods in estimating crop evapotranspiration in arid-zone agricultural development project. Agricultural Meteorology, 20: 237–247.
760. Hashemi, F., Habibian, M. T. (1979). Limitations of temperature based methods in estimating crop evapotranspiration in arid-zone agricultural development project. Agricultural Meteorology, 20: 237–247.
761. Hassan, Farouk A. (1999). Micro irrigation management and maintenance. Agro Industrial Management, 120–124.

762. Hatfield, J. L., Fuchs, M. (1990). Evapotranspiration models. In: *Management of farm irrigation systems*, eds. Hoffman, G. J., Howell, T. A., and Solomon, K. H., 33–59 pages. St. Joseph, MI: ASAE.
763. Hauck, R. D. (1984). Agrononic and technological approaches to improving the efficiency of nitrogen use by crop plants. Proc. Int. Symp. Nitrogen and the environment, Lahor, Pakistan, pages 317.
764. Haynes, R. J. (1988). Comparison of fertigation with broadcast applications of urea-N on levels of available soil nutrients and on growth and yield of trickle-irrigated peppers. *Scientia Horticulturae* 35: 189–198.
765. Haynes, R. J., and Swift, R. S. (1987). Effect of trickle fertigation with three forms of nitrogen on soil pH, levels of extractable nutrients below the emitter and plant growth. Plant and Soil, 102: 211–221.
766. Healy, M. G., Rodgers, M., Mulqueen, J. (2004). Recirculating sand filters for the treatment of synthetic dairy parlour washings. Journal of Environmental Quality, 33: 713–718.
767. Heldman R. H. (2003). Encyclopedia of agricultural, food and biological engineering: Drip irrigation. New York: Marcel Dekker. 206–211.
768. Heldman, R. H. (2003). Encyclopedia of agricultural, food and biological engineering. California Polytechnic State University. 4–20. New York: Marcel Dekker.
769. Hemond, Harold F., Fechner, L., Elizabeth, J. (1999). Chemical fate and transport in the environment. Academic Press, New York. Page 224.
770. Hensley, D., Deputy, J. (1999). Using tensiometers for measuring soil water and scheduling irrigation. College of Tropical Agriculture and Human Resources, University of Hawaii at Manoa. 644–667.
771. Henggeler, J. (1995). Use of drip irrigation on alfalfa. Proc. of Central Plains Irrigation Short Course and Equipment Exposition. February 7–8, 1995, Garden City, Kansas. Central Plains Irrigation Association, Colby, Kansas. pages 160–168.
772. Henggeler, J. C. (1995). A history of drip-irrigated cotton in Texas. In: *Micro irrigation for a Changing World: Conserving Resources/Preserving the Environment*, Proc. Fifth Intl. Micro irrigation Congress, ed. Lamm, F. R., 669–674. St. Joseph, MI: ASAE.
773. Henggeler, J., Kinnibrugh, J., Multer, W., Kight, D., Scott, R. (1996). Economic impact resulting from the adoption of drip irrigation cotton. College Station, TX: Texas Agricultural Extension Service, Texas A&M University.
774. Hensott, Blaine. *Drip Irrigation for Row Crops*. Division of Agricultural and Natural Resources, Publication No. 3376, University of California.
775. Heuvelink, E. (2005). Tomatoes: Irrigation. The Netherlands: Wageningen University. Page 171.
776. Helyar, K. R. (1976). Nitrogen cycling and soil acidification. Journal of the Australian Institute of Agricultural Science, 217–221 pages.
777. Helyar, K. R., Cregan, P. D., and Godyn, D. L. (1990). Soil acidity in New South Wales-current pH values and estimates of acidification rates. Australian Journal of Soil Research, 28: 523–537.
778. Hiaring, Peter E. (1987). Murphy -Goode vineyard in Alexander Valley: Going underground with drip. Wines and vines, January.
779. Higashitani, K., and Oshitani, J. (1998). Magnetic effects on thickness of adsorbed layer in aqueous solutions evaluated directly by atomic force microscope. J. Colloid Interf. Sci., 204: 363–368.
780. Higashitani, K., Kage, A., Katamura, S., Imai, K., and Hatade, S. (1993). Effects of magnetic field on formation of CaCO3 particles. J. Colloid Interf. Sci., 156: 90–95.
781. Hiler, E. A., Howell, T. A. (1973). Grain sorghum response to trickle and subsurface irrigation. Trans. of ASAE, 16(4): 799–803.
782. Hillel, D. (1982). *Introduction to soil physics*. New York: Academic Press.
783. Hills, D. J., and El-Ebaby, F. G. (1990). Evaluation of micro irrigation self-cleaning emitters. Journal of Applied Engineering in Agriculture. 6(4): 441–445.

784. Hills, D. J., Nawar, F. J., and Waller, P. M.,1989. Effect of chemical clogging on drip-tape irrigation uniformity. Trans. of ASAE, 32(4): 1202–1206.
785. Hills, D. J., Brenes, M. J. (2001). Micro irrigation of wastewater effluent using drip tape. Appl. Engr. Agric., 17(3): 303–308.
786. Hills, D. J., Tajrishy, M. A., Tchobanoglous, G. (2000). The influence of filtration on ultraviolet disinfection of secondary effluent for micro irrigation. Trans. ASAE 43(6)1499–1505.
787. Hills, D. J., Tajrishy, M. A. M., Gu, Y. (1989). Hydraulic considerations for compressed subsurface drip-tape. Trans. of ASAE, 32(4): 1197–1201.
788. Historical aspects of KSU-NWREC and agriculture plus tour of SDI facilities to Farm Bureau Masters Tour, Colby Kansas, Oct. 8, (2008).
789. Hla, Aung K., Thomas Scherer, F. (2003). Introduction to Micro-Irrigation. AE-1243, March.
790. Hochmuth, G. J., Locacio, S. J., Crocker, T. E., Stanley, C. D., Clark, G. A., and Parsons, L. R. (1993). Impact of micro irrigation on Florida Horticulture. HortTechnology, 3(2): 223–229.
791. Hoffman, G. J. (1986). Guidelines for reclamation of salt-affected soils. Appl. Agric. Res. 1(2): 65–72.
792. Hoffman, G. J., Jobes, J. A., Alves, W. J. (1983). Response to tall fescue to irrigation water salinity, leaching fraction, and irrigation frequency. Agricultural Water Management, 7: 439–456.
793. Hoffman, G. J., Howell, T. A., Solomon, K., H. (1990). *Management of farm irrigation systems*. ASAE Monograph. ASAE, St. Joseph MI. 1015 pages.
794. Hoffmann, C., Tarantino, A., Mongiovì, L. (2006). Thermal effects on response of high suction tensiometer. Proceedings 4th International Conference on Unsaturated Soils, April 2–5, Phoenix, USA, 423–456.
795. Hoitink, D. J., Burk, K. W., Ramsdell, J. V., Shaw, W. J. (2003). *Hanford Site Climatological Data Summary 2002 with Historical Data*. PNNL-14242. Pacific Northwest National Laboratory, Richland, WA. 645–665.
796. Holcombe, G. (2007). Subsurface drip irrigation offers water-saving alternatives. KSRE Report to Kansas Legislature. pages 26–27.
797. Holzapfel, E. A., C. Lopez and J. P.Joublan, (2001). Effect of water and fertigation on development and production of Thompson Navel (Spanish). Agricultura Técnica, 61(1): 51–60.
798. Hook, W. R., T.Ferre, P. A., Livingston, N. J. (2004). The effects of salinity on the accuracy and uncertainty of water content measurement. Soil Sci. Soc. of America J., 68: 47–56.
799. Hopen, H. J., Oebkar, N. F. (1976). Vegetable crop responses to synthetic mulches: An annotated bibligraphy. National Agricultural Plastics Association Tech. Bull. #1, U. S.A.
800. Hopen, J. H., Oebker, N. F. (1976). Vegetable crop responses to synthetic mulches. Special publication # 42. University of Illinois.
801. Horn, B. (2001). Annual industry survey, (2001). In: *Australian processing tomato grower*. 22: 4–7. The Australian Processing Tomato Research Council Inc., Shepparton, Vic.-AU.
802. Horne, D. W., Thorne, M. D. (1979). *Soil water and crop production*. Wesport, CT: AVI Publishing Company.
803. Horwatich, J. A., Corsi, S. R., Bannerman, R. T. (2004). Effectiveness of a pressurized stormwater, filtration system in green bay. U. S. Geological Survey Reston, VA.
804. Hosted Li, G. (2007). China Agricultural University, Beijing, China, July 12, for discussions and facility tours concerning SDI.
805. Hottel, H. C. (1976). A simple model for estimating the transmittance of direct solar radiation through clear atmospheres. Solar energy, 18: 129.
806. House, E. B. (1998). Irrigation by means of underground porous pipe. Bulletin 240. Fort Collins, Colo.: Colorado Agricultural Experiment Station.
807. Howell T. A., Meron, M. (2007). Irrigation Scheduling. In: *Micro irrigation for crop production,* (Lamm, F. R., Ayars, J. E., Nakayama, F. S., eds.). Elsevier.
808. Howell, T. A. (2000). Irrigation's role in enhancing water use efficiency. In Proc. 4th Decennial Natl. Irrigation Symp., 66–80. eds. Evans, R. G., Benham, B. L., T. P. Trooien. St. Joseph, MI: ASAE.

809. Howell, T. A., Schneider, A. D., Evett, S. R. (1997). Subsurface and surface micro irrigation of corn, Southern High Plains. Trans. ASAE, 40(3): 635–641.
810. Howell, T. A., Bucks, D. A., Chesness, J., L. (1980). Advances in trickle irrigation. Proc. 2nd National Irrigation Symposium, pages 69–94. St. Joseph, MI: ASAE.
811. Howell, T. A., Stevenson, D. S., Aljibury, F. K., Gitlin, H. M., I.-Wu, P., Warrick, W., Raats, P. A., C. (1983). Chapter 16: Design and operation of trickle (drip) systems. In: *Design and Operation of Farm Irrigation Systems*, 663–717 pages. ASAE Monograph No. 3. St. Joseph, MI.
812. Howell, T. A., Hiler, E., A. (1974). Trickle irrigation lateral design. Transactions American Society of Agricultural Engineers, 17(5): 902–908.
813. Howell, T. A., Hiler, E. A. (1974). Designing trickle irrigation laterals for uniformity. Journal of the Irrigation and Drainage Division, ASCE, 100 (IR4): 443–454, Paper 10983.
814. Howell, T. A. (2001). Enhancing water use efficiency in irrigated agriculture. Agronomy Journal, 93(4): 281–289.
815. Howell, T. A., Schneider, A. D., Dusek, D. A. (2002). Effects on furrow diking on corn response to limited and full sprinkler irrigation. Soil Sci. Soc. of America J., 66(1): 222–227.
816. Howell, T. A., Schneider, A. D., and Jensen, M. E. (1991). History of lysimeter design and use for evapotranspiration measurements. In: *Lysimeters for evapotranspiration and environmental measurements*, eds. Allen, R. G., Howell, T. A., Pruitt, W. O., Walter, L. A., Jensen, M. E., 1–9. New York, NY: ASCE.
817. Howell, T. A., Bucks, D. A., Goldhamer, D. A., Lima, J. M. (1986). Management Principles in Irrigation Scheduling. Chapter 4, In: *Trickle irrigation for crop production: Design, operation and management*, eds. Nakayama, F. S., Bucks, D. A., Elsevier Publisher: New York.
818. Howell, T. A., Hatfield, J. L., H. Yamada and Davis, K. R. (1984). Evaluation of cotton canopy temperature to detect crop water stress. Trans. of ASAE, 27(1): 84–88.
819. Howell, T. A., Meron, M., Davis, K. R., Phene, C. J., Yamada, H. (1987). Water management of trickle and furrow irrigated narrow row cotton in the San Joaquin Valley. Appl. Eng. Agric., 3: 222–227.
820. Howell, T. A., Evett, S. R., Tolk, J. A., Schneider, A. D., Steiner, J. L. (1996). Evapotranspiration of corn: Southern High Plains. In *Evapotranspiration and irrigation scheduling*, eds. Camp, C. R., Sadler, E. J., and Yoder, R. E., 158–166 pages by ASAE.
821. Huang, T. J., Chang, Y. H., Yang, P. C., Fang, Y. T. (1982). Studies on subsurface drip irrigation for coarse-texture sugarcane field in Taiwan. In Report of the Taiwan Sugar Research Institute, 13–32. Taiwan, Taiwan: Taiwan Sugar Research Institute.
822. Huang, Z, (1989). The research and application of plastic films in China. Beijing – China: Chinese Plastics Mulch Research Association.
823. Huang, Z. B., Assouline, S., Zilberman, J., Ben-Hur, M. (2000). Tillage and saline irrigation effects on water and salt distribution in a sloping field. Soil Sci. Soc. of America J., 64(1): 2096–2102.
824. Hubbell, J. M., Sisson, J., B. (2004). Comments on tensiometer modification for diminishing errors due to the fluctuating inner water column.Soil Sci. Soc. of America J., 68(2): 709.
825. Huisman, J. A., Weerts, A. H., Heimovaara, T. J., Bouten, W. (2002). Comparison of travel time analysis and inverse modeling for soil water content determination with time domain reflectometry. Water Resources Research, 38: 13–18.
826. Hunt, L. A., Reynolds, M. P., Sayre, K. D., Rajaram, S., White, J. W., Yan, W. (2003). Crop modeling and the identification of stable coefficients that may reflect significant groups of genes, Agronomy Journal, 95: 20–31.
827. Hunt, P. G., Bauer, P. J., Camp, C. R., Matheny, T. R. (1998). Nitrogen accumulation in cotton grown continuously or in rotation with peanut using subsurface micro-irrigation and GOSSYM/COMAX management. Crop Sci., 38: 410–415.
828. Hutchinson, P. A. (2006). Routine measurement of the soil water potential gradient near saturation using a pair of tube tensiometers. Australian Journal of Soil Research, 39(5): 1147–1156.

829. Hutmacher, R. B., Phene, C. J., Mead, R. M., Clark, D., Shouse, P., Vail, S. S., Swain, R., M. van Genuchten, Donovan, T., Jobes, J. (1992). Subsurface drip irrigation of alfalfa in the Imperial Valley. Proc. 22nd California/Arizona Alfalfa Symposium 22: 20–32.
830. Hutmacher, R. B., Phene, C. J., Davis, K. R., Kerby, T. A. (1993). Acala and pima cotton responses to subsurface drip irrigation: Water use, plant water relations, and yield. In Proceedings of the Beltwide Cotton Conference, 1221–1224 pages. Memphis, TN: National Cotton Council.
831. Hutmacher, R. B., Phene, C. J., Davis, K. R., Vail, S. S., Kerby, T. A., Peters, M., Hawk, C. A., Keeley, M., Clark, D. A., Ballard, D., Hudson, N. (1995). Evapotranspiration, fertility management for subsurface drip acala and pima cotton. In Proceedings of the 5th International Micro irrigation Congress, ed. Lamm, F. R., 147–154 pages. St. Joseph, MI: ASAE.
832. Hutmacher, R. B., Phene, C. J., Mead, R. M., Clark, D., Shouse, P., Vail, S. S., Swain, R., M. van Genuchten, Donovan, T., Jobes, J. (1992). Subsurface drip irrigation of alfalfa in the Imperial Valley. In Proceedings of the 22nd California/Arizona Alfalfa Symposium, 22: 20–32, University of California and University of Arizona Cooperative Extension, Holtville, CA, December 9–10.
833. Hutmacher, R. B., Vail, S. S., Muthamia, J. G., V. Mwaja and Liu, R. C. (1985). Effect of trickle irrigation frequency and installation depth on tomato growth and water status. In Proceedings of the 3rd International Drip/Trickle Irrigation Congress, 2: 798–804. St. Joseph, MI: ASAE.
834. Hydrological Processes, 14: 785–809.
835. IAEA- TECDOC-875. Nuclear methods for plant nutrients and water balance studies, (1996).
836. Ibáñez, M., Castellví, F. (2000). Simplifying daily evapotranspiration estimates over short full-canopy crops.Soil Sci. Soc. of America J., 64: 628–632.
837. Ibrahim, M. A.M., El-Sirafy, Z. M., El-Gohary, S. A., Willardson, L. S. (1997). Interactive effect of irrigation and nitrogen fertilization on cotton, soil, and groundwater nitrogen. Comm. Soil Sci. Plant Anal., 28: (1/2): 173–187.
838. Ibrahim, A. M., Abd El-Samad, G. A. (2009). Effect of different irrigation regimes and partial substitution of N-Mineral by organic manures on water use, growth and productivity of Pomegranate Trees. European Journal of Scientific Research, 38(2), pages 199–218.
839. Idso, S. B., Jackson, R. B. (1969). Thermal radiation from the atmosphere. Journal of Geophysical Research, 74: 5397–5403.
840. Improved irrigation water management with SDI. KSU Sedgewick County SDI Field Day, Wichita, Kansas, June 1, (2005).
841. Improving water productivity through micro-irrigation in arid punjab regions. Central Institute of Post Harvest Engineering and Technology(CIPHET) (ICAR), Malout Hanumangarh bye Pass, Abohar District: Fazilka(Punjab) 152116, India.
842. INCID, (1994). Drip Irrigation in India. Indian National Committee on Irrigation & Drainage (INCID), Ministry of Water Resources, Govt., of India, New Delhi.
843. INCID (1998). Sprinkler Irrigation in India. Indian National Committee on Irrigation & Drainage, Ministry of Water R esources, Govt., of India, New Delhi.
844. IPCC, (1996). Greenhouse gas emissions from agricultural soils. In: Revised IPCC guidelines for national greenhouse gas inventories: Reference manual. Intergovernmental Panel on Climate Change (IPCC), Geneva.
845. IPCC, (2007). *Climate Change 2007: Impacts, adaption and vulnerability*. Cambridge University Press, Cambridge, UK.
846. Irrigating through grain filling stage. In: Pioneer Hi-Bred Crop Insights, April, (2008). Pioneer Hi- Bred Crop Insights, April, (2008).
847. Irrigation Association, (2001). Glossary of Irrigation Terms.
848. Irrigation systems. The Journal of Agricultural Faculty of Ege University, 31(2): 177–184.
849. Irrigation: predominant bacteria in treated Colorado River water and biologically clogged emitters. Irrigation Science, 3: 123–132.
850. *Irrometer moisture indicator – reference book*. Riverside, CA: Irrometer Co.

851. Irtwange, S. V. (2006). Application of biological control agents in pre- and post harvest operations. Agricultural Engineering International: the CIGR Ejournal. Invited Overview No. 3, VIII: 1–10.
852. Isbell, R. F. (1996). *The Australian Soil Classification*. CSIRO, Melbourne, Vic. – AU.
853. Isbell, R. F., McDonald, W. S., Ashton, L. J. (1997). Concepts and Rationale of the Australian Soil Classification. CSIRO Land and Water, Canberra – AU.
854. Isobe, M. (1972). Agronomic research in subsurface and drip irrigation. In: 31st Report Hawaii Sugar Technology Annual Conference, 23–26 pages. Honolulu, Hawaii: Hawaiian Sugar Planters' Association.
855. Israelson, W., Hansen, V. E. (1965). Principles and applications of the irrigation. S. A. Barcelona-Madrid: Reverte Editorial.
856. Itier, B. (1996). *Measurement and estimation of evapotranspiration*. In: Sustainability of irrigated agriculture, eds. Pereira, L. S., Feddes, R. A., J. R.Gilley, Leseffr, B., 171–191 pages. Dordrecht: Kluwer Academic Publishers.
857. Itier, B., Perrier, A. (1976). Présentation d'une étude analytique de l'advection: I. Advection liée aux variations horizontales de concentration et de température. Annual Agronomy, 27(2): 111–140.
858. Itier, B., Brunet, Y., Mcaneney, K. J., Lagouarde, J. P. (1994). Downwind evolution of scalar fluxes and surface resistance under conditions of local adection. Part I: A reappraisal of boundary conditions. Agricultural and Forest Meteorology, 71: 211–255.
859. Ivahnenko, T., Bundash, J. (2004). Chloroform in the hydrologic system. Journal of Engineering Mechanics, pages 20–24.
860. Jackson, R. D. (1982). Canopy temperature and crop water stress. In: Advances in irrigation, ed. D. I. Hillel. I: 43–85. New York: Academic Press.
861. Jain, B. H. (2003). A Telling tale, Jain Irrigation Systems Ltd., Jalgaon, pages 1–112.
862. Jain, N., Chauhan, H. S., Singh, P. K., Shukla, K. N. (2000). Response of tomato under drip irrigation and plastic mulching. In: Proceeding of 6th International Micro-irrigation Congress, Micro-irrigation Technology for Developing Agriculture. 22–27 October, South Africa.
863. Janat, M., and Stroehlein, J. L. (1986). Response of drip irrigated cotton to phosphorus fertilizer. Fertilizer issues J., 3(4).
864. Jayasuriya, S. K., Shand, R. T. (1985). Technical change and labor absorption in Asian agriculture: some recent trends. World Development, 14(3): 415–428.
865. Jenkinson, B. J., Franzmeier, D. P., Lynn, W. C. (2002). Soil hydrology on an end moraine and a dissected till plain in West-Central Indiana.Soil Sci. Soc. of America J., 66: 1367–1376.
866. Jensen, J. R. (1988). Effect of asymmetric, daily air temperature and humidity waves on calculation of reference evapotranspiration. Proceedings of the European Economic Community Workshop.
867. Jensen, M. E. (1980). *Design and Operation of Farm Irrigation Systems*. ASAE Monograph #3, American Society of Agricultural Engineers, Chapter 6: 189–225.
868. Jensen, M. E., Burman, R. D., Allen, R., G. (1990). *Evapotranspiration and irrigation water requirements*. ASCE Manuals and Reports on Engineering Practice No. 70. American Society of CivilEngineers, New York 332 pages.
869. Jensen, M. E., Haise, H. R. (1963). Estimating evapotranspiration from solar radiation. Journal of Irrigation and Drainage Engineering Division, ASCE, 89: 15–41.
870. Jensen, M. E. (1968). Water consumption by agricultural plants. In: *Water deficits and plant growth*, ed. Kozlowski, T., 1–22. New York: Academic Press.
871. Jensen, M. E. (1974). Consumptive use of water and irrigation water requirements. A Report prepared by the Technical Committee on Irrigation Water Requirements. Irrigation and Drainage Division, ASCE, 227.
872. Jensen, M. H., Malter, A. J. (1994). Protected agriculture—A global review. World Bank technical paper no. 253. The World Bank, Washington, DC.
873. Jensen, M. H. (1965). Concluding results of air-supported row covers for early vegetable production. Proceedings of National Agriciulture Plastics Congress, 6: 100–112.

874. Jensen, M. H. (1967). A new approach to high yields. Am. Vegetable Grower, 15(2): 16, 23–24.
875. Jensen, M. H., M. Valenzuela and Fangmeier, D. D. (1998). Using non-woven floating covers on summer squash for exclusion of whitefly—transmited gemini viruses. Proceedings of National Agriciulture Plastics Congress, 27: 159–164.
876. Ji, S., Unger, P. W. (2001). Soil water accumulation under different precipitation, potential evaporation and straw mulch conditions. Soil Sci. Soc. of America J., 65(2): 442–448.
877. Jiménez, C. C., Tejedor, M., F. Diaz and Rodríguez, C. M. (2005). Effectiveness of sand mulch in soil and water conservation in an arid region, Lanzarote, Canary Islands, Spain. Journal of Soil and Water Conservation, 60(1): 63–67.
878. Jnad, I., Lesikar, B., Kenimer, A., Sabbagh, G. (2001). Subsurface drip dispersal of residential effluent: I. Soil chemical characteristics. Trans. ASAE, 44(5): 1149–1157.
879. Jnad, I., Lesikar, B., Kenimer, A., G. Sabbagh. 2001b. Subsurface drip dispersal of residential effluent: II. Soil hydraulic characteristics. Trans. ASAE, 44(5): 1159–1165.
880. Johnson, D. O., Arriaga, F. J., Lowery, B. (2005). Automation of a falling head permeameter for rapid determination of hydraulic conductivity of multiple samples. Soil Sci. Soc. of America J., 69: 828–833.
881. Johnson, J., Spears, F., Wells, R. (2004). Disease management in overhead sprinkler and subsurface drip irrigation systems for peanut. Agronomy Journal, 96: 1058–1065.
882. Jones, F. E. (1992). *Evaporation of water*. Chapter 9: 123–140. CRC Press.
883. Jones, J. W., Allen, L. H., Shih, S. F., Rogers, J. S., Hammond, L. C., Smajstrala, A. G., Martsolf, J. D. (1984). Estimated and measured evapotranspiration for Florida climatic, crops and soils. Agricultural Experiment Station, Institute of Food and Agricultural Sciences, University of Florida, Gainesville. Bulletin No. 840, 1–65.
884. Jones, S. B., Or, D. (2004). Frequency domain analysis for extending time domain reflectometry water content measurement in highly saline soils. Soil Sci. Soc. of America J., 68: 1568–1577.
885. Jordan, J. E., White, R. H., Victor, D. M., Hale, T. C., Thomas, J. C., Engelke, M. C. (2003). Effect of irrigation frequency on turf quality, shoot density and root length of five bentgrass cultivars. Crop Science, 43(1): 282–287.
886. Jordan, W. R., Ritchie, J. T. (1971). Influence of soil water stress on evaporation, root absortion and internal water status of cotton. Plant Physiology, 48: 783–788.
887. Jorgenson, G. S., Norum, K., N. (1992). Subsurface drip irrigation B theory, practices and application. Conference proceedings sponsored by California State University-Fresno and USDA ARS-Water Management Research Laboratory. CATI Publication No. 92–1001, CSUF, Fresno, CA. 212 pages.
888. Julich, W., Schubert, J. (2000). Proceedings of the International Riverbank Filtration. Dusseldorf-Germany. Boston: Kluwer Academic Publishers.
889. Kabat, P., B. J. van den Broek and Feddes, R. A. (1992). SWACROP: A water management and crop production simulation model. ICID Bulletins, 41(2): 61–84.
890. Kabutha, C, Blank, H., B. Van Koppen, (2000). Drip irrigation kits for small-holder farmers in Kenya: Experience and a way forward. Paper presented at the 2000 Micro Irrigation Conference, Cape Town, South Africa. 1–8.
891. Kadlec, R. H. (2006). Water temperature and evapotranspiration in surface flow wetlands in hot arid climate. pages 1–12. Elsevier.
892. Kaleita, A. L., Tian, L., Yao, H. (2003). Soil moisture estimation from remotely sensed hyperspectral data. ASAE Paper No: 031047. Presented at Annual International Mtg. of ASAE, Las Vegas, Nevada. July 27–30.
893. Kalfountzos, D., Alexiou, I., Kotsopoulos, S., Zavakos, G., Vyrlas, P. (2007). Effect of subsurface drip irrigation on cotton plantations. Water Resource Management 21: 1341–1351.
894. Kamal, G. Y., and Hwat, B. S. (2001). The relationship between leaf water potential and stem diameter in Sorghum. Agron. J., 93: 1341–1343.
895. Kang, Y., Jun, H. (2002). Effect of sprinkler irrigation on field micro-climate. Paper presentation at ASAE Annual Meeting. 1–15.

896. Kang, Y., Nishiyama, S. (1995). Hydraulic analysis of micro irrigation submain units. Trans. ASAE, 38(5): 1377–1384.
897. Kang, Y., Nishiyama, S. (1996). A simplified method for design of micro irrigation laterals. Trans. ASAE 39(5): 1681–1687.
898. Kang, Y., Nishiyama, S. (1996). Analysis of micro irrigation systems using a lateral discharge equation. Trans. ASAE, 39(3): 921–929.
899. Kangrang, A., Chaleeraktrakoon, C. (2007). A Fuzzy-GAs model for determining varied irrigation efficiency. American Journal of Applied Sciences 4(6): 339–345.
900. Karmeli, D. (1977). Classification and flow regime analysis of drippers. Journal of Agricultural Engineering Research. 22(2): 165–173.
901. Kasperbauer, M. J. (2000). Strawberry yield over red vs. black plastic mulch. Crop Science Journal, 40(1): 171–174.
902. Katerji, N., J. W. van Hoorn, Hamdy, A., Mastrorilli, M. (2003). Salinity effect on crop development and yield analysis of salt tolerance according to several classification methods. Agric. Water Manage., 62: 37–66.
903. Katerji, N., J. W. van Hoorn, Hamdy, A., Mastrorilli, M. (2004). Comparison of corn yield response to plant water stress caused by salinity and by drought. Agric. Water Manage., 65: 95–101.
904. Katerji, N., Perrier, A. (1983). Modélization de l'évapotranspiration réelle ETR d'une parcelle de luzerne: rôle d'un coefficient cultural. Agronomie, 3(6): 513–521 (in French).
905. Kay, M. (1988). *Sprinkler Irrigation: Equipment & Practice*. BT Batsford Limited, London.
906. Keller, J. (1990). Modern Irrigation in Developing Countries. Fourteenth Congress on Irrigation and Drainage of ICID, Brazil, 113–138.
907. Keller, J., and Bliesner, R. D. (1990). *Sprinkler and trickle irrigation*. An AVI book. New York.
908. Keller, J., Karmelli, D. (1974). Trickle irrigation design parameters. Trans. of ASAE.
909. Keller, J. (2000). Sprinkler and trickle irrigation. New York: Van Nostrand Reinhold'.
910. Keller, J., Bliesner, R. D. (2000). Sprinkler and trickle irrigation. Caldwell, NJ: Blackburn Press.
911. Keller, J., Bliesner, R. D. (2002). Sprinkler and trickle irrigation. Blackburn Press.
912. Keller, J., Bliesner, R. D. (2004). Sprinkler and trickle irrigation. New York (USA): Van Nostrand Rienhold.
913. Khasawneh, F. E., Sample, E. C., Hashimoto, I. (1974). Reactions of ammonium ortho- and polyphosphate fertilizers in soil: I. Mobility of phosphorus. Soil Science Society of America Proceedings 38: 446–450.
914. Khodabandeh, N. (2006). Agriculture Industry Plants. Tehran, Sepehr Press, pages 19–29.
915. Khurshid, K., Iqbal, M., Arif, M. S., Nawaz, A. (2006). Effect of tillage and mulch on soil physical properties and growth of maize. Int. J. Agric. Biol., 5: 593–596.
916. Kidder, G., Hanlon, E. A., Jr, (1997). Neutralizing excess bicarbonates from irrigation water. Univ. of Florida, Coop. Ext. Ser., SL-142. 7 pages.
917. Kijne, J. W., Barker, R., and Molden, D. (eds.). *Comprehensive Assessment of Water Management in Agriculture Series* 1. CABI/IWMI, Wallingford/Colombo; 251–271 pages.
918. Kincaid, D. E., Heerman, D. F. (1974). Scheduling irrigation using a programmable calculator. Publication ARS-NC-12. Washington D. C.: USDA Agricultural Research Service.
919. King, B. A., Wall, R. W., Kincaid, D. C., Westermann, D., T. (2005). Field testing of a variable rate sprinkler and control system for site-specific water and nutrient application. Applied Engineering in Agriculture, 21(5): 847–853.
920. Kite, G. W., Droogers, P. (2000). Comparing evapotranspiration estimates from satellites, hydrological models and field data. Journal of Hydrology, (Amsterdam), 229: 3–18.
921. Kiwifruit. Postharvest Biology and Technology, 16: 233–243.
922. Kizer, M. A., Elliott, R. L., Stone, J. F. (1990). Hourly ET model calibration with eddy flux and energy balance data. Journal of Irrigation and Drainage Engineering, ASCE, 116(2): 172–181.
923. Klein, I., Bar-Yosef, B., Assaf, R., Berkovitz, A. (1989). Drip nitrogen fertigation of "Starking Delicious" apple trees. Plant and Soil, 119: 305–314.

924. Klocke, N. L., Stone, L. R., Clark, G. A., Dumler, T. J., Briggeman, S. (2006). Water Allocation Model for Limited Irrigation. Applied Engineering in Agriculture, 22(3): 373–381.
925. Knapp, K. C. (1993). Economics of irrigation system investment. In: *Subsurface Drip Irrigation-Theory, Practices and Application*, 127–139 pages. CATI Pub. No. 92–1001. Fresno, CA: California State University.
926. Knox, J. W., Weatherhead, E. K. (2003). Trickle irrigation in England and Wales. Bristol-England: R&D Technical Report W6–070/TR Environment Agency.
927. Kolar, J. J., Kohl, R. A. (1976). Irrigating alfalfa for seed production. University of Idaho Agricultural Experiment Station Current Information Series 357, Moscow, Idaho. 3.
928. Koo, R. C.J., Smajstrla. A. G. (1985). Effects of trickle irrigation and fertigation on fruit production and juice porduction and juice quality of 'Valencia' orange. Proc. Fla. State Hort. Soc., 97: 8–10.
929. Keeney, D. R., and Nelson, D. W. (1982). Nitrogen: inorganic forms. 672–675 pages. In: *Agronomy Monograph No. 9* by ASA and SSSA: Madison, WI.
930. Korea Meteorological Administration (KMA), (2008). Understanding of climate change and application of climate change scenarios. Ministry of Construction and Transportation, Comprehensive Water Resources Plan-Water Vision, (2020).
931. Krebs, C. (2004). Researcher helps row crop farmers adopt drip irrigation. Ag. J., 57(25): 6, 19.
932. Kremer, S. A., Feigein, Mitchnick, Z. (1979). Management of nitrogen fertilization in trickl- irrigated onions grown in the Arava. Hassadeh. 59(10): 2021–2025.
933. Kruse E. G., Haise, H. R. (1974). Water use by native grasses in high altitude Colorado meadows. USDA Agricultural Research Service, Western Region report ARS-W-6–1974. 60.
934. Kruse, E. G., Israeli, I. (1987). Evaluation of a subsurface dripirrigation system. ASAE Paper No. 87–2034. St. Joseph, MI: ASAE.
935. Kuchaki, A. (1996). Agricultural Crops in Arid Regions. Ferdowsi University of Mashhad, Jihad Press, Mashhad,. 61 pages.
936. Kumar, Santosh, Singh, K. G., and Chetan Singla, (2013). Development and evaluation of low head and minimum pressure drop equipment for fertilizer application through drip irrigation system. The Indian Journal of Agricultural Sciences, 83(2).
937. Kumar, Satyendra Kumar, Rajbir Singh, Kaledhonkar, M. J., Nangare, D. D., and Ashwani Kumar, (2013). Improving water productivity through micro-irrigation in arid punjab regions. 62(3): 330–339.
938. Ladaniya, M. (2008). Fruit quality control, evaluation and analysis. In: *Citrus fruit: biology, technology and evaluation*, pages 475–500. Elsevier Academic Press., San Diego CA, USA.
939. Laegreid, M., Bockman, O. C., Kaarstad, O. (1999). Agriculture fertilizers and environment, CABI Publ., Wallingford and Norsk Hydro ASA, Oslo. 5–12.
940. Laflen, J. M., J. Tian and Huang, C. H. (2000). *Soil erosion and dryland farming.* Florida – USA: CRC Press. 1–744.
941. Lamm, F. R. (Ed.), (1995). Micro irrigation for a Changing World: Conserving Resources/Preserving the Environment. Proc. Fifth Int'l Micro irrigation Congress, Apr. 2–6, 1995, Orlando, FASAE, L., St. Joseph MI. 978 pages.
942. Lamm, F. R. (1995). Narrow corn row spacing for cost effective, water efficient subsurface drip irrigation. Kansas Water Resources Research Institute, Manhattan, KS. KWRRI Contribution No. 316. 20 pages.
943. Lamm, F. R. (1996). Narrow corn row spacing for cost effective, water efficient subsurface drip irrigation, 1994–1995. KS. Water Resources Research Institute, Manhattan, KS. KWRRI Contribution No. 323. 19 pages.
944. Lamm, F. R. (1996). Subsurface drip irrigation for corn. Presentation at the annual meeting of the Ground Water Management Districts Association, Las Vegas, NV, Dec. 9–11.
945. Lamm, F. R. (1997). Subsurface drip irrigation for corn and KSU's bed management system for SDI. Presentation at the semi-annual meeting of the Groundwater Management Districts Assn, Colby, KS, June 4, (1997).

946. Lamm, F. R. (1999). SDI for corn production in Kansas. Israeli Volcani Institute, Bet Dagan, Israel, March 22.
947. Lamm, F. R. (2000). Design and management considerations for SDI systems. Crop Modeling Workshop, Univ. of Nebraska, Lincoln, NE, Dec. 4–7.
948. Lamm, F. R. (2000). SDI for corn production in Kansas. USDA-RRF, W-128 Micro irrigation Committee, Mauii, Hawaii, Sep. 19–22.
949. Lamm, F. R. (2002). Advantages and disadvantages of subsurface drip irrigation. In Proc. International Meeting on Advances in Drip/Micro Irrigation, Puerto de La Cruz, Tenerife, Canary Islands, December 2–5. Instituto Canario de Investigaciones Agrarias, Canary Islands. 13 pages.
950. Lamm, F. R. (2003). KSU research for corn production using SDI: 14 years of progress. In proceedings of the Central Plains Irrigation Conference, Colby, KS, Feb. 4–5, (2003). CPIA, 760 N.Thompson, Colby, KS. pages 230–245.
951. Lamm, F. R. (2004). Comparison of SDI and Simulated LEPA Sprinkler Irrigation for Corn. In Proc. Irrigation Assn. Int'l. Irrigation Technical Conf., November 14–16, 2004, Tampa, FL. Irrigation Assn., Falls Church, VA. IA Paper No. IA04–1098. 475–485.
952. Lamm, F. R. (2005). SDI for conserving water in corn production. In Proc. ASCE-EWRI Water Congress, May 15–19, 2005, Anchorage, AK. 12 pages.
953. Lamm, F. R. (2006). Progress with SDI research at Kansas State University. In: Proc. Central Plains Irrigation Conference, Colby, KS, Feb. 21–22, (2006). CPIA, 760 N.Thompson, Colby, KS. pages 67–85.
954. Lamm, F. R. (2009). A look at twenty years of SDI research in Kansas. In: Proc. Central Plains Irrigation Conference, Colby, KS., Feb. 24–25, (2009). CPIA, 760 N.Thompson, Colby, KS. pages 152–176.
955. Lamm, F. R. (2009). Unique challenges with subsurface drip irrigation. ASAE paper no. 095927. ASABE, St. Joseph, MI. 25 pages.
956. Lamm, F. R., A. A. Abou Kheira, (2009). Corn irrigation macromanagement at the seasonal boundaries – Initiating and terminating the irrigation season. In: Proc. Central Plains Irrigation Conference, Colby, KS., Feb. 24–25, (2009). CPIA, 760 N.Thompson, Colby, KS. pages 40–60.
957. Lamm, F. R., A. A. Abou Kheira, (2008). Effects of early-season water stresses on corn production. Proc. 29th Annual Int'l. Irrigation Assoc. Tech. Conf., Anaheim, California, Nov. 2–4. Irrigation Association, Falls Church, VA. Paper No IA09–1052. 10 pages.
958. Lamm, F. R., A. A. Abou kheira, Trooien, T., P. (2010). Sunflower, soybean, and grain sorghum crop production as affected by dripline depth. Applied Engr. in Agric., 26(5): 873–882.
959. Lamm, F. R., Aiken, R., M. (2008). Comparison of Temperature-Time Threshold-and ET-based Irrigation Scheduling for Corn Production. ASABE paper no. 084202. ASABE, St. Joseph, MI. 12 pages.
960. Lamm, F. R., Aiken, R., M. (2005). Effect of irrigation frequency for limited subsurface drip irrigation of corn. Proc. Irrig. Assoc. Int'l. Tech. Conf., Nov. 6–8, 2005, Phoenix, AZ. Irrigation Assoc., Falls Church, VA. IA Paper No. IA05–1264. 10 pages.
961. Lamm, F. R., Ayars, J. E., F. S. Nakayama (Eds.), (2007). *Micro irrigation for Crop Production: Design, Operation and Management.* Elsevier Publications. 608 pages.
962. Lamm, F. R., Bordovsky, J. M., Schwankl, L. J., Grabow, G. L., Enciso-Medina, J., Peters, T. R., Colaizzi, P. D., Trooien, T. P., Porter, D., O. (2010). Subsurface Drip Irrigation: Status of the Technology in, (2010). Proc Fifth Nat'l. Irrigation Symp., Phoenix, Dec. 5–8. ASABE, St Joseph Michigan. 14 pages.
963. Lamm, F. R., Camp, C., R. (2007). Chapter 13: Subsurface drip irrigation. In *Micro irrigation for Crop Production: Design, Operation, and Management*, 473–551. Lamm, F. R., Ayars, J. E., Nakayama, F. S., eds. Amsterdam, The Netherlands: Elsevier.
964. Lamm, F. R., Clark, G. A., Yitayew, M., Shoneman, R. A., Mead, R. M., and Schneider, A. D. (1995). Installation issues for SDI systems. In proceedings of the International Exposition and Technical Conference, Phoenix, AZ, Nov. 12–14. IA, Fairfax, VA. pages 29–35.

965. Lamm, F. R., Clark, G. A., Yitayew, M., Shoneman, R. A., Mead, R. M., and Schneider, A. D. (1997). Installation issues for SDI systems. Presented at the 1997 Int'l. ASAE meeting. Paper no. 972074, ASAE, St. Joseph, MI. 6 pages.
966. Lamm, F. R., Harmoney, K. R., Aboukheira, A. A., Johnson, S., K. (2012). Alfalfa production with subsurface drip irrigation in the Central Great Plains. Trans. of the ASABE, 55(4): 1203–1212.
967. Lamm, F. R., Lawless, J. R., and Sunderman, H. D. (1990). Agricultural Research. Progress Report 602, June, (1990). KAES, Manhattan, KS. 78 pages.
968. Lamm, F. R., and Manges, H. L. (1991). Nitrogen fertilization for drip irrigated corn in northwest Kansas. The international winter meeting of the ASAE, Chicago, December 17–20. ASAE paper no. 912596, St Joseph, MI, USA.
969. Lamm, F. R., Manges, H. L., Rogers, D. H., Spurgeon, W. E., and Farmer, M. H. (1990). Design and installation of a drip irrigation system for research purposes. Presented at the international winter meeting of the ASAE, Chicago, December 18–21. ASAE paper no. 902530, St. Joseph, MI, USA. 21 pages.
970. Lamm, F. R., Manges, H. L., Stone, L. R., and Khan, A. H. (1991). Water requirement of drip-irrigated corn in northwest Kansas. Presented at the 1991 Int'l. winter ASAE meeting. Paper no. 902591, ASAE, St. Joseph, MI. 20 pages.
971. Lamm, F. R., Manges, H. L., Stone, L. R., Khan, A. H., and Rogers, D. H. (1995). Water requirement of subsurface drip-irrigated corn in north- west Kansas. Trans ASAE 38(2): 441–448.
972. Lamm, F. R., D. M. O'Brien, Rogers, D. H., Dumler, T., J. (2002). Sensitivity of center pivot sprinkler and SDI comparisons. Proc. Irrigation Assn. Int'l. Irrigation Technical Conf., Oct. 24–26, New Orleans, LA. Irrigation Assn., Falls Church, VA.
973. Lamm, F. R., D. M. O'Brien, Rogers, D. H., Dumler, T., J. (2009). Using the K-State center pivot sprinkler and SDI economic comparison spreadsheet. Proc. Central Plains Irrigation Conference, Colby, KS., Feb. 24–25, (2009). CPIA, 760 N.Thompson, Colby, KS. pages 177–187.
974. Lamm, F. R., D. M. O'Brien, Rogers, D. H., and Dumler, T. J. (2003). Center pivot sprinkler and SDI economic comparisons. In proceedings of the Central Plains Irrigation Conference, Colby, KS, Feb. 4–5, (2003). CPIA, 760 N.Thompson, Colby, KS. pages 199–211.
975. Lamm, F. R., D. M. O'Brien, Rogers, D. H., and Dumler, T. J. (2006). Using the K-State center pivot sprinkler and SDI economic comparison spreadsheet. In: Proc. Central Plains Irrigation Conf., Colby, KS, Feb. 21-.
976. Lamm, F. R., Pacey, D. A., Manges, H., L. (1987). Spreadsheet templates for the calculation of Penman reference evapotranspiration. ASAE Mid-Central Regional Meeting. ASAE Paper No. MCR-87106. St. Joseph, MI: ASAE.
977. Lamm, F. R., and Rogers, D. H. (1983). Scheduling irrigation using computed evapotranspiration. ASAE paper no. MCR 83–109, St Joseph, MI, USA.
978. Lamm, F. R., and D. H., Rogers, (1985). Corn yield response to different irrigation regimes. ASAE paper no. MCR 85–131, St Joseph, MI, USA.
979. Lamm, F. R., Rogers, D. H., Alam, M., D. M. O'Brien, Trooien, T., P. (2009). Summary of twenty years of Kansas SDI research. Proc. Irrigation Association Technical Conference, San Antonio, Texas, December 2–5, (2009). IA, Falls Church, VA. 24 pages .
980. Lamm, F. R., Rogers, D. H., Alam, M., D. M. O'Brien, Trooien, T., P. (2009). Twenty years of progress with SDI in Kansas. ASABE Paper No. 095923. St. Joseph, MI: ASABE.
981. Lamm, F. R., Rogers, D. H., M. Alam and Clark, G. A. (2003). Management considerations for operating a subsurface drip irrigation (SDI) system. KSU Cooperative Ext. Irrigation Mgmt. Series, MF-2590. 8 pages.
982. Lamm, F. R., Rogers, D. H., and Spurgeon, W. E. (1994). Design and management considerations for subsurface drip irrigation systems. In Proc. Central Plains Irrigation Short Course, Kearney, NE, Feb. 7–8, (1994). KSU Ext. Agric. Engr., Manhattan, KS. pages 98–105.
983. Lamm, F. R., Rogers, D. H., and Spurgeon, W. E. (1995). Design and management considerations for subsurface drip irrigation systems. In Proc. Central Plains Irrigation Short Course,

Garden City, KS, Feb. 7–8, (1995). KSU Extension Agricultural Engineering, Manhattan, KS. pages 148–155.

984. Lamm, F. R., Rogers, D. H., and Spurgeon, W. E. (1997). Design and management considerations for subsurface drip irrigation systems. In Proc. Central Plains Irrig. Short Course, Colby, KS, Feb. 4, (1997). KSU Ext. Ag Engr., Manhattan, KS. pages 65–71.
985. Lamm, F. R., Rogers, D. H., and Spurgeon, W. E. (1998). Design and management considerations for subsurface drip irrigation systems. In Proc. Central Plains Irr. Short Course, North Platte, NE, Feb. 17–18, (1998). CPIA, 970 Fifth, W., Colby, KS. pages 6–12.
986. Lamm, F. R., Rogers, D. H., and Spurgeon, W. E. (2000). Design and management considerations for subsurface drip irrigation systems. In Proc. of the Central Plains Irrigation Short Course, Garden City, KS, Feb. 9–10, (2000). CPIA, 970 Fifth, W., Colby, KS. pages 72–80.
987. Lamm, F. R., Rogers, D. H., and Spurgeon, W. E. (2001). Design and management considerations for subsurface drip irrigation systems. In Proc. of the 2001 Central Plains Irrigation Short Course, Kearney, NE, Feb. 5–6, (2001). CPIA, 760 Thompson, N., Colby, KS. pages 29–37.
988. Lamm, F. R., Rogers, D. H., and Spurgeon, W. E. (2003). Design and management considerations for subsurface drip irrigation systems. In proceedings of the Central Plains Irrigation Conference, Colby, KS, Feb. 4–5, (2003). CPIA, 760 N.Thompson, Colby, KS. pages 189–198.
989. Lamm, F. R., Schlegel, A. J., and Clark, G. A. (1997). Nitrogen fertigation for corn using SDI: A BMP. The international meeting of the ASAE, Minneapolis, Minn., August 10–14. ASAE paper no. 972174, St Joseph, MI, USA.
990. Lamm, F. R., Schlegel, A. J., Clark, G., A. (2004). Development of a best management practice for nitrogen fertigation of corn using SDI. Applied Engineering in Agriculture, 20(2): 211–220.
991. Lamm, F. R., Schlegel, A. J., Clark, G. A. (1997). Nitrogen fertigation for corn using SDI: A BMP. ASAE Paper No. 97–2174. St. Joseph, MI: ASAE.
992. Lamm, F. R., Schlegel, A. J., and Clark, G. A. (1997). Optimum nitrogen fertigation for corn using SDI. Proc. Int'l. Technical Conference, Nashville, TN, Nov. 2–4, (1997). IA, Fairfax, VA. pages 251–258.
993. Lamm, F. R., Schlegel, A. J., and Clark, G. A. (1998). Optimum nitrogen fertigation for corn using SDI. International Water and Irrigation Review, 18(2): 13–14, 16–18.
994. Lamm, F. R., Schegel, A. J., and Clark, G. A. (1999). Optimum nitrogen fertigation for corn using SDI. In Proc. of the Central Plains Irrigation Short Course, Sterling, CO, Feb. 9–10, (1999). CPIA, 970 Fifth, W., Colby, KS. pp 52–61.
995. Lamm, F. R., Schlegel, A. J., and Clark, G. A. (2001). Optimum nitrogen fertigation of corn with SDI. SDI workshop, Alma, NE, Jan. 31, (2001).
996. Lamm, F. R., Schlegel, A. J., and Clark, G. A. (2004). Development of a Best Management Practice for Nitrogen Fertigation of Corn Using SDI. Appl. Engr in Agric. 20(2): 211–220.
997. Lamm, F. R., Spurgeon, W. E., Manges, H. L., and Rogers, D. H. (1991). Drip irrigation for corn production–the research approach in western Kansas. Poster presentation at the 1991 Irrigation Assn. Int'l. Expo. and Tech. Conf., San Antonio, TX, Nov. 9–13, (1991).
998. Lamm, F. R., Spurgeon, W. E., Manges, H. L., Rogers, D., H. (1991). Drip irrigation for corn production -the research approach in western Kansas. In Proc. Central Plains Irrigation Short Course, Goodland, KS, Feb. 4–5, (1992). KSU Ext. Ag. Engr., Manhattan, KS, pages 87–92.
999. Lamm, F. R., Spurgeon, W. E., Manges, H. L., and Rogers, D. H. (1992). Drip irrigation for corn production–the research approach in western Kansas. Poster presentation at the American Farm Bureau Annual Meeting, Kansas City, MO, Jan. 13, (1992).
1000. Lamm, F. R., Spurgeon, W. E., Manges, H. L., and Rogers, D. H. (1992). Drip irrigation for corn–a promising prospect. Irrigation Journal, Mar, (1992). pp 12, 14–16.
1001. Lamm, F. R., Spurgeon, W. E., Manges, H. L., and Rogers, D. H. (1993). Using drip irrigation for a non traditional crop – evaluating the practice for irrigated corn. In Proc. 24th National Agricultural Plastics Congress, Overland Park, KS, June 4–8, (1993). American Society for Plasticulture, St. Augustine, FL. pp 136–141.
1002. Lamm, F. R., Spurgeon, W. E., Manges, H. L., and Rogers, D. H. (1993). Using drip irrigation for a nontraditional crop. Arab Water World, March/June, 17: 2–3. pp 5–6.

1003. Lamm, F. R., Spurgeon, W. E., Manges, H. L., and Rogers, D. H. (1994). Subsurface drip irrigation research for corn. In Proc. Central Plains Irrigation Short Course, Kearney, NE, Feb. 7–8, (1994). KSU Extension Ag. Engr., Manhattan, KS. pages 107–110.
1004. Lamm, F. R., Spurgeon, W. E., Manges, H. L., and Rogers, D. H. (1995). Subsurface drip irrigation research for corn. In Proc. Central Plains Irrigation Short Course, Garden City, KS, Feb. 7–8, (1995). KSU Extension Ag Engr, Manhattan, KS. pages 156–159.
1005. Lamm, F. R., Spurgeon, W. E., Rogers, D. H., and Manges, H. L. (1995). Corn production using subsurface drip irrigation. Proceedings of the 5th international micro irrigation congress, Orlando, Florida, April 2–6. ASAE, St Joseph, MI, USA, pages 388–394.
1006. Lamm, F. R., Spurgeon, W. E., Rogers, D. H., and Manges, H. L. (1997). KSU Research for corn production using SDI. In Proc. Central Plains Irrigation Short Course, Colby, KS, Feb. 4, (1997). KSU Ext. Agric. Engr., Manhattan, KS. pages 72–80.
1007. Lamm, F. R., Spurgeon, W. E., Rogers, D. H., and Manges, H. L. (1998). KSU research for corn production using SDI. In Proc. Central Plains Irrigation Short Course, North Platte, NE, Feb. 17–18, (1998). CPIA, 970 Fifth, W., Colby, KS. pages 13–21.
1008. Lamm, F. R., Stone, L. R., Manges, H. L., D. M. O'Brien, (1997). Optimum lateral spacing for subsurface drip-irrigated corn. Trans. ASAE, 40(4): 1021–1027.
1009. Lamm, F. R., Stone, L. R., and Manges, H. L. (1992). Optimum lateral spacing for drip-irrigated corn. Presented at the 1992 Int'l. Winter ASAE meeting. Paper no. 922575, ASAE, St. Joseph, MI. 15 pages.
1010. Lamm, F. R., Storlie, C. A., and Pitts, D. J. (1997). Revision of EP-458: Field evaluation of micro irrigation systems. Presented at the 1997 Int'l. ASAE meeting. Paper no. 972170, ASAE, St. Joseph, MI. 19 pages.
1011. Lamm, F. R., and Sunderman, H. D. (1995). Agricultural Research. Report of Progress 741, Nov, (1995). KAES, Manhattan, KS. 66 pages.
1012. Lamm, F. R., Sunderman, H. D., and Lawless, J. R. (1993). Agricultural Research. Rep. of Progress 688, Aug, (1993). KAES, Manhattan, KS. 56 pages.
1013. Lamm, F. R., Trooien, T., P. (1998). SDI and the declining Ogallala. Proc. 15th Annual Water and the Future of Kansas Conference, Manhattan, KS, March 3, (1998). pages 12–15.
1014. Lamm, F. R., Trooien, T., P. (1999). SDI research in Kansas after ten years. Proc. Irrigation Assoc. International Irrigation Show and Conf., pages 1–8., Fairfax, Va.: Irrigation Assoc.
1015. Lamm, F. R., and Trooien, T. P. (2000). SDI for corn production: A ten year summary of research. In Proceedings of the 6th Int'l Micro irrigation Congress (Micro 2000), Capetown, South Africa, Oct. 22–27, (2000). 9 pages.
1016. Lamm, F. R., and Trooien, T. P. (2001). Irrigation capacity and plant population effects on corn production using SDI. In Proc. Int'l. Irrigation Technical Conf., Nov. 4–6, 2001, San Antonio, TX. 73–80. Irrigation Assn., Falls Church, VA.
1017. Lamm, F. R., and Trooien, T. P. (2001). SDI in the Great Plains. SDI workshop, Alma, NE, Jan. 31, (2001).
1018. Lamm, F. R., and Trooien, T. P. (2002). Irrigation capacity and plant population effects on corn production using SDI. In Proc. of the Central Plains Irrigation Short Course, Lamar, CO, Feb. 5–6, (2002). CPIA, 760 N.Thompson, Colby, KS. pages 122–130.
1019. Lamm, F. R., and Trooien, T. P. (2003). Effect of dripline depth on field corn production in Kansas. In Proc. Irrigation Assn. Int'l. Irrigation Technical Conf., November 18–20, 2003, San Diego, CA. Irrigation Assn., Falls Church, VA.
1020. Lamm, Freddie R., and Todd Trooien, P. (2003). Subsurface drip irrigation for corn production: a review of 10 years of research in Kansas. Irrig Sci., 22 (3–4): 195–200.
1021. Lamm, F. R., Trooien, T. P. (2005). Dripline depth effects on corn production when crop establishment is nonlimiting. Appl. Engr in Agric., 21(5): 835–840.
1022. Lamm, F. R., Trooien, T. P., Clark, G. A., Rogers, D. H., Alam, M. (1997). SDI and electrotechnologies. In proceedings of the EPRI-Agricultural Technology Alliance semi-annual meeting, Boise, ID, May 28–30, (1997). EPRI-ATA, Palo Alto, CA. pages 40–42.

1023. Lamm, F. R., Trooien, T. P., Clark, G. A., Stone, L. R., Alam, M., Rogers, D. H., Schlegel, A., J. (2002). Using beef lagoon wastewater with SDI. Proc. Irrig. Assoc. Int'l. Irrigation Technical Conf., October 24–26, 2002, New Orleans, LA. Irrig. Assoc., Falls Church, VA, 8 pages.
1024. Lamm, F. R., Trooien, T. P., Manges, H. L., and Sunderman, H. D. (2001). Nitrogen fertilization for subsurface drip-irrigated corn. Trans ASAE, 44(3): 533–542.
1025. Lamm, Trooien, F. R. T., P., Schlegel, A., J. (2007). Application and utilization of livestock effluent through SDI systems. Int. Symposium on Air Quality and Waste Management for Agriculture, Broomfield, CO. Sept. 15–19. ASABE. 8 pages.
1026. Lamm, F. R., Trooien, T. P., and Schlegel, A. J. (2006). Applying swine effluent with SDI and LEPA sprinkler irrigation. Proc. 27th Annual Int. Irrigation Assoc. Tech. Conf., San Antonio, Texas, Nov. 5–7. Irrigation Association, Falls Church, VA. Paper No. IA06–1517.
1027. Lamm, F. R., Trooien, T. P., Stone, L. R., Alam, M., Rogers, D. H., Clark, G. A., and Schlegel, A. J. (2001). Using livestock wastewater with SDI – A status report after three seasons. In Proc. of the 2001 Central Plains Irrigation Short Course, Kearney, NE, Feb. 5–6, (2001). CPIA, 760 Thompson, N., Colby, KS. pages 97–104.
1028. Lamm, F. R., Trooien, T. P., Stone, L. R., Alam, M., Rogers, D. H., Clark, G. A., and Schlegel, A. J. (2002). Using livestock wastewater with SDI: A status report after four seasons. In proceedings of the Central Plains Irrigation Short Course, Lamar, CO, Feb. 5–6, (2002). CPIA, 760 N.Thompson, Colby, KS. pages 8–15.
1029. Lamont, W. J., Sorensen, K. A., Averre, C. W. (1990). Painting aluminum strips on black plastic mulch reduces mosaic symptoms on summer squash. Horticulture Science, 25: 1305.
1030. Lampinen B. D., Shackel, K. A., Southwick, S. M., Olson, B., and Yeager, J. T. (1995). Sensitivity of yield and fruit quality of French prune to water deprivation at different fruit growth stages. J. Amer. Soc. Hort. Sci., 120(2): 139–147.
1031. Lang, S. (2002). Hillside landscaping. It is all the planning. Sunset Books Inc., New York. 79–90.
1032. Lanier, J. E., Jordan, D. L., J. Stephen Barnes, Mathews, J., Gary Grabow, L., William J. Griffin Jr., Jack Bailey, E., P. Dewayne Johnson, Janet F. Spears and Randy Wells, (2004). Disease management in overhead sprinkler and subsurface drip irrigation systems for peanut. Agronomy Journal, 96(4): 1058–1065.
1033. Lanting, S. (1975). Subsurface irrigation-Engineering research. In 34th Report Hawaii Sugar Technology Annual Conference, 57–62. Honolulu, Hawaii: Hawaiian Sugar Planters' Association.
1034. Laosheng, W. U. (2000). Drip irrigation using low-quality water. Irrigation Journal, 50(3): 18–20.
1035. Larhrafi, M., Nishiyama, S. (1996). Bypass flow across a bend for chemical injection into micro irrigation systems. Trans. ASAE, 39(4): 1321–1327.
1036. Larson, D., Peterson, D. (2006). Larson Irrigation, Inc., Haxtun, Colorado and SDI system owner, Fort Collins, Colorado.
1037. Larson, R. A. (1993). Impact of plastics in the floriculture industry. HorticultureTechnology, 3: 28–34.
1038. Lascano, Robert J. (2000). A general system to measure and calculate daily crop water use. Agronomy Journal, 92: 821–832.
1039. Latif, M. (1990). Sprinkler irrigation to harness potential benefits of water scarcity areas in Pakistan. National Seminar on Water Resources Development & Management in Arid Areas organised by PCR-WAR at Quetta, August 11–12.
1040. Lawless, J. R., Sunderman, H. D., and Lamm, F. R. (1991). Agricultural Research. Rep. Progress 635, Aug, (1991). KAES, Manhattan, KS. 87 pages.
1041. Lazarovitch, N., Ben-Gal, A., Shani, U. (2006). An automated rotating lysimeter system for greenhouse evapotranspiration studies.Soil Sci. Soc. of America J., 70: 801–804.
1042. Lazarovitch, N., Warrick, A. W., Furman, A., Šimůnek, J. (2007). Subsurface water distribution from drip irrigation described by moment analyses. Vadose Zone J., 6: 116–123.

1043. Lazorovitch, N., U. Shani. 2004, November. Subsurface drip irrigation in gravel-filled cavities. Vadose Zone Journal, 3: 1407–1413.
1044. Leaflet, (2280). Division of Agricultural Sciences, University of California, Berkley, CA.
1045. Lebron, I., Robinson, D. A., S. Goldberg and Lesch, S. M. (2004). The dielectric permittivity of calcite and arid zone soils with carbonate minerals. Soil Sci. Soc. of America J., 68: 1549–1559.
1046. Lee, C. O. (1920). Irrigation tile. U. S. Patent No. 1,350, 229.
1047. Lee, J., Horton, R., and Jaynes, D. B. (2000). A time domain reflectometry method to measure immobile water content and mass exchange coefficient. Soil Sci. Soc. of America J., 64: 1911–1917.
1048. Lee, R. (1980). *Forest hydrology.* New York: Columbia University Press.
1049. Leffert, M., G. A. Clark., Hutchinson, S. L., and Barden, C. L. (2008). Evaluation of poplar trees irrigated with livestock lagoon wastewater. Trans. of the ASABE, 51(6): 2051–2060.
1050. Legaz, F., and Primo-Millo, E. (1988). In: *Guidelines for Citrus fertilization.* Technical Report, Department of the Agriculture, Fish and Food of the Valencian Government, Valencia, Spain, No 5.88 (In Spanish).
1051. Leib, B. G., Jarrett, A. R., Orzolek, M. D., Mumma, R., O. (2000). Drip chemigation of Imidacloprid under plastic mulch increased yield and decreased leaching caused by rainfall. Trans. of ASAE, 43(3): 615–622.
1052. Leib, B. G., Jabro, J. D., Jarrett, A. R. (2003). Pesticide movement under drip chemigation: model simulations and field measurements. ASAE Paper No. 032022. International Mtg. of the ASAE, Las Vegas, NV. July 27–30.
1053. Leslie Long, F. (1982). A new solid – state device for reading tensiometers. Soil Science, 133(3): 131–132.
1054. Letey, J., Dinar, A. (1986). Simulated crop-water production functions for several crops when irrigated with saline waters. Hilgardia, 54(1): 1–32.
1055. Letey, J., Dinar, A., and Knapp, K. C. (1985). Crop-water production function model for saline irrigation waters. Soil Sci. Soc. of America J., 49: 1005–1009.
1056. Levin, I., F. C. van Rooyen, (1977). Soil water flow and distribution in horizontal and vertical directions as influenced by intermittent application rates. Soil Sci. 124(6): 355–365.
1057. Levin, I., P. C. van Rooyen, F. C. van Rooyen, (1979). The effect of discharge rate and intermittent water application by point-source irrigation on the soil moisture distribution pattern. Soil Sci. Soc. Am. J., 43: 8–16.
1058. Levitt, D. G., Hartmann, M. J., Kisiel, K. C., Criswell, C. W., Dwain-Farley, P., Christensen, C. (2005). Comparison of the water balance of an asphalt cover and an evapotranspiration cover at technical area 49 at the Los Alamos National Laboratory. Soil Sci. Soc. of America J., 69: 789–797.
1059. Ley, T. W., Stevens, R. G., Topielec, R. R., W. H Neibling, (1994). Soil water monitoring and measurement. Pacific NW Pub. PNW475. 35 pages.
1060. Li, G. Y. (2006). Influence of geometrical parameters of labyrinth flow path of drip emitters on hydraulic and anti-clogging performance. Trans. of ASAE, 49(3): 637–643.
1061. Li, J., Zhang, J., Rao, M. (2005). Modeling of water flows and nitrate transport under surface drip fertigation. Trans. of ASAE, 48(2): 627–637.
1062. Lilleboe, D. (2005). SDI primer. The Sunflower 31(2): 7. February.
1063. Linacre, E. T. (1977). A simple formula for estimating evapotranspiration rates in various climates, using temperature data along. Agricultural Meteorology, 18: 409–424.
1064. Lindeburg, M. R. (2003). Environmental engineering reference manual for the PE Exam.California, CA: Professional Publications, 302.
1065. Lindsay, C. A., Sutton, B. G., Collis-George, N. (1989). Irrigation scheduling of subsurface drip-irrigated salad tomatoes. *Acta Horticulturae*, 247: 229–232.
1066. Lindsay, W. L. (1979). *Chemical equilibria in soils.* NY: John Wiley & Sons Press, 449 pages.
1067. List, R. (1984). *Smithsonian meteorological tables*, 6th rev. ed., Washington, DC: Smithsonian Institution, 539.

1068. Littell, R. C., Milliken, G. A., Stroup, W. W., Wolfinger, R. D., Schabenberger, O. (2006). *SAS for Mixed Models.* 2nd Ed. Cary, N. C.: SAS Institute, Inc.
1069. Liu, Shi-Ping, (2002). Irrigation system in agriculture. Paper number 022285, ASAE Annual Meeting, St. Joseph – MI.
1070. Liu, B. Y.H., Jorden, R. C. (1960). The interrelationship and characteristic distribution of direct diffuse and total solar radiation. Solar Energy, 4(3): 1–19.
1071. Liu, L. C., Goyal, M. R. (1989). Selective herbicides to control grass weeds in transplanted tomatoes and peppers. J. Agric. U. P.R., 73(3): 231–238.
1072. Liu, L. C., Padilla, M. A., Goyal, M. R., Ibanez, J. G. (1987). Integrated weed management in transplanted tomatoes and peppers under drip irrigation. Journal of Agriculture of the University of Puerto Rico, 71(4): 349–358.
1073. Liu, Y., Teixeira, J. L., Zhang, H. J., Pereira, L. S. (1998). Model validation and crop coefficients for irrigation scheduling in the North China. Agricultural Water Management, 36: 233–246.
1074. Livestock Wastewater use through SDI – A Pilot Project. Victory Electrical Coop. Annual Irrigation Meeting, Dodge City, Kansas, February 8, (2008).
1075. Lobell, D. B., Asner, G. P. (2002). Moisture effects on soil reflectance. Soil Sci. Soc. of America J., 66: 722–727.
1076. Locascio, S. J., Clark, G. A., Csizinszky, A. A., Stanley, C. D., Olson, F. M., Rhoads, F., Smajstrla, A. G., Vellidis, G., Edling, R. J., Hanna, H. Y., Goyal, M. R., Crossman, S., and Navarro, A. A. (1992). Water and nutrient requirements for drip irrigated vegetables in humid regions. Southern Cooperative Series Bulletin 363. Southern Regional Research Project S-143. 17 pages.
1077. Lomen, D. O., and Warrick, A. W. (1978). Linearized moisture flow with loss at the soil surface. Soil Sci. Soc. Am. J., 42: 396–400.
1078. Long, D. S., Wraith, J. M., Kegel, G. (2002). A heavy-duty time domain reflectometry soil moisture probe for use in intensive field sampling. Soil Sci. Soc. of America J., 66: 396–401.
1079. Loommis, R. S., Williams, W. A. (1969). Productivity and the morphology of crop stand patterns with leaves. In: *Physiological aspects of crop yield*, ed. Eastin, J. D., 27–47 pages. ASA, CSSA and SSSA, Madison, WI.
1080. López, G., Rabones, A., Del Campo, J., Mata, M., Vallverdú, X., Girona, J., Marsal, J. (2008). Response of peach trees to regulated deficit irrigation during stage II of fruit development and summer pruning. Spanish Journal of Agricultural Research, 6: 479–491.
1081. Head loss in effluent filtration in micro irrigation systems using dimensional analysis. Biosystems Eng., 92(3): 383–390.
1082. Loy, B., Lindstrom, J., Gordon, S., Rudd, D., Wells, O. (1989). Theory and development of wavelength selective mulches. Proc. of National Agriciulture Plastics Congress, 21: 193–197.
1083. Loy, J. B. and. Wells, O. S. (1982). A comparison of slitted polyethylene and spun-bonded polyester for plant row covers. Horticulture Science. 17: 405–407.
1084. Lubana, P. P.S., Narda, N. K. (2001). Modelling soil water dynamics under trickle emitters – a review. J. Agric. Eng. Res., 78: 217–232.
1085. Lugo López, M. A. (1953). Moisture relationships of Puerto Rico soils. Technical Paper No. 9. Agricultural Experiment Station, University of Puerto Rico, Río Piedras, Puerto Rico.
1086. Luthra, S. K., Kaledhonkar, M. J., Singh, O. P., Tyagi, N. K. (2000). Design and development of an auto irrigation system. Agricultural Water Management, 53(2,): 169–181.
1087. Lyannoi, A. D., and Shevchenko, I. V. (1985). Effectiveness of fertilizer aplication with drip irrigation in vineyards. Vinodelie Vinogadarstvo, (1): 35–37.
1088. Ma, L., Selim, H. M. (2005). Pesticide transport in mulch amended soils: A two-compartment model. Soil Sci. Soc. of America J., 69(2): 318–327.
1089. Maarten, P., Raangs, R., Verjeij, H. (2000). Response of the osmotic tensiometer to varying temperature. Laboratory for Inorganic Materials Science, Department of Chemical Technology and MESA Research Institute, University of Twente. 231–234.

1090. Maas, E. V. (1990). Crop salt tolerance. In: *Agricultural salinity assessment and management*, ed. Tanji, K. K., 262–304 pages, ASCE Manuals and Reports on Engineering Practice No. 71. New York: ASCE.
1091. Magazine "Agriculture of the Americas." 1983 and 1984 July and February of, (1988).
1092. Mahmood, N., Shah, N., Ross, M., Vomacka, J. (2005). Evapotranspiration of two vegetation covers in a shallow water table environment.Soil Sci. Soc. of America J., 69: 492–499.
1093. Mahmood, R., Hubbard, K. G. (2003). Simulating sensitivity of soil moisture and evapotranspiration under heterogeneous soils and land uses. Journal of Hydrology (Amsterdam), 280: 72–90.
1094. Makens, T. P., Spurgeon, W. E., Lamm, F. R., and Manges, H. L. (1992). Subsurface dripline length for corn production. Presented at the 1992 Mid-central meeting of the ASAE. Paper no. MC92–100, ASAE, St. Joseph, MI. 12 pages.
1095. Makkink, G. F. (1957). Testing the Penman formula by means of lysimeters. Journal of the Institution of Water Engineers, 11(3): 277–288.
1096. Malhotra, A. N. (1984). Sprinkler irrigation case study. Proceedings of the Seminar on Sprinkler and Drip Irrigations, Delhi.
1097. Maman, N., Lyon, D. J., Mason, S. C., Galusha, T. D., Higgins, R. (2003). Pearl millet and grain sorghum yield response to water supply in Nebraska. Agronomy Journal, 95: 1618–1624.
1098. Mamedov, A. I., I. Shainberg and Levy, G. J. (2000). Irrigation with Influent water: effects of rain fall energy on soil infiltration. Soil Science Society of American Journal, 64(2): 732–737.
1099. Mamedov, A. I., C. Huang and Levy, G. J. (2006). Antecedent moisture Content and aging duration effects on seal formation and erosion in specific soils.Soil Sci. Soc. of America J., 70: 832–843.
1100. Management of Water Resources in Cash Crops and in Alternative Production Systems. Brussels, Belgium. 12.
1101. Mane, M. S. (2006). *Principles of drip irrigation*. New Delhi, India: Asiatic Books. 1–168.
1102. Manges, H. L., Spurgeon, W. E., Huang, Z. M., Tomsicek, D. J. (1995). Subsurface dripline spacing and plant population for corn production. Proceedings of the 5th International Micro irrigation Congress, ed. Lamm, F. R., ASABE.
1103. Manoliadis, O. (2006). Analysis of irrigation systems using sustainability-related criteria. Environment Journal, 30: 1150–1153.
1104. Manoliadis, O. G. (2002). Analysis of irrigation systems using sustainability-related criteria. Journal of Environmental Quality, 30: 1150–1153.
1105. Mansell, M. G.,2003. *Rural and urban hydrology*. London: Thomas Telford. 52–179.
1106. Mansour, N. S. (1991). The use of field covers in vegetable production. Proceedings of International Workshops on Implied Vegetable Production Through the use of fertilizers, Mulching and Irrigation, Chaing Mai University, Thailand.
1107. Marouelli, W. A., Waller, P. M. (1999). Oil drop generator for foliar chemigation: Theory and laboratory evaluation. Trans. of ASAE, 42(5): 1289–1302.
1108. Marsh, A. W., Strohman, R. A., Spaulding, S., Younger, V., Gibeault, V. (1980). Turf grass irrigation research at the University of California: warm and cool season grasses tested for water needs. Irrigation Journal, 20–21, 32–33.
1109. Marshall, T. J., and Holmes, J. W. (1988). Management of soil water. In: *Soil Physics*. Cambridge Univ. Press: Cambridge, UK.
1110. Marthaler, H. P., Vogelsanger, W., F. Richard and Wirenga, P. J. (1983). A pressure transducer for field tensiometers. Soil Sci. Soc. of America J., 47: 634–637.
1111. Martin, D. L., Gilley, J. R. (1993). Irrigation Water Requirements. Chapter 2, In: *The SCS national engineering handbook*. Washington, DC: USDA Soil Conservation Service.
1112. Martin, E. C., Slack, D. C., Pegelow, E. J. (1996). Crop coefficients for vegetables in Central Arizona. Proceedings of the International Conference on Evapotranspiration and Irrigation Scheduling, eds. C. R. Camp.
1113. Martínez-Fernández, J., Ceballos, A. (2003).Temporal stability of oilmoisture in a large-field experiment in spain.Soil Sci. Soc. of America J., 67: 1647–1656.

1114. Martius, C., H. Tiessen and Vlek, P. L. G. (2001). *Managing organic matter in tropical soils: Scope and developments in plant and soil sciences*. New York: Springer Press. 1–248.
1115. Mastrorilli, M., Katerji, N., and Rana, G. (1995). Water efficiency and stress on grain sorghum at different reproductive stage. Agric. Water Manage., 28: 23–34.
1116. Masumoto, Mas David, (1999). *Harvest son: Planting roots in American soil*. W. W. Norton & Company. 175.
1117. Matesi, S. W. (2003). Organic byproducts of drinking water chlorination. Global Nest, 1: 143–156.
1118. Matías, L. R., Bollero, G. A., Hoeft, R. G., Bullock, D. G. (2005). Spatial variability of the Illinois soil nitrogen test: Implications for soil sampling. Agronomy Journal, 97: 1485–1492.
1119. Matias, P. G.M. (1992). SWATCHP: A model for a continuous simulation of hydrologic processes in a system vegetation – soil – aquifer – river. PhD dissertation, Technical University of Lisbon.
1120. Mavi, H. S. (2004). *Agrometeorology: Principles and applications of climate studies in agriculture*. New York: Haworth Press. pages 76–83.
1121. Mazor, I., Mazor, E. (2003). Chemical and isotopic groundwater hydrology. Marcel Dekker. Page 20–25.
1122. McGee, E. A., Thompson, T. L., McGreary, T. W. (1995). An apparatus for application of nitrogen-15 fertilizers through buried drip tubing. Soil Sci. Soc. Am. J., 59: 1495–1498.
1123. McAlavy, T. (1997). K-State's drip irrigation research holds promise for farmers. KSU Ag Report, fall, (1997). 5.
1124. McCabe, D. (1998). Plunging into drip. Feature article about Nebraska farmer using SDI. Nebraska Farmer, Mid-Feb. pages 8–9.
1125. McCabe, D. (2002). Uncover Water. Nebraska Farmer, March. pages 12–13.
1126. McCarthy, M. G., Loveys, B. R., Dry, P. R., and Stoll, M. (2002). Regulated deficit irrigation and partial rootzone drying as irrigation management techniques for grapevines. In: *Deficit irrigation practices*, FAO Water Reports No. 22. Rome, Italy, pages 79–87.
1127. McElroy, D. L., Hubbell, J. M. (2004). Evaluation of the conceptual flow model for a deep vadose zone system using advanced tensiometers. Vadose Zone Journal, 55: 170–182.
1128. McGill, S. (1993). Buried drip for alfalfa. The Furrow, 98(7): 26–27.
1129. McKenzie, D. C. (1998). *SOILPAK for Cotton Growers*. 3rd Edition, NSW Agriculture, Orange-NSW-AU.
1130. McMichael, B., Gitz, D. C., Lascano, R., Mahan, J. R., and Wanjura, D. F. (2006). The Growth and development of cotton under sub-surface drip irrigation. ASA-CSSA-SSSA Annual Meeting, Indianapolis, Indiana. Paper No. 165–2.
1131. McNamara, J. B. (1970). Subirrigation-The basis of tomorrow's agriculture. Proceedings of the National Irrigation Symposium, C1-C13. St. Joseph, MI: ASAE.
1132. McNaughton, K. G., Jarvis, P. G. (1984). Penman-Monteith equation predictively. Agricultural Water Management, 8: 263–278.
1133. McWilliams, D. (2002). Drought strategies for alfalfa. Circular 581. Las Cruces, N. M.: New Mexico State University Cooperative Extension.
1134. Mead, R. M., Hutmacher, R. B., Phene, C. J. (1993). Subsurface drip irrigation of alfalfa. In Subsurface Drip Irrigation-Theory, Practices and Application, 145–146. CATI Pub. No. 92–1001. Fresno, CA: California State University.
1135. Meek, D. W., Singer, J., W. (2004). Estimation of duration indices for repeated tensiometer reading. Agronomy Journal, 96: 1787–1790.
1136. Mehmet, O., H.A Biçak, (2002). Modern and traditional irrigation technologies in the eastern mediterranean. IDRC. *Drip Irrigation in Northern Cyprus*, 73–83.
1137. Mehmet, O. (2002). Modern and traditional irrigation technologies in the eastern mediterranean. International Development Research Center (IDRC), 130–132.
1138. Meiri, A., Letey, J. (2003). Evaluation of a model for irrigation management under saline conditions: I. Effects on plant growth. Soil Sci. Soc. of America J., 67: 71–76.

1139. Melby, P. (2003). Simplified irrigation design. New York: Van Nostrand Reinhold Publishers. 14–30.
1140. Mergos, G. J. (2000). Sustainability issues and technology choice in irrigation investment. Water Resource Management Journal, 5: 244–251.
1141. Merriam, J. L. (1966). A management control concept for determining the economical depth and frequency of irrigation. Trans. ASAE, 9: 492–498.
1142. Meshkat, M., Warner, R. C., Workman, S., R. (1998). Comparison of water and temperature distribution profiles under sand tube irrigation. Trans. ASAE, 41(6): 1657–1663.
1143. Meshkat, M., Warner, R. C., Workman, S., R. (2000). Evaporation reduction potential in an undisturbed soil irrigated with surface drip and sand tube irrigation. Trans. ASAE, 43(1): 79–86.
1144. Methods of estimating potential evapotranspiration in southern Italy. L'Irrigazione XXXII (4): 23–28 (in Italian).
1145. Metochis, C. (1980). Irrigation of lucerne under semi-arid conditions in Cyprus. Irrig. Sci., 1(4): 247–252.
1146. Meyer, J. L. (1985). Cleaning drip irrigation systems. Proc. Third International Drip/Trickle Irrigation Congress, Drip/Trickle Irrigation in Action, Fresno, California, November 18–21. 41–44.
1147. Meyer, W. S., Mateos, L. (1990). Effects of soil type on soybean crop water use in weighing lysimeters. II: Effect of lysimeter canopy height discontinuity on evaporation. Irrigation Science, 11: 233–237.
1148. Mhammed, L., Nishiyama, S. (1996). Characteristics of bypass flow across pvc pipe elbows. Trans. ASAE, 39(2): 505–510.
1149. Michael, A. M. (1978). Irrigation theory and practice. Chapter 7: 448–584. New Delhi: Vikas Publishing.
1150. Mieri, A. (1984). Plant response to salinity: experimental methodology and application to the field. In: *Soil salinity under irrigation: Processes and management*, eds. Shainberg, I., Shalhevet, J., Section 8.3: 284–297. Berlin: Springer-Verlag.
1151. Mikkelsen, R. L. (1989). Phosphorus fertilization through drip irrigation. Journal of Production Agriculture, 2(3): 279–286.
1152. Miller, D. E., and Aarstad, J. S. (1974). Calculation of the drainage component of soil water depletion. Soil Sci., 118: 11–15.
1153. MILLER, R. S. et al, (1981). Labelled nitrogen uptake by drip-irrigated tomatoes. Agronomy J., 73: 265–270.
1154. Millera, S. D., Vanceb, G. F., Zhangb, R. (2001).Agroeconomic analyses of drip irrigation for sugarbeet production.Agronomy Journal, 93(3): 517–523.
1155. Minimum System Requirements with SDI. KSU NWREC Field Day, Colby, Kansas, August 29, (2007).
1156. Ministry of Agriculture, (1997). Current Statistics on Agriculture. Beirut, Lebanon. Number 77, Table 1.
1157. Mitchell, P. D., Chalmers, D. J. (1982). The effect of reduced water supply on peach tree growth and yields. Journal of the American Society of Horticultural Science, 107: 853–856.
1158. Mitchell, P. D., van den Ende, B., Jerie, P. H., Chalmers, D. J. (1989). Response of 'Bartlett' pear to withholding irrigation, regulated deficit irrigation, and tree spacing. Journal of the American Society of Horticultural Science, 114: 15–19.
1159. Mitchell, Peter D., Ian Goodwin, (1996). Micro-Irrigation of Vines and Fruit Trees. Agriculture Victoria, AU.
1160. Mitchell, W. H. (1981). Subsurface irrigation and fertilization of field corn. Agronomy J., 73: 913–916.
1161. Mitchell, W. H., Tilmon, H., D. (1982). Underground trickle irrigation: the best system for small farms? Crops and Soils, 34: 9–13.
1162. Mitchell, W. H., Sparks, D. L. (1982). Influence of subsurface irrigation and organic additions on top and root growth of field corn. Agronomy Journal, 74(6): 1084–1088.

1163. Miyamoto, K., T.Yamamoto, (2005). The effect of chlorine on emitter clogging induced by algae and protozoa and the performance of drip irrigation. Trans. of ASAE, 48(2): 519–527.

1164. Mizyed, N., Kruse, E. G. (1989). Emitter discharge evaluation of subsurface trickle irrigation systems. Trans. of ASAE, 32(4): 1223–1228.

1165. Mmolawa, K., Or, D. (2003). Experimental and numerical evaluation of analytical volume balance model for soil water dynamics under drip irrigation. Soil Science Society of American Journal, 67(6): 1657–1671.

1166. Mmolawa, K., Or, D. (2000). Water and solute dynamics under a drip-irrigated crop: Experiments and analytical model. Trans. ASAE, 43(6): 1597–1608.

1167. Mmolawa, K., Or, D. (2000). Root zone dynamics under drip irrigation: A review. Plant and Soil, 222: 163–190.

1168. Mmolowa, K., Or, D. (2003). Experimental and numerical evaluation of analytical volume balance model for soil water dynamics under drip irrigation.Soil Sci. Soc. of America J., 67: 1657–1671.

1169. Mo, X., Liu, S., Lin, Z., Zhao, W. (2004). Simulating temporal and spatial variation of evapotranspiration over the Lushi basin. Journal of Hydrology (Amsterdam), 285: 125–142.

1170. Modern methods of irrigation, (2004). Earth and Environmental Science, Geo Journal, 35 (1): 59–63.

1171. Mofoke, A. L.E. (2006). Yield of tomato grown under continuous-flow drip irrigation in Bauchi state of Nigeria. Agricultural Water Management, 84(1/2): 166–172.

1172. Mohd Ekhwan Toriman, Mazlin Mokhtar, Muhamad Barzani Gasim, Sharifah Mastura Syed Abdullah, Othman Jaafar and Nor Azlina Abd Aziz, (2009). Water resources study and modeling at North Kedah: A Case of Kubang Pasu and Padang Terap water supply schemes. Research Journal of Earth Sciences, 1(2): 35–42.

1173. Moines, D. (2006). *All About Sprinklers & Drip Systems*. Meredith Books, Iowa. 1–127.

1174. Monteith, J. L. (1965). Evaporation and Environment. 19th Symposia of the Society for Experimental Biology, Cambridge: Cambridge University Press, 19: 205–234.

1175. Monteith, J. L. (1981). Evaporation and surface temperature. Quarterly Journal of the Royal Meteorological Society, 107: 1–27.

1176. Monteith, J. L. (1985). Evaporation from land surfaces: progress in analysis and prediction since, (1948). In Advances in Evapotranspiration: Proceedings of the ASAE Conference on Evapotranspiration, Decenmber 16–17, Chicago, Ill. St. Joseph, MI: ASAE. 4–12.

1177. Monteith, J. L., Unsworth, M. H. (1990). *Principles of environmental physics*, 2nd ed., London: Edward Arnold.

1178. Montero, J., Tarjuelo, J. M., J. F.Ortega, (2000). Heterogeneity analysis of the irrigation in fields with medium size sprinklers. Agricultural Engineering International: CIGR Journal of Scientific Research and Development. Volume II.

1179. Moore, R. C., Fitschen, J. C. (1990). The drip irrigation revolution in the Hawaiian sugarcane industry. Proceedings of the 3rd National Irrigation Symruposium, 223–227 pages. St. Joseph, MI: ASAE.

1180. Moore, S., Han, Y. J., A.Khalilian, Owino, T. O., B.Niyazi, (2005). Instrumentation for variable-rate lateral irrigation system. Paper number 052184 at ASAE Annual Meeting.

1181. Morari, F., Giardini, L. (2002). Irrigation automation with heterogeneous vegetation: the case of the Padova botanical garden. Agricultural Water Management, 55(3): 183–201.

1182. Morgan, K. T., Parsons, L. R., Wheaton, T. A., Pitts, D. J., and Obreza, T. A. (1999). Field calibration of a capacitance water content probe in fine sand soils. Soil Sci. Soc. of America J., 63: 987–989.

1183. Morgan, K. T., Obreza, T. A., Scholberg, J. M. S., Parsons, L. R., Wheaton, T. A. (2006). Citrus water uptake dynamics on a sandy Florida entisol. Soil Sci. Soc. of America J., 70: 90–97.

1184. Moriana, A., Orgaz, F., Pastor, M., Fereres, E. (2003). Yield responses of mature olive orchard to water deficits. Journal of the American Society for Horticultural Science, 123: 425–431.

1185.Morris, M., Schwankl, L. (2008). *The California micro irrigation pocket guide* – System management and maintenance. National Center for Appropriate Technology. 92 pages.

1186. Morrison, C. T., Green, W. T., Hadley, P. (1989). Energy exchange by plastic row covers. Proceedings of National Agriciulture Plastics Congress, 21: 269–275.
1187. Morrow, M. R., Krieg, D. R. (1990). Cotton management strategies for s short growing season environment: Water-nitrogen considerations. Agron. J., 82(1): 52–56.
1188. Moshabbir, PM., Ahmad, S., Yasin, M., and Ahmad, M. M. (1993). Indigenization of Trickle Irrigation Technology. Proc., USAID- ISMR Symposium, IIMI, Lahore.
1189. Moynihan, M. J., and Haman, D. Z. (2001). Micro irrigation systems for small-scale farms in the Rio Cobre basin area of Jamaica. Applied Engineering in Agriculture Journal, 8: 617–623.
1190. Muney, J. R., Henderix, D. L. (1988). Responses of glasshouse-grown cotton to irrigation with CO_{2-}saturated water. Crop Sci., 28: 835–838.
1191. Muñoz, O., Rodríguez, V. R. (1986). Soil Classification in Puert Rico (Spanish). Colegio de Ciencias Agrícolas, Universidad de Puerto Rico.
1192. Murray, F. W. (1967). On the computation of saturation vapor pressure. Journal of Applied Meteorology, 6: 203–204.
1193. Mustafa, O. S., Arshad, M., Sattar, I., Ali, S. (2003). Adoption of kostiakof model to determine the soil infiltration for surface irrigation methods under local conditions. Int. J. Agric. Biol., 1: 40–42.
1194. Nachabe, M., Shah, N., Ross, M., Vomacka, J. (2005). Evapotranspiration of two vegetation covers in a shallow water table environment. Soil Sci. Soc. of America J., 69(1): 492–499.
1195. Naden, P. S., Blyth, E. M., Broadhurst, P., Watts, C. D., Wright, I. R. (2000). Modeling the spatial variation in soil moisture at the landscape scale: An application to five areas of ecological interest in the UK. Hydrological Processes, 14: 785–809.
1196. Nadler, A., Gamliel, A., Peret, I. (1999). Practical aspects of salinity effect on TDR-measured water content: A field study. Soil Sci. Soc. of America J., 63: 1070–1076.
1197. Nadler, A., Green, S. R., I. Vogeler and Clothier, B. E. (2002). Horizontal and vertical TDR measurements of soil water content and electrical conductivity. Soil Sci. Soc. of America J., 66: 735–743.
1198. Nadler, Arie, Eran Raveh, Uri Yermiyahu and Green, S. R. (2003). Evaluation of TDR to monitor water content in stem of lemon trees, and soil and their response to water stress. Soil Sci. Soc. of America J., 67: 437–448.
1199. Nakayama, F. S., D. A. Bucks (Eds.). (1986). *Trickle irrigation for crop production*. Elsevier, Netherlands. 383 pages.
1200. Nakayama, F. S., Bucks, D. A. (2002). *Trickle irrigation for crop production*. Maryland Heights – MO: Elsevier Publishing Company. 1–383.
1201. Nakayama, F. S., and Bucks, D. A. (1981). Emitter clogging effects on trickle irrigation uniformity. Trans. of ASAE, 24(1): 77–80.
1202. Nakayama, F. S., and Bucks, D. A. (1991). Water quality in drip/trickle irrigation: a review. Journal of Irrigation Science, 12: 187–192.
1203. Nakhjavani, M. M., Sadrghaen S. H., Dehghanisanij, H. (2008). The role of micro irrigation systems on water productivity improvement of row crops. Proceeding of the Workshop on Pressurized Irrigation and Sustainable Development in Iran. 21 Feb. Karaj. Iran. (In Farsi).
1204. Nakhjavani-Moghadam, M. M., Sadr-Cain, H., Dehghani-Sanych, H. (2007). Use of drip irrigation system on improving water use efficiency of crops. National Seminar on Science and Sustainable Development of Irrigation under Pressure, pages 591–597.
1205. Narayanamoorthy, A. (1997). Beneficial impact of drip irrigation: A study based on Western India. Water Resources Journal, 1997: 17–31.
1206. National Research Council, (2002). *A New Era For Irrigation*. New York, USA.
1207. National Task Force on Micro irrigation, (2004). Ministry of Agric., Government of India Report 330 pages.
1208. Natural Resource Conservation Service. *National Engineering Handbook*: Section 15. Chapter 1, Soil-Plant-Water Relationships; Chapter 2, Irrigation Water Requirements; Chapter 4, Border Irrigation; Chapter 5, Furrow Irrigation; Chapter 7, Trickle Irrigation; Chapter 11, Sprinkler Irrigation. United States Department of Agriculture, U. S.A.

1209. Neale, C. M.U. (1987). Development of reflectance based crop coefficients for corn. Unpublished PhD Dissertation, Agricultural Engineering, Colorado State University, Fort Collins, USA.
1210. Neale, C. M.U., Kruse, E. G., Yoder, R. E. (1991). Field experience with hydraulic weighing lysimeters. In: *Lysimeters for evapotranspiration and environmental measurements*, eds. R. G. Allen et al., 160–169 pages. New York, NY: ASCE.
1211. Neale, C. M.U., Bausch, W. C., Heerman, D. F. (1989). Development of reflectance-based crop coefficients for corn. Trans. of ASAE, 32(6): 1891–1899.
1212. Neibling, H., Brooks, R. (1995). Potato production using subsurface drip irrigation-Water and nitrogen management. Proceedings of the 5th International Micro irrigation Congress, ed. Lamm, F. R., 656–663 pages. St. Joseph, MI: ASAE.
1213. Neibling, W. H., Weihing, W. (2004). Using the IDWR "Keep-up and Catch-up" slide rule. University of Idaho CIS 1117. 8 pages.
1214. Neitzel, D. A. (2003). Hanford site national environmental policy act (NEPA) characterization. PNL-6415, Rev. 15. Pacific Northwest National Laboratory, Richland, WA. 44–56.
1215. Nemery, Hoet, B., P. H. M., Nowak, D. (2002). Indoors swimming pools, water chlorination and respiratory health, Eur. Respir. J., 19 (May): 790–793.
1216. New strategy for optimizing water application under trickle irrigation. Journal of Irrigation and Drainage Engineering, 128(5): 287–297. September/ October, (2002).
1217. Newham, D. (1992). Reclaimed Water in the United States with a Worldwide Overview. Proceedings of the Recycled Water Seminar, Wagga Wagga, NSW, Australia, May 19–20. Australian Water and Waste Water Assoc.
1218. Newport, A. (2000). Pivots or drip? Irrigation Extra, pages IE10- IE13, Kansas Farmer, Mid-January, (2000).
1219. Ngigi, S. N., Thome, J. N., Waweru, D. W., Blank, H. G. (2001). Low-cost irrigation for poverty reduction: An evaluation of low-head drip irrigation technologies in Kenya. In IWMI. Annual report 2000–2001.
1220. Nieuwenhuijsen, M., Toledano, M., Eaton, N. (2000). Chlorination disinfection by-products in water and their association with adverse reproductive outcomes: A review. Occupational and Environmental Medicine, 57: 73–85.
1221. Nightingale, H. I. (1972). Nitrates in soil and groundwater beneath irrigated and fertilized crops. Soil Science, 114: 300–311.
1222. Nightingale, H. I., Phene, C. J., Patton, S. H. (1985). Trickle irrigation effects on soil chemical properties. Proceedings of the 3rd International Drip/Trickle Irrigation Congress, 2: 730–735. St. Joseph, MI: ASAE.
1223. Nikolaou, A. D., Golfinopoulos, S. K. (2002). Investigation of the formation of organic byproducts during chlorination of surface waters. International Conference Protection and Restoration of the Environment VI, Skiathos-Greece, 1: 115–122.
1224. Nikolaou, A. D., Golfinopoulos, S. K., Lekkas, T. D., Kostopoulou, M. N. (2004). DBP levels in chlorinated drinking water: Effect of humic substances. Journal of Environmental Monitoring and Assessment, 93: 301–319.
1225. Nimah, M. (1992). Needs in irrigation water in Lebanon. UNESCO National Seminar on water in Lebanon, 27–28 November.
1226. Nissen, H., Moldrup, P., Olesen, T. (1999). Printed circuit board time domain reflectometry probe: Measurements of soil water content. Soil Sci. Soc. of America J., 64: 454–466.
1227. Nissen, H., Moldrup, P., L. de Jonge, Jacobsen, O. (1999). Time domain reflectometry coil probe measurements of water content during fingered flow. Soil Sci. Soc. of America J., 63: 493–500.
1228. Noborio, K., R. Horton and Tan, C. S. (1999). Time domain reflectometry probe for simultaneous measurement of soil meatric potential and water content. Soil Sci. Soc. of America J., 63: 1500–1505.
1229. Noi, Y. (1967). Soil-Water Relations (Spanish). Servicio de Extensión Agrícola, Israel, Departamento de Capacitación para el Extranjero, Israel.

1230. Norsida, Man and Sami Ismaila Sadiya, (2009). Off-farm employment participation among paddy farmers in the MUDA agricultural development authority and Kemasin Semerak granary areas of Malaysia. Asia-Pacific Development Journal.16 (2): 141–153.
1231. Northcote, K. H. (1979). *A Factual Key for the Recognition of Australian Soils*. 4th Edition, Rellim Technical Publications, Adelaide SA.
1232. Nujamudeen, M. S., Dharmasena, P. B. (2002). Performance of chillie under drip irrigation with mulch. Annual of the Sri Lanka Department of Agriculture, 4: 89–94.
1233. Nyhan, J. W. (2005). A seven-year water balance study of an evapotranspiration landfill cover varying in slope for semiarid regions.Soil Sci. Soc. of America J., 69: 466–480.
1234. O'Brien, D. M., Dumler, T. J., Rogers, D., H. (2011). Irrigation capital requirements and energy costs. KSU Farm Management Guide, MF-836. Manhattan, Kansas. 4 pages.
1235. O'Brien, J. J., Oberbauser, S. F. (2001). An inexpensive portable meter for measuring soil moisture. Soil Sci. Soc. of America J., 65: 1081–1083.
1236. O'Brien, D. M., Rogers, D. H., Lamm, F. R., Clark, G. A. (1998). An economic comparison of subsurface drip and center pivot sprinkler irrigation systems. Applied Engineering in Agriculture. 14(4): 391–398.
1237. O'Brien, D. M., Rogers, D. H., and Lamm, F. R. (1996). An economic comparison of alternative irrigation systems on small and irregularly shaped fields. Proc. Central Plains Irrigation Short Course, Burlington, CO, Feb. 6–7, (1996). CSU Dept. of Chemical and Bioresource Engineering, Ft. Collins, CO. pages 74–88.
1238. O'Brien, D. M., Rogers, D. H., Lamm, E. R., Clark, G. A. (1997). Economics of SDI for corn in western Kansas. ASAE Paper No. 97–2072. St. Joseph, MI: ASAE.
1239. O'Brien, D. M., Rogers, D. H., Lamm, F. R., and Clark, G. A. (1997). Irrigation system economics as affected by field size. Proc. Central Plains Irrigation Short Course, Colby, KS, Feb. 4, (1997). KSU Extension Ag. Engineering, Manhattan, KS. pages 81–93.
1240. Ochsner, T. E., Horton, R., Ren, T. (2001). Simultaneous water content, air-filled porosity and bulk density measurements with thermo-time domain reflectometry. Soil Sci. Soc. of America J., 65: 1618–1622.
1241. Ogallala Aquifer Project progress report. Can Subsurface Drip Irrigation (SDI) be a Competitive Irrigation System in the Great Plains Region for Commodity Crops, F. R. Lamm. Amarillo, Texas, March.
1242. Ogallala Aquifer Project written report. Evaluating TTT Technology for Irrigation of Corn in the Central Great Plains, Lamm, F. R., Amarillo, Texas, March 11–13, (2008).
1243. Ogallala Aquifer Project written report. Strategies for improved crop germination with SDI in Kansas and Texas, F. R. Lamm. Amarillo, Texas, March 11–13, (2008).
1244. Ogallala Aquifer Project written report. Comparison of LEPA and SDI for corn, sunflower, soybean and sunflower, F. R. Lamm. Amarillo, Texas, March 11–13, (2008).
1245. O'Halloran, T. F. (1997). Reported crop acreages by month for the imperial irrigation district. Imperial Irrigation District, Imperial, CA, USA.
1246. Ollier, C, Cardoso, F., DrInu, M. (2009). Summary results of the EU-27 orchard survey. In: *Statistics in Focus, Agriculture and Fisheries*. Eurostat. European Commission.
1247. Onder, D., Akiscan, Y., Onder, S., Mert, M. (2009). Effect of different irrigation water level on cotton yield and yield components. African Journal of Biotechnology, 8: 1536–1544.
1248. On-Farm Irrigation Committee of Irrigation and Drainage Division, (1987). Selection of irrigation for methods agriculture. New York, NY: American Society of Civil Engineers, 95.
1249. Onken, A. B., Wendt, C. W., Wilke, O. C., Hargrove, R. S., Bausch, W., Barnes, L. (1979). Irrigation system effects on applied fertilizer nitrogen movement in soil. Soil Sci. Soc. of America J., 43(2): 367–372.
1250. Oosthuizen, L. K., Botha, P. W., B. Groové and Meiring, J. A. (2005). Cost-estimating procedures for drip-micro and furrow irrigation systems. J. Water South Africa, 31(3): 403–406.
1251. *Operating instructions–soil moisture tester*. Delmhorst Instrument Co.
1252. Optimum Fertigation for Corn Using SDI at Central Plains Irrigation Conference, February 27–28, 2007, Kearney, Nebraska.

1253. Or, D. (2001). Who invented the tensiometer? Soil Sci. Soc. of America J., 65: 1–3.
1254. Or, D. (1995). Soil water sensor placement and interpretation for drip irrigation management in heterogeneous soil. Proceedings of the 5th International Micro irrigation Congress, ed. Lamm, E. R., 214–221 pages. St. Joseph, MI: ASAE.
1255. Orloff, S. B., Carlson, H. L., Hanson, B. R. (1997). Chapter 4: Irrigation. 25–40. In: *Intermountain Alfalfa Management* by Orloff, S. B., Carlson, H. L., L. R. Teuber (eds.). Pub, (3366). Davis, CA: University of California, Division of Agriculture and Natural Resources.
1256. Oron, G, DeMalach Y, Hoffman Z, Keren Y, Hartman H, and Plazner N. (1991). Wastewater disposal by subsurface drip irrigation. Water Sci. Tech., 23: 2149–2158.
1257. Oron, G., DeMalach, J., Hoffman, Z., Cibotaru, R. (1991). Subsurface micro irrigation with effluent. Journal of Irrigation and Drainage Engineering, 117(1): 25–36.
1258. Oron, G., Goemans, M., Manor, Y., Feyan, J. (1995). Polio virus distribution in the soil-plant system under reuse of secondary wastewater. Water Res., 29(4): 1069–1078.
1259. Oron, G., Y DeMalach, Hoffman, Z., Manor, Y. (1992). Effect of effluent quality and application method on agricultural productivity and env. Control. Wat. Sci. Tech., 26(7–8): 1593–1601.
1260. Oron, G., DeMalach, Y., Gillerman, L., David, I. (1995). Pear response to saline water application under subsurface drip irrigation. Proceedings of the 5th Int'l. Micro irrigation Congress, ed. Lamm, F. R., 97–103 pages. St. Joseph, MI: ASAE.
1261. Oron, G., DeMalach, Y., Gillerman, L., David, I., Rao, V. P. (1999). Improved saline-water use under subsurface drip irrigation. Agric. Water Manage. 39(1): 19–33.
1262. Oron, G., DeMalach, Y., Hoffman, Z., Keren, Y., Hartman, H., Plazner, N. (1991). Wastewater disposal by sub-surface trickle irrigation. Water Sci. Tech., 23: 2149–2158.
1263. Oron, G., DeMalach, Y., Hoffman, Z. (1989). Subsurface trickle irrigation of alfalfa with treated wastewater. In: Progress Report. Israel: Ben Gurion Univ., Institute of Desert Research.
1264. Orzolek, M. D., Murphy, J. H. (1993). The effect of colored polyethylene mulch on the yield of squash and pepper. Proceedings of National Agriciulture Plastics Congress, 24: 157–161.
1265. Osmond, D. L., Hardy, D. H. (2004). Characterization of turf practices in five North Carolina communities. Journal of Environmental Quality, 33: 565–575.
1266. Oster, J. D., Shainberg, I., Abrol, I., P. (1999). Reclamation of salt-affected soils. In: *Agricultural Drainage*, 669–672. ASA Monograph No. 38. Skaggs, R. W., J. Van Schilfgaarde, eds. Madison, Wisc.: ASA, CSSA, SSSA.
1267. Oster, J. D., Willardson, L. S., Hoffman, G. J. (1973). Sprinkling and ponding techniques for reclaiming saline soils. Trans. ASAE, 16(1): 115–117.
1268. Oster, J. D. (1992). Water Quality Impacts: issues, information sources and mitigation strategies. Proc. of Reclaimed Wastewater: Practical Approaches to Developing an Alternative Water Supply, Davis, Ca. May 18. Davis, U. C., CA 95616.
1269. Oster, J. D. (1994). Irrigation with poor quality water: Review article. Agricultural water management, 25: 271–297.
1270. Oster, J. D., I. Shainberg and Abrol, I. P. (1996). Reclamation of salt-affected soil. In: *Soil erosion, conservation, and rehabilitation*, ed. Agassi, M., Chapter 14: 315–335. New York: Marcel Dekker, Inc.
1271. Ould Ahmed, B. A., Yamamoto, T., Inoue, M., and Anyoji, H. (2006). Drip irrigation scheduling for sorghum under greenhouse condition. Trans. JSIDRE J., 244: 133–141.
1272. Ould Ahmed, B. A., Yamamoto, T., Rasiah, V., Inoue, M., and Anyoji, H. (2007). The impact of saline water irrigation management options in a dune sand on available soil water and its salinity. Agricultural Water Management Journal, **88**: 63–72.
1273. Ould-Ahmed, B. A., Yamamoto, T., Inoue M., and Dehghanisanij, H. (2007). Evaluation of sorghum density as affected by two water qualities under drip irrigation system. American Journal of Environmental Sciences. 3(4): 241–246.
1274. Overrein, L. N., and Mac, P. G. (1967). Factors effecting urea hydrolysis and ammonia volatilization in soil. Soil Science Society of America Journal, 31: 57–61.

1275. Ozay, M., Bicak, H. A. (2002). Modern and traditional irrigation technologies in the eastern Mediterranean. Chapter 9, 90–97.
1276. Özekici, Sneed, R., E. (1995). Manufacturing variation for various trickle irrigation on-line emitters. Applied Engr. in Agric., 11(2): 235–240.
1277. Oztekin T., Brown, L. C., Holdsworth, P. M., Kurunc, A. (1999). DRAINMOD: Modeling, runoff, subsurface drainage, wastewater irrigation. Applied Engineering in Agriculture, 15(5): 449–455.
1278. Oztekin, S., Soysal, Y. (2000). Comparison of adsorption and desorption isoteric heats for some grains. Agricultural Engineering International: CIGR Journal of Scientific Research and Development, 2(4): 2–16.
1279. Kramer, P. J. (Editor), (1984). *Crop Water Requirements*. INRA Conf., 11–14 September, INRA, Paris, pages 221–234.
1280. Padhye, A. H. (1990). Micro and Sprinkler Irrigation in India. Proc. Int. Congress on the Use of Plastics in Agriculture, New Delhi.
1281. Page, A. L., Miller, R. H., and D. R. Keeney (eds). *Agronomy Monograph No. 9*, ASA and SSSA: Madison, WI.
1282. Pair, Claude H. (1983). Irrigation. 5th Edition. Irrigation Association. 10–45.
1283. Paknejad, Farzad, Fatemeh Majidi Fakhr and Seied Mehdi Mirtaheri, (2012). Validation of the CERES-wheat for prediction of wheat varieties in irrigation and terminal drought stress. American Journal of Agricultural and Biological Sciences, 7(2): 180–185.
1284. Palacios, M. P., Haman, D. Z., Del-Nero, E., Pardo, A., Pavon, N. (2000). Banana production irrigated with treated effluent in the Canary Islands. Trans. ASAE, 43(2): 309–314.
1285. Palacios, M., Pampillon, J. F., Rodriguez, M. E. (2002). Organohalogenated compounds levels in chlorinated drinking waters and current compliance with quality standards throughout the European Union. Water Research, 34(3): 1002–1016.
1286. Panagiotis, V., Sakellariou-Makrantonaki, M. (2005). Intermittent water application through surface and subsurface drip irrigation. ASAE Annual Meeting. 22–78.
1287. Papadopoulos, I. (1996). Use of saline and brackish waters for irrigation in Cyprus. Mediterranean colloquim on protected cultivation, October 6–9. Marocco: A6.1 to A6.22.
1288. Papadopoulos, I. (1985). Constant feeding of field-grown tomatoes irrigated with sulfate water. Plant and Soil, 88: 213–236.
1289. Papadopoulos, I. (1987). Nitrogen fertigation of greenhouse-grown tomato. Communication in Soil Science and Plant Analysis, 18: 897–907.
1290. Papadopoulos, I. (1988). Nitrogen fertigation of trickle irrigated potato. Fertilizer Research, 16: 157–167.
1291. Papadopoulos, I. (1989). Report on fertigation consultancy mission in Egypt. FAO of the United Nations, Rome, 31 pages.
1292. Papadopoulos, I. (1990). The role of fertigation and chemigation in increasing productivity and efficient use of inputs. FAO Proceedings Regional Consultation Meeting on Efficient Resource Use In Near East Agriculture. Amman, Jordan.
1293. Papadopoulos, I. (1993). Environmentally sound water management of protected agriculture under Mediterranean and arid climates. Bari, Italy, (1993).
1294. Papadopoulos, I. (1993). Fertigation of vegetables under protected conditions. Regional consultative meeting on Greenhouse production in the Mediterranian Region. 15–18 November. Agadir, Morocco.
1295. Papadopoulos, I. (1993). Regional Middle East and Europe projection Nitrogen Fixation and water balance studies. Amman, Jordan, 18–27 October. RER /5/004–62 pages.
1296. Papadopoulos, I. (1994). Irrigation/fertigation research and application at farmers level in Cyprus. Expert consultation on research and extension in effective water use at farm level in the near east region. Cairo, Egypt.
1297. Papadopoulos, I. (1995). Regional Workshop on water Balance and Fertigation for Crop Management. 24 September–10 October, Annually-Turkey, (1995).

1298. Papadopoulos, I. (1997). Fertigation of irrigated Crops: Step by Step Approach. Seminar on Fertigation Management. Beirut, April 12, (1996).
1299. Papadopoulos, I., and Eliades, G. (1987). A fertigation system for experimental purposes. Plant and Soil, 102: 141–143.
1300. Papadopoulos, I., and Stylianou, Y. (1988). Trickle irrigation of cotton with treated sewage effluent. J. Environm. Qual. 17 (1988) 574–580.
1301. Parchomchuk, P., Neilsen, G. H., and Hague, E. J. (1993). Effects of drip fertigation of NH4-N and P on soil pH and cation leaching. Canadian Journal of Soil Science, 73: 157–164.
1302. Park, H. S., Hwang, T. M., Kang, J. W. H., Choi and Oh, H. J. (2001). Characterization of raw water for the ozone application measuring ozone consumption rate. Water Research, 35(11): 2607–2614.
1303. Parsons, L. R., Wheaton, T. A., Cross, P. (1995). Reclaimed municipal water for citrus irrigation in Florida. In: Micro irrigation for a changing world. Proc 5th Int. Micro irrigation Congress. Lamm, F., R. (Ed.), St. Joseph, MI: ASAE. pages 262–268.
1304. Parsons, L. R., Wheaton, T. A., Castle, W., S. (2001). High application rates of reclaimed water benefit citrus tree growth and production. HortScience, 36(7): 1273–1277.
1305. Parsons, S., Judd, S. J., Stephenson, T., Udol, S., and Wang, B. L., 1997). Magnetically augmented water treatment. Trans. Inst. Chem. Eng, 75: 98–104.
1306. Pastor, M., Orgaz, F. (1994). Riego deficitario del olivar: Los programas de recorte de riego en olivar. Agricultura, 746: 768–776 (in Spanish).
1307. Patterson, H. D., and Thompson, R. (1971). Recovery of inter-block information when block sizes, are unequal. Biometrika, 58: 701–714.
1308. Pathak, B. K., Kazama, F., Iida, T. (2004). Monitoring of nitrogen leaching from a tropical paddy field in Thailand. Agricultural Engineering International: CIGR Journal of Scientific Research and Development, 6(4): 2–9.
1309. Payne, W. A. (1999). Shallow tillage with a traditional West African hoe to conserve oil water. Soil Sci. Soc. of America J., 63: 972–976.
1310. Payne, W. A. (1999). Shallow tillage with a traditional West African hoe to conserve oil water. Soil Sci. Soc. of America J., 63: 972–976.
1311. Payne, W. A., B. Gerard and Klaij, M. C. (1995). Subsurface drip irrigation to evaluate transpiration ratios of pearl millet. Proceedings of the 5th International Micro irrigation Congress, ed. Lamm, E. R., 923–931. St. Joseph, MI: ASAE.
1312. Payne, W. A., C. Chen and Ball, D. A. (2004). Agronomic potential of narrow-leafed and white lupins in the Inland Pacific Northwest.Agronomy Journal, 96: 1501–1508.
1313. Pearce, A. J., J.Gash, H. C., Stewart, J. B. (1980). Rainfall interception in a forest stand estimated from grassland meteorological data. Journal of Hydrology, 46: 147–163.
1314. Pelton, W. L., King, K. M., Tanner, C. B. (1960). An evaluation of the Thornthwaite and mean temperature methods for determining potential evapotranspiration. Agronomy Journal, 52: 387–395.
1315. Peng Y. H., and Rabe. E. (1998). Effect of differing irrigations regimes on fruit quality, yield, fruit size and net CO_2 assimiliation on "Mihbowase" satsuma. J. Hort. Sci. Biotecnology, 73(2): 229–234.
1316. Peng, S., Krieg, D. R., Hicks, S. K. (1989). Cotton response to accumulated heat units and soil water supply. Field Crops Res., 19(4): 253–262.
1317. Penman, H. L. (1948). Natural evaporation from open water, bare soil and grass. Proceedigs of the Royal Society London, 193: 120–146.
1318. Penman, H. L. (1949). The dependence of transpiring on climate and soil conditions. Journal of Soil Sciences, 1: 74–89.
1319. Penman, H. L. (1963). *Vegetation and hydrology*. Tech. Comm. No. 53, Commonwealth Bureau of Soils, Harpenden, England. 125.
1320. Pereira, J. J., Mohamad Barzani Gasim, Sharifah Mastura, S. A., and Nor Azlina Abdul Aziz, (2010). Issues of Climate Change and Water Resources in Peninsular Malaysia: The Case of Northern Kedah. The Arab World Geographer, 12(1–2): 87–94.

1321. Pereira, L. S. (1989). Mitigation of droughts: Irrigation. ICID Bull. 38, pages 16–34.
1322. Pereira, L. S. (1998). Water and soil management for sustainable agriculture in the Huang-Huai-Hai rivers plain (North China). Final Report of EC Research Contract CT93–250, Instituto Superior de Agronomia, Lisbon.
1323. Pereira, L. S. (1999). Higher performances through combined improvements in irrigation methods and scheduling: a discussion. Agric. Water Manage. 40, 153–169.
1324. Pereira, L. S., Oweis, T., Zairi, A. (2002). Irrigation management under water scarcity. Agric. Water Manage. 57(3): 175–206.
1325. Pereira, L. S., Smith, M. (1989). Proposed procedures for revision of guidelines for predicting crop water requirements.Land and Water Use Division, FAO Rome. 36.
1326. Pereira, L. S., Allen, R. G. (1998). Crop water requirements. In: *Handbook of agricultural engineering*, Chapter 1.5.1. CIGR and ASABE.
1327. Pereira, L. S., Perrier, A., Allen, R. G., Alves, I. (1996). Evapotranspiration: Review of concepts and future trends. In: *Evapotranspiration and irrigation scheduling*, eds. Camp, C. R., Sadler, E. J., R. E. Yoder. 109–115. Journal of Irrigation and Drainage Engineering, ASCE 25. New York: ASAE.
1328. Pereira, L. S., B. J. van den Broek, P. Kabat and Allen, R. G. (1995). *Crop-water simulation models in practice*. Wageningen: Wageningen Academic Publishers, 339.
1329. Pereira, L. S., Teixeira, J. L., Pereira, L. A., Ferreira, M. I., and Fernando, R. M. (1987). Simulation models of crop response to irrigation management: research approaches and needs. In: *Simulation models for cropping systems in relation to water management*, ed. Feyen, J., pages 19–36. Luxembourg.
1330. Pereira, L. S., M. A. Ait Kadi Perrier, Kabat, P. (1992). *Crop water models*. Special issue of the ICID Bulletin, 200.
1331. Pereira, L. S., Oweis, T., Zairi, A. (2002). Irrigation management under water scarcity. Agricultural Water Management, 57(3): 175–206.
1332. Pérez-Sarmiento F., Alcobendas, R., Mounzer, O., Alarcón, J., Nicolás, E. (2010). Effects of regulated deficit irrigation on physiology and fruit quality in apricot trees. Spanish Journal of Agricultural Research, 8(S2): S86-S94.
1333. Performance of drip system. Appl. Engr. Agric., 12(3): 307–313.
1334. Perrier, A., Tuzet, A. (1991). Land surface processes: Description, theoretical approaches, and physical laws underlying their measurements. In: *Land surface evaporation: measurement and parameterization*, eds. Schmugge, T. J., André, J. C., 145–155. Berlin: Springer-Verlag.
1335. Perrier, A. (1978). Importance des définitions de l'évapotranspiration dans le domaine pratique de la mesure, de l'estimation of de la notion de coefficients culturaux. XV' Journal of Hydraulics, Société Hydrotechnique de France, Question IV, Rapport 1: 1–7 (in French).
1336. Perrier, A. (1982). Land surface processes: vegetation. In: *Land surface processes in atmospheric general circulation models*, ed. Eagleson, P. S., 395–448 pages. Cambridge, MA: Cambridge University Press,.
1337. Perrier, A. (1985). Updated evapotranspiration and crop water requirement definitions. In: Perrier, A., Riou, C. (eds) *Crop water requirements* (ICID Int. Conf., Paris, Sept. 1984). INRA, Paris. 885–887.
1338. Perrier, A., Itier, B., Bertolini, J. M., Katerji, N. (1976). A new device for continuous recording of the energy balance of natural surfaces. Agricultural Meteorology, 16(1): 71–85.
1339. Perrier, A., Pereira, L., Segeren, A. (1992). *Report of the expert consultation on procedures for revision of FAO guidelines for prediction of crop water requirements*. UN-FAO, Rome, Italy. 54.
1340. Perrier, A., Katerji, N., Gosse, G., Itier, B. (1980). In situ study of evapotranspiration rates for a wheat crop. Agricultural Meterology, 21: 295–311. (in French).
1341. Perrier, A., Archer, P., B. de Pablos, (1974). Etude de l'évapotranspiration réelle et maximele de diverses cultures. I: Dispositif et mesure. Annual Agronomy, 25(3): 229–243.
1342. Persaud, T., Goyal, M. R., Bellerive, P. (1988). Cost of drip irrigation system for vegetables farming in Puerto Rico. J. Agric. U. P.R., 72(1): 31–40.

1343. Persson, M., Wraith, J. (2002). Shaft mounted time domain reflectometry probe for water content and electrical conductivity measurements. Vadose Zone Journal, 1: 316–319.
1344. Persson, M., Sivakumar, B., Berndtsson, R., Jacobsen, O. H., Schjonning, P. (2002). Predicting the dielectric constant-water content relationship using artificial neural networks. Soil Sci. Soc. of America J., 66: 1424–1429.
1345. Peterson Jr., C. (1998). Subsurface drip irrigation saves water. Feature article about SW Kansas farmers using SDI. Successful Farming, March, (1998).
1346. Pettygrove, S. (1992). Nitrogen in Reclaimed Wastewater: Asset or Liability? Proc. of Reclaimed Waste Water: Practical Approaches to Developing an Alternative Water Supply, Davis Ca. May 18. Davis, U. C., CA 95616.
1347. Phansalkar, S. J. (2000). Appropriate dripirrigation technologies: A socio-economic assessment. New Delhi, India: International Development Enterprises (IDE).
1348. Phene, C. J. (1990). Drip Irrigation Saves Water. Proc. of Conserve 90, August, Phoenix, Arizona, pages 645–650.
1349. Phene, C. J. (1974). High-frequency porous tube irrigation for water-nitrogen management in humid regions. Proceedings of the 2nd International Drip Irrigation Congress, 166–171 pages. Riverside, CA: University of California.
1350. Phene, C. J. (1986). Fertilization of high yilding subsurface trickle irrigated tomatoes. Proc. of the 34th Fertilizer Conference. Califorina Fertilizer Assoc. Fresno, California. pages 33–34.
1351. Phene, C. J. (1995). The sustainability and potential of subsurface drip irrigation. Proceedings of the 5th International Micro irrigation Congress. pages 359–367, ASAE: St Joseph, MI.
1352. Phene, C. J. (1995). Research trends in micro irrigation. Proc. 5th Int. Micro irrigation Congress, Orlando, FL. Ed. Lamm, F. R., pages 6–24. St. Joseph, MI.: ASAE.
1353. Phene, R. C. (1996). Real-time irrigation scheduling with automated evaporation pan systems. Proceedings of the International Conf on Evapotrans-piration and Irrigation Scheduling, eds. Camp, C. R., Sadler, E. J., and Yoder, R. E., 1093–1098. St. Joseph, MI: ASAE.
1354. Phene, C. J., Bar-Yosef, B., Hutmacher, R. B., Patton, S. H., Davis, K. R., McCormick, R. L. (1986). Fertilization of high-yielding subsurface trickle irrigated tomatoes. Proceedings of the 34th Ann. Calif. Fertilizer Confernce, 33–43 pages, Fresno, CA.
1355. Phene, C. J., Beale, O., W. (1976). High-frequency irrigation for water nutrient management in humid regions. Soil Sci. Soc. Am. J., 40(3): 430–436.
1356. Phene, C. J., Beale, O. W. (1979). Influence of twin-row spacing and nitrogen rates on high-frequency trickle-irrigated sweet corn. Soil Sci. Soc. of America J., 43(6): 1216–1221.
1357. Phene, C. J., Dark, D. A., Cardon, G. E. (1996). Real time calculation of crop evapotranspiration using an automated pan evaporation system. In: Evaporation and irrigation scheduling, eds. Camp, C. R., Sadler, E. J., Yoder, R. E., 189–194 pages. ASCE.
1358. Phene, C. J., Davis, K. R., McCormick, R. L. (1987). Evapotranspiration and irrigation scheduling of drip irrigated cantaloupes. ASAE Paper No. 87–2526. St. Joseph, MI: ASAE.
1359. Phene, C. J., Davis, K. R., Hutmacher, R. B., and McCormick, R. L. (1987). Advantages of subsurface irrigation for processing tomatoes. Acta Horticulturae, 200: 101–114.
1360. Phene, C. J., Davis, K. R., Hutmacher, R. B., Bar-Yosef, B., Meek, D. W., Misaki, J. (1991). Effect of high frequency surface and subsurface drip irrigation on root distribution of sweet corn. Irrgaton Science, 12(2): 135–140.
1361. Phene, C. J., DeTar, W. R., Clark, D. A. (1992). Real-time irrigation scheduling of cotton with an automated pan evaporation system. Applied Engineering in Agriculture, 8(6): 787–793.
1362. Phene, C. J., Howell, T. A. (1984). Soil sensor control of high frequency irrigation.Trans. of ASAE, 27(2): 386–391, 396.
1363. Phene, C. J., Hutmacher, R. B., Ayars, J. E. (1992). Subsurface drip irrigation: Realizing the full potential. In: Proc. of Conference on Subsurface Drip Irrigation, pages 137–158. CATI Publication 921001. Fresno, Calif.: California State Univ. USA.
1364. Phene, C. J., Hutmacher, R. B., and Davis, K. R. (1992). Two hundred tons per hectare of processing tomatoes-Can we reach it? HortTechnology, 2(1): 16–22.

1365. Phene, C. J., Hutmacher, R. B., Davis, K. R., McCormick, R. L. (1990). Water- fertilizer management of processing tomatoes. Acta Horticulturae, 277: 137–143.
1366. Phene, C. J., Fouss, J. L., Sanders, D. C. (1979). Water – nutrient – herbicide management of potatoes with trickle irrigation. Am. Potato J., 56: 51–59.
1367. Phene, C. J., McCormick, R. L., Davis, K. R., Pierro, J. D., and Meek, D. W. (1989). A lysimeter feedback irrigation controller system for evapotranspiration measurements and real time irrigation scheduling. Trans. of ASAE, 32(2): 477–484.
1368. Phene, C. J., Ruskin, R. (1995). Potential of subsurface drip irrigation for management of nitrate in wastewater. In Proceedings of the 5th International Micro irrigation Congress, ed. Lamm, F. R., 155–167 pages. St. Joseph, MI: ASAE.
1369. Phene, C. J., Sanders, D. C. (1976). High-frequency trickle irrigation and row spacing effects on yield and quality of potatoes. Agron. J., 68(4): 602–607.
1370. Phene, C. J., Itier, B., Reginato, R. J. (1990). Sensing irrigation needs. Proceedings of the Third National Irrigation Symposium, Oct. 28- Nov. 1, Phoenix, AASAE, Z., St Joseph, MI. 761 pages.
1371. Phene, C. J., Yue, R., I Pai Wu, Ayars, J. E., Schoneman, R. A., Meso, B. (1992). Distribution uniformity of subsurface drip irrigation systems. ASAE Paper No. 92–2569, 14 pages, St. Joseph, MI: ASAE.
1372. Philip, J. R. (1968). Steady infiltration from buried point sources and spherical cavities. Water Resources Research, 4(5): 1039–1047.
1373. Phillip, D., Brown, C. P., Allen, V. G., Webster, D. B. (2006). Influence of irrigation on mineral concentration in three bluestem species. Crop Science Journal, 46(5): 2033–2040.
1374. Phocaides, A. (2000). *Technical Handbook on Pressurized Irrigation Techniques*, FAO, Rome.
1375. Pielou, E. C. (2000). *Fresh water.*University of Chicago Press. 67–68.
1376. Pier, J. W. (1997). Adoption of subsurface drip irrigation for cotton in west Texas. Proceedings of the Beltwide Cotton Conference. 1: 655–656. National Cotton Council: Memphis, TN.
1377. Pier, J. W., Doerge, T. A. (1992). What happens near a quasi-linear point source? Water Resources. Res., 28(1): 47–52.
1378. Pier, J. W., Doerge, T. A. (1995). Nitrogen and water interactions in trickle-irrigated watermelon. Soil Sci. Soc. of America J., 59(1): 145–150.
1379. Pier, J. W., Doerge, T. A.,1995. Concurrent evaluation of agronomic, economic, and environmental aspects of trickle-irrigated watermelon production. Journal of Environmental Quality, 24(1): 79–86.
1380. Pierzgalski, E. (1995). Application of subsurface irrigation on a hop plantation. Proc. of the 5th International Micro irrigation Congress, ed. Lamm, F. R., 729–734 pages. St. Joseph, MI: ASAE.
1381. Pikul, J. L. Jr., Aase, J. K., Cochran, V. L. (2004). Water use and biomass production of oat–pea hay and lentil in a semiarid climate.Agronomy Journal, 96: 298–304.
1382. Pimentel, D., Pimentel, M., Karpenstein-Machan, M. (1999). Energy use in agriculture: An Overview. Agricultural Engineering International: CIGR Journal of Scientific Research and Development. Volume I.
1383. Piper, B. S. (1989). Sensitivity of Penman estimates of evaporation to errors in input data. Agricultural Water Management, 15: 279–300.
1384. Pitts, D. J., Clark, G. A. (1991). Comparison of drip irrigation to subsurface for tomato production in southwest Florida. Applied Engineering In Agriculture, 7(2): 177–184.
1385. Pitts, D. J., Haman, D. Z., Smajstrla, A. G. (1990). Causes and prevention of emitter plugging in micro irrigation systems. Univ. of Florida, Cooperative Extension Service, Bulletin 258. 12 pages.
1386. Pitts, D., Peterson, K., and Gilbert, G., and Fastenau, R. (1996). Field assessment of irrigation system performance. Appl. Eng. Agric., 12: 307–313.
1387. Plant, Z., Carmi, A., Grava, A. (1996). Cotton root and shoot responses to subsurface drip irrigation and partial wetting of the upper soil profile. Irrigation Science, 16(3): 107–113.

1388. Plant, Z., Rom, M., Meiri, A. (1985). Cotton response to subsurface trickle irrigation. In Proceedings of the 3rd International Drip/Trickle Irrigation Congress, Volume II: 916–920. St. Joseph, MI: ASAE.
1389. Plaut, Z., Ben-Hur, M. (2005). Irrigation management of peanut with a moving sprinkler system: Runoff, yield and water use efficiency. Agronomy Journal, 97(1): 1202–1209.
1390. Playan, E., Mateos, L. (2006). Modernization and optimization of irrigation systems to increase water productivity. Agricultural Water Management, 80(1–3): 100–116.
1391. Playán, Ortiz, R., Royo, A., 1999, September. A new drip-injection irrigation system for crop salt tolerance evaluation.Soil Sci. Soc. of America J., 63: 1397–1403.
1392. Pochi, D., Vannucci, D. (2002). Prediction of pesticide distribution on the ground based on sprayer boom movements. Agricultural Engineering International: CIGR Journal of Scientific Research and Development, IV(19): 12–22.
1393. Pocock, Richard. *Irrigation System Design Manual*. Bauer GmBH-Australia & New Zealand.
1394. Polomski, Bob, (2001). *Month-by-month gardening in the Carolinas*. Franklin, TN: Cool Springs Press. 241.
1395. Postel, S., Polak, P. (2005). Drip irrigation for small farmers: A new initiative to alleviate hunger and poverty. New Delhi, India: International Development Enterprises. 1–21.
1396. Postel, S. (1999). Pillars of sand: Can the irrigation miracle last? W. W. Norton & Co, 313.
1397. Postel, S., Polak, P., Gonzales, F., Keller, J. (2001). Drip irrigation for small farmers: A new initiative to alleviate hunger and poverty. Water International, 26(1): 3–13.
1398. Postel, Sandra and Paul Polak, (2002). Drip irrigation for small farmers: a new initiative to alleviate hunger and poverty. New Delhi: International Development Enterprises.
1399. Powell, N. L., Wright, E. S. (1993). Grain yield of subsurface microirrigated corn as affected by irrigation line spacing. Agronomy Journal, 85(6): 1164–1169.
1400. Pratt, P. F., Lund, L. J., and Rible, J. M. (1978). An approach to measuring leaching of nitrate from freely drained irrigated fields. In: *Nitrogen in the environment*, (Eds Nielsen, D. R., J. G. MacDonald) pages 223–256. Academic Press: New York.
1401. Preece, J. E., Read, P. E. (2005). *The biology of horticulture: An introductory textbook*. John Wiley & Sons. 221–222.
1402. Prevatt, J. W., Clark, G. A., and Stanley, C. D. (1992). A comparative cost analysis of vegetable irrigation systems. HortTechnology, 2(1): 91–94.
1403. Prevatt, J. W., Stanley, C. D., Gilreath, P. R., Clark, G. A. (1992). Return-risk analysis of adopting drip irrigation. Applied Engineering in Agriculture, 8(1): 47–52.
1404. Price, J. F. (2003). Dimensional analysis of models and data sets. Am J Phys, 71: 437–447.
1405. Priestley, C. H.B., Taylor, R. J. (1972). On the assessment of surface heat flux and evaporation using large scale parameters. Monthly Weather Review, 100: 81–92.
1406. Prinz, D. (1996). Water harvesting – past and future. In: Pereira LS, Feddes RA, Giley JR, Lessafre B (Eds), *Sustainability of Irrigated Agriculture*. Kluwer Academic Publishers, Dordrecht, pages 137–168.
1407. Priyanjith, K. J.K. T., D.Kuruppuarachchi, S. P., Gunathilaka, H. A. (2002). Field evaluation of drip irrigation system for small scale banana (musa) orchards. Makandura, Sri Lanka: Wayamba University of Sri Lanka. Faculty of Agriculture and Plantation Management. 185–193.
1408. Provenzano, G., Pumo, D. (2006). Experimental analysis of local pressure losses for micro irrigation laterals. Journal of Irrigation and Drainage Engineering, 132(2): 193–194.
1409. Pruitt, W. O. (1991). Development of crop coefficients using lysimeters. Proceedigs of the ASCE International Symposium on Lysimetry, eds. Allen, R. G., et al., *Lysimeters for Evapotranspiration and Environmental Measurements*, Honolulu, HA. New York, NY: ASCE, 182–190.
1410. Pruitt, W. O. (1996). Empirical method of estimating evapotranspiration using primarily evaporation pans. Proceedings of the Conference on Evapotranspiration and its Role in Water Resources Management. December, Chicago. New York: ASAE, 57–61.
1411. Pruitt, W. O., Doorenbos, J. (1977). Background and development of methods to predict reference crop evapotranspiration (ETo). Appendix 11, In: FAO-ID-24, 108–119.

1412. Pruitt, W. O., Fereres, E., Martin, P. E., Singh, H., Henderson, D. W., Hagan, R. M., Tarantino, E., Chandio, B. (1984). Microclimate, evapotranspiration, and water-use efficiency for drip- and furrow-irrigated tomatoes. Proceedings of the 12th Congress, International Commission on Irrigation and Drainage, Ft. Collins, CO., 367–394.
1413. Pruitt, W. O., Lourence, F. J. (1985). Experiences in lysimetry for ET and surface drag measurements. *Advances in Evapotranspiration*. St. Joseph, MI: ASAE, 51–69.
1414. Pruitt, W. O. (1986). Traditional methods. Evapotranspiration research priorities for the next decade. ASAE Paper No. 86–2629. 23.
1415. Pruitt, W. O., Morgan, D. L., Lourence, F. J. (1973). Momentum and mass transfers in the surface boundary layer. Quarterly Journal of the Royal Meteorological Society, 99: 370–386.
1416. Pruitt, W. O., Swann, B. D. (1986). Evapotranspiration studies. In: *N. S.W.: Daily vs. hourly meteorological data*. Irrigation 1986, Darling downs Institute of Advanced Education, Toowoomba, Queensland, Australia, 29.
1417. Ptacek, L. R. (1986). Subsurface irrigation and the use of chemicals. Proc. of the Irrigation Association Annual Conference. 225–234.
1418. Puckett, W. E., Dane, J. H. (1981). Testing tensiometers by a vacuum method. Soil Science, 132: 444–445.
1419. Puig-Bargués, J., Arbat, G., Barragán, J., and Cartagena, F. R. (2005). Effluent particle removal by micro irrigation system filters. Spanish Journal of Agricultural Research, 3(2): 182–191.
1420. Puig-Bargués, J., Barragán, J., and Cartagena, F. R. (2005). Development of equations for calculating the head loss. Spanish Journal of Agricultural Research, 3(2).
1421. Puig-bargués, J., Arbat, G., Barragán, J., and Cartagena F. R. (2003). Hydraulic performance of drip irrigation subunits using WWTP effluents. Proc. VI Inter-Regional Conference on Environment – Water, Albacete, Spain. September 3–5, pages 97–98.
1422. Puig-Bargués, J., Lamm, F. R., Trooien, T. P., Clark, G. A. (2010). Effect of dripline flushing on subsurface drip irrigation systems. Trans. of the ASABE, 53(1): 147–155.
1423. Pulver, E. L., and Nguyen, V. N. (1998). Sustainable rice production issues for the third millennium. In: Proceedings of the 19th Session of the International Rice Commission, FAO.
1424. Punmia, B. C., Pande, B. B. L. (2005). Irrigation and water power engineering. New Delhi: Laxmin Publications Pvt. Ltd. Page 89.
1425. Qassim, A. (2003). Micro-irrigation: A situation analysis. 1–5.
1426. Quiñones, A., Martínez-Alcántara, B., and Legaz, F. (2007). Influence of irrigation system and fertilization management on seasonal Distribution of N in the soil profile and on N- uptake by citrus trees. Agri. Ecosyst. Environ., 122: 399–409.
1427. Quiñones, A., Martínez-Alcántara, B., San-Francisco, S., García-Mina, J. M., Legaz, F. (2011). Methyl Xanthine as a potential alternative to Gibberellic acid in enhancing fruit set and quality in clementine citrus trees in Spain. Experimental Agricultural, 47(1): 159–171.
1428. Quinones, H., Ruelle, P. (2001). Operative calibration methodology of a TDR sensor for soil moisture monitoring under irrigated crops. Subsurface Sensing Technologies and Applications, 2: 31–45.
1429. Raats, P. A.C. (1971). Steady infiltration from point sources, cavities, and basins. Soil Sci. Soc. Am. Proc., 35: 689–694.
1430. Raats, P. A.C. (1972). Steady infiltration from sources at arbitrary depth. Soil Sci. Soc. Am. Proc., 36: 399–401.
1431. Radam, A., Ismail, A. L. (1995). Off farm labor decisions by farmers in Northwest Selangor Integrated Agricultural Development Project (IADP) in Malaysia. Bangladesh Journal of Agricultural Economics, 18(2): 51–61.
1432. Radin, J. W., Reaves, L. L., Mauney, J. R., and French, O. F. (1992). Yield enhancement in cotton by frequent irrigation during fruiting. Agron. J., 48: 551–557.
1433. Raes, D., Lemmens, H., P. Van Aelst, M. Vanden Bulcke, Smith, M. (1988). *IRSIS: Irrigation scheduling information system*. Reference Manual No. 3. Institute for Land and Water Management, Belgium.

1434. Rafiq, M., M. M.Alam, (2004). Annual Report 2003–04, International Waterlogging and Salinity Research Institute, Lahore, December, (2004).
1435. Rahimiyan, M. H., Kakhaki, A. (2007). Calculate the water requirements of cotton plants by lysimeter method in the region Kashmir. Journal of Soil and Water Sciences, 21(1): 141–145.
1436. Raine, S. R., Curran, N. (2007). *IPART Irrigation Performance Audit and Reporting Tool, User Manual*. National Center for Engineering in Agriculture Publication 1002475/1, USQ, Toowoomba-AU.
1437. Raine, S. R., Foley, J. P., Henkel, C. R. (2000). Drip irrigation in the Australian cotton industry: a scoping study. National Center for Engineering in Agriculture Publication 179757/2, USQ, Toowoomba-AU.
1438. Raine, S. R., Purcell, J., Schmidt, E. (2005). Improving whole farm and infield irrigation efficiencies using Irrimate™ Tools. Irrigation Assoc. of Australia National Conference, Townsville, QLD-AU.
1439. Raine, S. R., Smith, R. J. (2006). Simulation modelling for surface irrigation evaluation. training in using SIRMOD, InfiltV5 and IPARM. National Center for Engineering in Agriculture Publication 1000008/5, USQ, Toowoomba – AU.
1440. Raleigh, M. (2005). Development of Micro irrigation Systems. Irrigation Science, 10: 47–50.
1441. Ramos, C., Agut, A., and Lidon A. L. (2002). Nitrate leaching in important crops of the Valencian community region (Spain). Environmental Pollution 118: 215–223.
1442. Rangel, V. (2000). Lagoon wastewater re-use project in third season. Irrigation J., 50(5): 17.
1443. Ravalo, E. J., Goyal, M. R. (1988). Water requirements of rice in Lajas Valley of Puerto Rico. Dimension CIAPR, 2(8): 33–35. Jan-March issue.
1444. Ravalo, E. J., Goyal, M. R., Almoldóvar, C. R. (1986). Average monthly and annual rainfall distribution in Puerto Rico. J. Agric. U. P.R., 70 (4): 267–275.
1445. Ravina, I., Sagi, G., Paz, E., Schischa, A., Yechiely, Z., Sofer, Z., and Lev, Y. (1993). Filter performance and emitter clogging in conditions of effluent reservoir water. Proc. 6th Int. Conf. on Irrigation, Agritech, Tel-Aviv, Israel, pages 19.
1446. Rawlins, S. L., P. A.Raats, C. (1975). Prospects for high-frequency irrigation. Science, 88: 604–610.
1447. Ray, C. (2001). Managing nitrate problems for domestic wells in irrigated alluvial, aquifers. Journal of Irrigation and Drainage Engineering, 127(1): 49–53.
1448. Ray, C., Dordrecht, D. (2002). *Riverbank filtration: Understanding contaminant bio-geochemistry and pathogen removal*. Boston: Kluwer Academic Publishers.
1449. Rayment, G. E., and Higginson, F. R. (1992). Australian laboratory handbook of soil and water chemical method. Inkata Press: Melbourne – AU.
1450. Reckhow, D. A., Singer, P. C., Malcolm, R. L. (1999). Chlorination of humic materials: by-product formation and chemical interpretations. Environmental Science & Technology, 24(11): 1655–1664.
1451. Reddy, S., Neufeld, J., Gallian, J., Neibling, H., Ellsworth, J., Gortsema, S. (2007). Sugarbeet irrigation management using Watermark moisture sensors. University of Idaho CIS 1140. 7 pages.
1452. Reed, A. D. (1980). Irrigation costs. Leaflet 2875 October by Division of Agricultural Science, Berkley-CA: University of California, 1–10.
1453. Reich, L. (2001). *Weedless gardening.* Workman Publishing Company, New York. Page 48.
1454. Reichenberger, L. (2003). Drip irrigation makes a splash. The Furrow, 108(1): 34–36.
1455. Reichenberger, L. (2004). Outlook for the Ogallala. The Furrow, 109(2): 20–21.
1456. Ren, T., Noborio, K., Horton, R. (1999). Measuring soil water content, electrical conductivity and thermal properties with a thermo-time domain reflectometry probe. Soil Sci. Soc. of America J., 63: 450–457.
1457. Resh, H. M. (1985). *Hydroponics food production*, 3rd ed., Santa Barbarian, CA: Word Bridge Press.
1458. Ress, Steve, (2001). Four States Irrigation Council tour visits western Kansas in July. Water Current, 33(5): 5.

1459. Reston: Virginia, Publication 23136. 1–216.
1460. Review of concepts and future trends. Journal of Irrigation and Drainage Engineering, ASCE 25.
1461. Reyes-Pérez, L. A., Liu, L. A., Almodovar, Goyal, M. R., González, J. (1986). Integrated weed control in transplanted peppers. 22nd meeting of Caribbean Food Crops Society, St. Lucia. Vol. 22: 324–330.
1462. Rhodes, J. D., A. Kandiah and Mashali, A. M. (1992). The use of saline waters for crop production. Irrigation and Drainage Paper 48. Food and Agriculture Organization of the United Nations, Rome, 133.
1463. Richardson, S. D., Thruston, A. D., Caughran, T. V., Collette T. W., Patterson K. S., Lykins, B. W. (1999). Chemical by-products of chlorine and alternative disinfectants. Food Technology, 52(4): 58–66.
1464. Riddell, P. J. (1992). Omaha Effluent Treatment Scheme. Proceedings of the Recycled Water Seminar, Wagga Wagga, NSW, Australia, May 19–20, (1992). Australian Water and Wastewater Association.
1465. Rijtema, P. E. (1965). *Analysis of actual evapotranspiration*. Agricultural Research Report No. 69, Center for Agricultural Publishing and Documentation, Wageningen.
1466. Ritchie, J. T., Johnson, B. S. (1990). Soil and plant factors affecting evaporation. In: *Irrigation of agricultural crops*, eds. Stewart, B. A., Nielsen, D. R., Chapter 13: 363–390, Agronomy Series 30. American Society of Agronomy.
1467. Ritchie, J. T. (1991). Wheat phasic development. In: *Modeling plant and soil systems*, eds. Hanks, R. J., Ritchie, J. T., Chapter 3: 31–54, Agronomy Series No. 31. Madison, WI: American Society Agronomy.
1468. Ritchie, J. T., NeSmith, D. S. (1991). Temperature and crop development. In: *Modeling plant and soil systems*, eds. Hanks, R. J., Ritchie, J. T., Chapter 2: 5–29, Agronomy Series No. 31, Madison, WI: American Society of Agronomy.
1469. Rivera, L. E., Goyal, M. R. (1985). Mulch types on soil temperature at varying depths of drip irrigated summer and winter peppers. Journal of Agriculture of the University of Puerto Rico, 69(1): 121–123.
1470. Rivera, L. E., Goyal, M. R. (1986). Filtration system (Spanish). Agricultural Extension Service UPR-RUM, Mayagüez. Serie VIII, IA67. 1–72.
1471. Rivera, L. E., Goyal, M. R. (1986). Mulch types for soil moisture retention in drip irrigated summer and winter peppers. J. Agric. U. P.R., 70(4): 303–305.
1472. Rivera, L. E., Goyal, M. R. (1987). Drip irrigation: Methods to measure soil moisture (Spanish). Agricultural Extension Service, UPR-RUM, Mayagüez. Serie XIII, IA71. 1–48.
1473. Robert, W. J., Mears, D. R. (1969). Double covering a film greenhouse using air to separate film layers. Trans. of ASAE, 12: 32–33, 38.
1474. Robert, W. J., Simpkins, C., Kendall, P. (1976). Using solar energy to heat plastic film greenhouses. Proceedings of Solar Energy Fuel-Food Workshop. University of Arizona, Tucson. 142–159.
1475. Roberts Irrigation Products, (2003). RO-DRIP designer for Windows. Software version 2.1. San Marcos, California. Version. 2.1. (*Roberts Irrigation is now part of John Deere Water*).
1476. Robinson, D. A. (2004). Measurement of the solid dielectric permittivity of clay minerals and granular samples using a time domain reflectometry immersion method. Vadose Zone Journal, 3: 705–713.
1477. Robock, A., Vinnokov, K. Y., Srinivasan, G., Entin, J. K., Hollinger, S., Spernskaya, N. A., Liu, S., Namkhai, A. (2000). The global soil moisture data bank. Bulletin of the American Meteorological Society, 81: 1281–1299.
1478. Rodgers, M., J. Mulqueen and Healy, M. G. (2004). Surface clogging in an intermittentintermittent stratified sand. Soil Sci. Soc. of America J., 68: 1827–1832.
1479. Rodríguez de Miranda, Fabio, (2003). Automated irrigation control, micro irrigation, spatially-variable irrigation, distributed control. Paper number 031129, ASABE Annual Meeting.

1480. Rodríguez, L., Leal, M. (1997). Automation of the localized irrigation. Short communication. Agrotecnia de Cuba, 27(1): 22–26.
1481. Rogers, D. H. (2012). Summarization of alfalfa yields from western Kansas using Kansas Farm Facts as a data source.
1482. Rogers, D. H., F. R. Lamm. 2005, Key considerations for a successful subsurface drip irrigation (SDI) system. In proceedings of the Central Plains Irrigation Conference, Sterling, CO, Feb. 16–17, (2005). CPIA, 760 N.Thompson, Colby, KS. pages 113–118.
1483. Rogers, D. H., Lamm, F., R. (2007). Criteria for successful adoption of SDI systems. Proc. Central Plains Irrigation Conference, Kearney, NE., Feb. 27–28, (2007). CPIA, 760 N.Thompson, Colby, KS. pages 62–71.
1484. Rogers, D. H., Lamm, F., R. (2009). Keys to successful adoption of SDI: Minimizing problems and ensuring longevity. Proc. Central Plains Irrigation Conference, Colby, KS., Feb. 24–25, (2009). CPIA, 760 N.Thompson, Colby, KS. pages 140–151.
1485. Rogers, D. H., Lamm, F. R., Alam, M. (2003). Subsurface drip irrigation (SDI) components: Minimum requirements. KSU Cooperative Ext. Irrigation Mgmt. Series, MF-2576. 4 pages.
1486. Rogers, D. H., Lamm, F. R., Alam, M. (2003). Subsurface drip irrigation systems (SDI) water quality assessment guidelines. proceedings of the Central Plains Irrigation Conference, Colby, KS, Feb. 4–5, (2003). CPIA, 760 N.Thompson, Colby, KS. pages 220–229.
1487. Rogers, D., Lamm, F., Trooien, T., Alam, M. (2006). Subsurface drip irrigation (SDI) with livestock wastewater. KSU Cooperative Ext. Irrigation Mgmt. Series, MF2727. 4 pages.
1488. Rogers, D. H., Lamm, F. R., and Clark, G. A. (2004). Subsurface drip irrigation (SDI) system remediation – A case study. Presented at the Mid-Central ASAE mtg., St. Joseph, MO, March 26–27, (2004). ASAE Paper No. MC04–307. ASAE, St. Joseph, MI. 12 pages.
1489. Rogers, D. H., Lamm, F. R., Alam, M., and Powell, G. M. (2003). Shock chlorination for treatment for irrigation wells. KSU Cooperative Ext. Irrigation Mgmt. Series, MF-2589. 4 pages.
1490. Rogers, J. S., Allen, L. H., Calvert, D. J. (1983). Evapotranspiration for humid regions: Developing citrus grove, grass cover. Trans. of ASAE, 26(6): 1778–1783, (1792).
1491. Rolston, D. E., Rauschkolb, R. S., Phene, C. J., Miller, R. J., Uriu, K., Carlson, R. M., Henderson, D. W. (1979). *Applying nutrients and other chemicals to trickle-irrigated crops*. University of California Bulletin No, (1893). Davis, CA: University of California.
1492. Romero, P., Botia, P., and Garcia, F. (2004). Effects of regulated deficit irrigation under subsurface drip irrigation conditions on water relations of mature almond trees. Plant and Soil, 260: 155–168.
1493. Roose, E. J., Lal, R., Feller, C., B. Barthes and Stewart, B. A. (2005). *Soil erosion and carbon dynamics* (advances in soil science). Florida – USA: CRC Press. 1–376.
1494. Rose, J. (1992). Health aspects of recycled water use. Proceedings of the Recycled Water Seminar, Wagga Wagga, NSW, Australia, May 19–20, (1992). Australian Water and Waste Water Association.
1495. Rose, J. L., Chavez, R. L., Phene, C. J., Hile, M. M. S. (1982). Subsurface drip irrigation of processing tomatoes. Proceedings of the Specialty Conference on Environmentally Sound Water and Soil Management, eds. Kruse, E. G., Burdick, C. R., and Yousef, Y. A., 369–376. New York, NY: ASCE.
1496. Rosegrant, M., Ringler, C. (1999). Impact on food security and rural development of reallocating water from agriculture. IFPRWashington, I., DC.
1497. Rosenberg, N. J., Blad, B. L., Verma, S. B. (1983). *Microclimate: The biological environment*. A Wiley- Interscience. Chapter 7: 209–287.
1498. Rosner, M., Assouline, S., Cohen, S., Meerbach, D., Harodi, T. (2002). Microdrip irrigation of field crops: Effect on yield, water uptake, and drainage in sweet corn. Soil Sci. Soc. of America J., 66: 228–235.
1499. Ross, D. S., Funt, R. C., Reynolds, C. W., Coston, D. C., Fries, H. H., Smith, N. J. (2006). Trickle irrigation—an introduction. Technical Bulletin by The Northeast Regional Agricultural Engineering Service, Cornell University, Ithaca, N. Y.

1500. Rossi, G., A. Cancelliere and Pereira, L. S. (2003). Tools for drought mitigation in mediterranean regions. Netherlands: Kluwer Academic Publisher. Part III: pages 162–259.
1501. Roth, R. L., Sanchez, C. A., Gardner, B. R. (1995). Growth and yield of mature 'Valencia' oranges converted to pressurized irrigation systems. Applied Engr. in Agric., 11(1): 101–105.
1502. Royo, Aragüés, R., Playán, E., Ortiz, R. (2000). Salinity–grain yield response functions of barley cultivars assessed with a drip-injection irrigation system. Soil Sci. Soc. of America J. 64: 359–365.
1503. Rubeiz, I. G. (1990). Response of greenhouse cucumber to mineral fertilizers on a high phosphorus and potassium soil. J. Plant Nutrition, 13: 269–273.
1504. Rubeiz, I. G., Oebker, N. F., Strohlein, J., L. (1989). Surface drip irrigation and urea phosphate fertigation for vegetables on calcareous soils. J. Plant Nutrition, 12: 1457–1465.
1505. Rubeiz, I. G., Stroehlein, J. L., Oebker, N. F. (1991). Effect ofirrigation methods on urea phosphate reactions in calcareoussoils. Communications in Soil Science and Plant Analysis, 22(5&6): 431–435.
1506. Ruiz-Sánchez, M. C., Torrecillas, A., Pérez-Pastor, A., and Domingo, R. (2000). Regulated deficit irrigation in apricot trees. Acta Horticulturae, 537: 759–766.
1507. Running, S. W., Coughlan, J. C. (1988). A general model of forest ecosystem processes for regional applications: I. Hydrologic balance, canopy gas exchange and primary production processes. Ecological Modeling, 42: 125–154.
1508. Ruskin, R. (1992). Reclaimed Water and Subsurface Irrigation. ASAE Paper No. 92–2578 ASAE., St. Joseph, MI 49085.
1509. Ruskin, R., Ferguson, K. R. (1998). Protection of subsurface drip irrigation systems from root intrusion. In Proc. 19th Annual Irrigation Assoc. Int'l. Tech. Conf., 41–48. Falls Church, Va.: Irrigation Association.
1510. Ruskin, R., P. Van Voris and Cataldo, D. A. (1990). Root intrusionprotection of buried drip irrigation devices with slow-release herbicides. Proceedings of the 3rd National Irrigation Symposium, 211–216. St.Joseph, MI: ASAE.
1511. Ruth, B. (1999). A capacitance sensor with planar sensitivity for monitoring soil water content. Soil Sci. Soc. of America J., 63: 48–54.
1512. Ryszkowski, L. (2002). *Landscape ecology in agroecosystems management.* CRC Press. 76–81.
1513. Saad, F. (1995). Available water in Lebanon. Training Course on the use of modern irrigation system. Beirut, October 19–22.
1514. Sabra, M. (1995). Modern irrigation system in Lebanon. Training course on modern irrigation system. Beirut, October 19–22.
1515. Sadler, E. J., Camp, C. R., Busscher, W. J. (1995). Emitter flowrate changes caused by excavating subsurface micro irrigation tubing. Proceedings of the 5th International Micro irrigation Congress, ed. Lamm, F. R., 763–768 pages. St. Joseph, MI: ASAE.
1516. Sadras, V. O., Roget, D. K. (2004). Production and environmental aspects of cropping intensification in a semiarid environment of southeastern Australia.Agronomy Journal, 96: 236–243.
1517. Saeed, I. A.M., El-Nadi, A., H. (1997). Irrigation effects on the growth, yield, and water use efficiency of alfalfa. Irrig. Sci., 17(2): 63.
1518. Safely delaying the first irrigation of corn. Pioneer Hi- Bred Crop Insights, April, (2008).
1519. Safontas, J. E., J. C. di Paola, (1985). Drip irrigation of maize. Third international drip/trickle irrigation congress, Fresno, Calif., USA. American Society of Agricultural Engineers, St Joseph MI, pages 575–578.
1520. Saif, U., Maqsood, M., Farooq, M., Hussain, S., Habib, A. (2003). Effect of planting patterns and different irrigation levels on yield and yield component of maize. Int. J. Agric. Biol., 1: 64–66.
1521. Sakai, K. (2001). Nonlinear Dynamics and Chaos in Agricultural Systems.Schwankl, L., Prichard, T. (2001). Chemigation in tree and vine micro irrigation systems. Agriculture and Natural Resources Publication 21599. University of California, Davis – m CA.

1522. Salih, A. M., Sendil, U. (1984). Evapotranspiration under extremely arid climates. Journal of Irrigation and Drainage Division of ASCE, 110(IR3): 289–303.
1523. Sammis, T. W. (1980). Comparison of sprinkler, trickle, subsurface and furrow irrigation methods for row crops. Agron. J., 72(5): 701–704.
1524. Sánchez-Blanco, M. J., Torrecillas, A., León, A. and del Amor, F. (1989). The effect of different irrigation treatments on yield and quality of Verna lemon. Plant and Soil, 120: 299–302.
1525. Sanden, Douglas C. (2006). Drip irrigation systems: an operation and trouble shooting checklist. College of Agriculture and Life Sciences, North Carolina State University.
1526. Sando, S. K. (2001). *Irrigation of the angostura reclamation unit.* 4th ed. U. S. Dept. of the Interior, U. S. Geological Survey, 1–65.
1527. Sanewe, A., Plessis, M., Backeberg, G. (2003). Water utilization in agriculture. Water research Comission, 156–187.
1528. Sanjines, A, Ruskin, R. (1991). Root intrusion protection for subsurface drip emitters. ASAE Paper No. 91–2047 Am. Soc. Agr. Engr., St. Joseph, MI.
1529. Santhi, C, and Pundarikanthan, N. V. (2000). A new planning model for canal scheduling of rotational irrigation. Agric. Water Manage., 43: 327–343.
1530. Santiago, C. L., Goyal, M. R. (1985). Nutrient uptake and solute movement in drip irrigated summer peppers. Journal of Agriculture of the University of Puerto Rico, 69(1): 63–68.
1531. Sarwar A, and Bastiaanssen, W. G.M. (2001). Long-term effects of irrigation water conservation on crop production and environment in semi-arid areas. J. Irrig. Drain. Eng., 127: 331–338.
1532. Sau, F., Boote, K. J., McNair-Bostick, W., Jones, J. W., Inés-Mínguez, M. (2004). Testing and improving evapotranspiration and soil water balance of the DSSAT crop models. Agronomy Journal, 96: 1243–1257.
1533. Saxton, K. E., Johnson, H. P., Shaw, R. H. (1974). Modeling evapotranspiration and soil moisture. Trans. of ASAE, 17(4): 673–677.
1534. Scandella, D., Kraeutler, E., and Venien, S. (1997). Anticiper la qualit'e gustative des p^eches et nectarines. Infos CTIFL, 129: 16–19 (In French).
1535. Schabenberger, O., Pierce, F. J. (2001). *Contemporary statistical models for the plant and soil sciences*. Florida – USA: CRC Press. 1–738.
1536. Scherer, T. F. (2002). System installation: Design and control of sprinkler systems for crop disease research. Trans. ASAE, 45: 45–56.
1537. Scherer, T. F. (2005). Design and control of sprinkler systems for crop disease research. St. Joseph, MI: The American Society of Agricultural and Biological Engineers. Paper number 052182, ASAE Annual Meeting.
1538. Scherer, Thomas F. (1999). *Sprinkler Irrigation Systems*. Mid West Plan Service (MWPS).
1539. Scherm, H., A. H.C. van Bruggen, (1995). Comparative study of microclimate and downy mildew development in subsurface drip- and furrow-irrigated lettuce fields in California. Plant Disease, 79(6): 620–625.
1540. Scholberg, J., McNeal, B. L., Boote, K. J., Jones, W. W., Locascio, S. J., and Olson S. M. (2000). Nitrogen stress effects on growth and nitrogen accumulation by field-grown tomato. Agronomy J., 92: 152–167.
1541. Schneider, A. D., Evett, S. R. (1995). Evapotranspiration of irrigated winter wheat: Southern high plains. Trans. of ASAE, 38(3): 745–759.
1542. Schneider, A. D., Howell, T. A. (1999). LEPA and spray irrigation for grain crops. Journal of Irrigation and Drainage Engineering, 125(4): 167–172.
1543. Scholander, P. F., Hammel, L. H., Bradsteet, E. D., and Hemmingsen, E. A. (1965). Sap pressure in vascular plants. Science, 148: 339–346.
1544. Schulz, M. A. (2000). Subsurface drip irrigation for broad acre crops. Department of Natural Resources and Environment, State of Victoria, Australia.
1545. Schulze, K. (1995). Report of expert meeting for the preparation of an intercomparison of instruments and procedures for measurement and estimation of evaporation and evapotranspiration. World Meteorological Organization, Commission for Instruments and Methods of Observation. Geneva, Switzerland. 30.

1546. Schumann, A. W., Syvertsen, J. P., and Morgan, K. T. (2009). Implementing advanced citrus production systems in florida-early results. Proc. Fl. State Hort. Soc., 122: 108–113.
1547. Schwankel, L. J., Prichard, T. L. (1990). Clogging of buried drip irrigation systems. California Agriculture, 44(1): 16–17.
1548. Schwankel, L. J., Grattan, S. R., and Miyao, E. M. (1990). Drip irrigation burial depth and seed planting depth effects on tomato germination. Proceedings of the 3rd Nat. Irrigation Symposium, 682–687. St. Joseph, MI: ASAE.
1549. Schwankl, L., Prichard, T. (2001). Chemigation in tree and vine micro irrigation systems. Agriculture and Natural Resources Publication 21599. University of California, Davis, CA.
1550. Schwankl, L., Hanson, B., Prichard, T. (1998). Micro-irrigation of trees and vines. Pub, (3378). Davis, CA: Coop. Ext., Dept. Land, Air, and Water Res., Univ. of Calif.
1551. Schwankl, L., Hanson, B., Prichard, T. (2008). Maintaining micro irrigation systems. Univ. of California Agriculture and Natural Resources Pub. 21637. 53 pages.
1552. Science and Technology. 19(12): 323–331.
1553. SDI and the Great Plains CD-Rom. KSU, February, (2008).
1554. SDI facilities and other irrigation topics. KSU Biological and Agricultural Engineering Department Chair, Gary Clark, May 13, (2008).
1555. SDI facilities with Michael Dowgert, Netafim USA, August 20, (2008).
1556. SDI in the Great Plains at Central Plains Irrigation Conference, February 19–20, 2008, Greeley, Colorado.
1557. SDI research at Kansas State University. Victory Electrical Coop annual irrigation meeting, Dodge City, Kansas, February 8, (2008).
1558. SDI research for corn production in Kansas, 16 years of progress. SDI meeting, Sublette, Kansas, March 3, (2005).
1559. SDI research results at KSUSDA-RRF, U., W1128 regional project concerning micro irrigation, Tampa, FL, November 17–19, (2004).
1560. SDI research with USDA RRF Regional Project W1128, Honolulu, Hawaii, October 9–11, (2007).
1561. SDI research with USDA RRF Regional Project W1128, Portland, Oregon, October 22–24, (2008).
1562. Seckler, D., Upali, A., Molden, D., de Silva, R., and Barker R. (1998). *World Water Demand and Supply, 1990 to 2025: Scenarios and Issues*. Research Report 19, International Water Management Institute, Colombo, Sri Lanka, 40 pages.
1563. Seemann, J., Chirkov, Y. I., Lomas, J., Primault, B. (1979). *Agrometeorology*. Berlin, Heidelberg: Springer Verlag.
1564. Segal, E., Ben-Gal, A., Shani, U. (2000). Water availability and yield response to high-frequency micro-irrigation in sunflowers [CD-ROM]. Proc 6th Int. Micro-Irrigation Congr. 22–27 October, Int. Council Irr. Drainage. Cape Town, South Africa.
1565. Segarra, E., Almas, L., Bordovsky, J. P. (1999). Adoption of advanced irrigation technology: LEPA vs. drip in the Texas High Plains. Proc. Beltwide Cotton Conf., 1: 324–328. Memphis, TN: National Cotton Council.
1566. Seginer I, (1987). Spatial water distribution in sprinkler irrigation. In: Hillel, D. (Ed.), *Advances in Irrigation*, Vol. 4. Academic Press, Orlando, pages 119–168.
1567. Seginer, I. (1978). A note on the economic significance of uniform water application. Irrig Sci 1: 19–25.
1568. Seginer, I. (1979). Irrigation uniformity related to horizontal extent of root zone. Irrig. Sci. 1: 89–96.
1569. Seguin, B., Brunet, Y., Perrier, A. (1982). Estimation of evaporation: A review of existing methods and recent developments. European Geologic Society Symposium on Evaporation. Leeds, U. K., 21.
1570. Selim, H. M., Zhou, L., Zhu, H. (2003). Herbicide retention in soil as affected by sugarcane mulch residue. Journal of Environmental Quality, 32(4): 1445–1454.

1571. Selker, J. (2004). Irrigation system maintenance, groundwater quality and improved production. Oregon State University Extension Service. 1–56.
1572. Senock, R. S., Ham, J. M., Loughin, T. M., Kimball, B. A., Hunsaker, D. J., Pinter, P. J., Wall, G. W., Garcia, R. L., and LaMorte, R. L. (1996). Sap flow in wheat under free-air CO_2 enrichment. Plant, Cell & Environment, 19(2): 147–158.
1573. Seyfi, K., Rashidi, M. (2007). Effect of drip irrigation and plastic mulch on crop yield and yield components of cantaloupe. International Journal of Agriculture & Biology, 9(2): 246–249.
1574. Seyfried, M. S., Murdock, M. D. (2001). Response of a new soil water sensor to variable soil, water content and temperature. Soil Sci. Soc. of America J., 65: 28–34.
1575. Seyfried, M. S., Murdock, M. D. (2004). Measurement of soil water content with a 50-MHz soil dielectric sensor. Soil Sci. Soc. of America J., 68: 394–403.
1576. Shackel, K. A., Lampinen, B., Southwick, S., Olson, W., Sibbett, S., Krueger, W., Yeager, J., Goldhamer, D. (2000). Deficit irrigation in Prunes: Maintaining productivity with less water. HortScience, 35: 1063–1066.
1577. Shahidian, S. (2009). Drip irrigation using a PLC based adaptive irrigation system WSEAS transactions on environment and development (WSEAS), 2(5): 209–219.
1578. Shahin, M. A. (2002)., Hydrology and water resources of Africa.Springer Press, New York – USA. 157–212.
1579. Shahin, M. A. (2002). *Hydrology and water resources of Africa*. Springer. 157–212.
1580. Shalhevet, J. (1984). Management of irrigation with brackish water. In: *Soil salinity under irrigation: Processes and management*, eds. Shainberg, I., Shalhevet, J., Section 8.4: 298–318. Berlin: Springer-Verlag.
1581. Shalhevet, J. (1994). Using water of marginal quality for crop production: Major issues – review article. Agricultural Water Management, 25: 233–269.
1582. Shangguan, Z., Shao, M., Horton, R., Lei, T., Qui, L., and Ma, J. (2001). A model for regional optimal allocation of irrigation water under deficit irrigation and its applications. Agric. Water Manage., 52: 139–154.
1583. Shani, U., Ben-Gal A. (1999). Irrigation of vineyards with low-quality effluent: development of safe application methods. In: Annual Report 1998–1999, Arava Research and Development. pages 104–110.
1584. Shani, U., Or, D. (1995). In-situ method for estimating subsurface unsaturated hydraulic conductivity. Water Resour. Res., 21: 1863–1870.
1585. Shani, U., Xue, S., R. Gordin-Katz and Warrick, A. W. (1996). Soil-limiting flow from subsurface emitters. I. Pressure measurements. Journal of Irrigation and Drainage Engineering, 122(5): 291–295.
1586. Shani, U., Xue, S., Gordin-Katz, R., and Warrick, A. W. (1996). Soil-limiting flow from subsurface emitters: 1, Pressure measurements. J. Irrig. Drain. Eng. Am. Soc. Civ. Eng., 122: 291–295.
1587. Shani, U., Waisel, Y., and Eshel, A. (1995). The development of melon roots under trickle irrigation: Effects of the location of emitters. F. Baluska et al. (eds), Netherlands, 223–225 pages.
1588. Sharma, K. D. (2001). Rainwater harvesting and recycling. In: Goosen, MFA, Shayya WH (Eds.), *Water Management, Purification and Conservation in Arid Climates*. Technomic Publishing Company, Lancaster, PA, pages 59–86.
1589. Sharma, M. L. (1985). Estimating evapotranspiration. In *Advances in irrigation*, ed. Hillel, D., 3: 213–281. New York: Academic Press.
1590. Sharmasarkar, F. C., Sharmasarkar, S., Held, L. J., Miller, S. D., Vance, G. F., Zhang, R. (2001). Agroeconomic analyses of drip irrigation for sugarbeet production. Agronomy Journal, 93(1): 517–523.
1591. Shaw, R. H., Pereira, A. R. (1982). Aerodynamic roughness of a plant canopy: A numerical experiment. Agricultural Meteorology, 26: 51–65.
1592. Shedd, M., Dukes, M. D., and Miller, G. L. (2007). Evaluation of evapotranspiration and soil moisture-based irrigation control on turfgrass. Proceedings ASCE-EWRI World Environmental & Water Resources Congress.

1593. Sheldrake, R., Langhans, R. (1961). Heating study with plastic greenhouses. Proceedings of National Horticultural Plastics Congress, 2: 16–17.
1594. Sherrod, L. A., Peterson, G. A., Westfall, D. G., Ahuja, L. R. (2005). Soil organic carbon pools after 12 years in no-till dry land agroecosystems.Soil Sci. Soc. of America J., 69: 1600–1608.
1595. Shih, S. F. (1984). Data requirement for evapotranspiration estimation. Journal of Irrigation and Drainage Division of ASCE, 110(IR3): 263–274.
1596. Shock, C. C., Feibert, E. B., Saunders, M. (1998). SDI irrigation scheduling for profit and environmental protection. In Proc. Irrig. Assoc. Int'l. Tech. Conf., 33–39. Falls Church, Va.: Irrigation Association.
1597. Shock, C. C. (2001). Drip irrigation: An introduction. Oregon State University, Extension Service. 25–32.
1598. Shock, C. C., Flock, R. J., Feibert, E. B. G., Shock, C. A., Pereira, A. B., Jensen, L. B. (2005). Irrigation monitoring using soil water tension. Oregon State University Extension Service. EM 8900. 1–6.
1599. Shock, Clinton C., Feibert, B. G., Lamont D. Saunders and Eric Eldredge, P. (2002). Water Management: Allium Cepa, (onion), Solanum Tuberosum, (potato), watermark, granular matrix sensor, soil water potential, and SDI. Proceedings of the World Congress of Computers in Agriculture and Natural Resources, (13–15 March, at Iguacu Falls-Brazil) 701P0301, 809–816.
1600. Shrive, S. C., McBride, R. A., Gordon, A. M. (1994). Photosynthetic and growth responses of two broad-leaf tree species to irrigation with municipal landfill leachate. Journal of Environmental Quality, 23(3): 534–542.
1601. Shrock, C. C. (2001). Drip Irrigation: An Introduction. Oregon State University, Extension Service, 67–89.
1602. Shuttleworth, W. J. (1993). Evaporation. In: *Handbook of hydrology*, ed. Maidment, D. R., 4.1–4.53. New York: McGraw Hill.
1603. Shuttleworth, W. J., Wallace, J. S. (1985). Evaporation from sparse crops – an energy combination theory. Quarterly Journal of the Royal Meteorological Society, 111: 839–853.
1604. Siemens, J., Kaupenjohann, M. (2004). Comparison of three methods for field measurement of solute leaching in a sandy soil. Soil Sci. Soc. of America J., 68: 1191–1196.
1605. Sijali, I. V. (2001). Drip irrigation: Options for smallholder farmers in eastern and Southern Africa. Regional Land Management Unit, Nairobi-Kenya.
1606. Simmers, I. (2003). Understanding water in a dry environment: Hydrological processes in arid and semi-arid zones. Taylor Francis: UK. 92–99.
1607. Simunek, J., Sejna, M., M. T. van Genuchten, (1999). The HYDRUS-2D software package for simulating two-dimensional movement of water, heat, and multiple solutes in variably saturated media, Version 2.0, Rep. IGWMC- TPS-53, 251 pages, IGWMC, Colorado School of Mines, Golden, Co.
1608. Sinai, G., Haramati, A., Gorbonos, V. (2005). Attenuation with soil depth of dynamic fluctuations in the ratio of chemicals to irrigation water in chemigation. Applied Engineering in Agriculture, 21(3): 357–370.
1609. Sinclair, T. R. (1984). Leaf area development in field-grown soybeans. Agronomy Journal, 76: 141–146.
1610. Singer, J. S., Munns, D. N. (1999). *Soils – An Introduction.* 4th Edition, Prentice Hall -New Jersey.
1611. Singh, D. K., Rajput, T. B.S., Singh, D. K., Sikarwar, H. S., Sahoo, R. N., and Ahmad, T. (2006). Simulation of soil wetting pattern with subsurface drip irrigation from line source. Agric. Water Manage., 83: 130–134.
1612. Singh, H. P. (2001). Emerging scenario of micro irrigation in India. Proceedings of International Conference on Micro and Sprinkler Irrigation Systems, 8–10 Feb., CBIP publication. 282, pages 18–30.
1613. Singh, H. P. (2000). Micro irrigation. International Conference on Micro and Sprinkler Irrigation Systems. New Delhi-India: Central Board of Irrigation and Power.

1614. Sisson, J. B., Gee, G. W., Hubbell, J. M., Bratton, W. L., Ritter, J. C., Ward, A. L., Caldwell, T. G. (2002). Advances in tensiometry for long-term monitoring of soil water pressures. Geo Sciences Research Department, Idaho National Engineering and Environmental Laboratory, Bechter, Idaho. 34–45.

1615. Sivanappan, R. K., Padmakumari, O. (1980). Drip irrigation. TNAU, Coimbatore, India.

1616. Sivanappan, R. K. (1994). Prospects of micro-irrigation in India. Irrigation and Drainage Engineering, 8: 49–58.

1617. Sivanappan, R. K., Rajgopal, A., Paliniswami, D. (1974). Response of vegetable to the drip irrigation. Madras Agric. J., 65: 576–579.

1618. Siyal, A. A., and Skaggs, T. H. (2009). Measured and simulated soil wetting patterns under porous clay pipe sub-surface irrigation Agricultural Water Management, 96: 893–904.

1619. Skaggs, R. K. (2001). Predicting drip irrigation use and adoption in a desert region.Agricultural Water Management Journal, 51(2): 125–142.

1620. Skaggs, T. H., Trout, T. J., Simunek, J., and Shouse, P. J. (2004). Comparison of HYDRUS-2D simulations of drip irrigation with experimental observations. J. Irrig. Drain.Eng., 30: 304–310.

1621. Skaggs, T. H., T. J.Trout, Rothfuss, Y. (2010). Drip irrigation water distribution patterns: Effects of emitter rate, pulsing, and antecedent water. Soil Sci. Soc. Amer. J., 74(6): 1886–1896.

1622. Slattery, W. J., Conyers M. K., and Aitken R. L. (1999). Soil pH, aluminum, manganese and lime requirement. In: *Soil analysis: an interpretation manual* (Eds Peverill, K. I., Sparrow, L. A., D. J. Reuter) pages 103–129. CSIRO Publishing: Collingwood, Vic. – AU.

1623. Slatyer, R. O., McIlroy, I. C. (1961). Evaporation and the principle of its measurement. In: *Practical meteorology* by C B Garretty.Paris: CSIRO (Australia) and UNESCO.

1624. Slima, M. B., Morawski, R. Z., Kraszewski, A. W., Barwicz, A. (1999). Calibration of a microwave system for measuring grain moisture content. IEEE Transactions on Instrumentation and Measurement, 48: 778–782.

1625. Smajstrla, A. G. (1985). Design and management of drip irrigation systems for tomatoes. Agricultural Engineering Extension Mimeo Report 85–13. University of Florida.

1626. SmajstrlaA. G., Boman, B. J., Haman, D. Z., Pitts, D. J., Zazueta, F. S. (2002). Field evaluation of micro irrigation water application uniformity. Cooperative Extension Service, Department of Agricultural Engineering, Gainesville, FL: University of Florida.

1627. Smajstrla, A. G., Clark, G., A. (1982). Water stress effects on water use and yield of soybeans. Soil Crop Sci. Soc. Fla. Proc., 41: 179–181.

1628. Smajstrla, A. G., Dalton, D. S., F. X. Duran. Tensiometers for soil moisture measurement and irrigation scheduling. Circular 487 Florida Cooperative Extension Service, Gainesville – FL.

1629. Smajstrla, A. G., Harrison, D. S., and Clark, G. A. (1985). Trickle irrigation scheduling 1: Durations of water applications. IFAS Bulletin 204. University of Florida.

1630. Smajstrla, A. G., Harrison, D. S., and Becker, W. J. (1986). Chemigation safety. Agricultural Engineering Fact Sheet AE-58. University of Florida.

1631. Smajstrla, A. G., Harrison, D. S., Becker, W. J., Zazueta, F. S., and Haman, D. Z. (1991). Backflow prevention requirements for florida irrigation systems. IFAS Bulletin 217. University of Florida.

1632. Smajstrla, A. G., Haman, D. Z., and Zazueta, F. S. (1992). Calibration of fertilizer injectors for agricultural irrigation systems. Circular, (1033). Fla. Coop. Ext. Ser., Univ. of Florida.

1633. Smajstrla, A. G., Locascio, S., J. (1996). Tensiometer-controlled, drip-irrigation scheduling for tomato. Applied Engr. in Agric., 12(3): 315–319.

1634. Smajstrla, A. G., Locascio, S. J., Weingartner, D. P., Hensel, D. R. (2000). Subsurface drip irrigation for water table control and potato production. Applied Engr. in Agric., 16(3): 225–229.

1635. Smajstrla, A. G., and Zazueta, F. S. (1985). Design and management of drip irrigation systems for strawberries. Agricultural Engineering Extension Mimeo Report 85–14. University of Florida.

1636. Smajstrla, A. G., Zazueta, F. S., and Haman, D. Z. (1985). Soil characteristics affecting irrigation in Florida. Agricultural Engineering Extension Report 85–2 (revised). University of Florida.

1637. Smith, M. (1992). *CROPWAT: A computer program for irrigation planning and management.* FAO Irrigation and Drainage Paper 46, FAO, Rome.
1638. Smith, M., Allen, R. G., Monteith, J. L., Perrier, A., Pereira, L., Segeren, A. (1992). Report of the expert consultation on procedures for revision of FAO guidelines for prediction of crop water requirements. UN-FAO, Rome, Italy, 54.
1639. Smith, R. B., Oster, J. D., Phene, J. C. (1991). Subsurface drip irrigation produced highest net return in wasteland area study. Calif. Agric., 45(2): 8–10.
1640. Smith, Steve, (2002). Converting to drip – The basics of converting from furrow to drip irrigation. Irrigation Business and Technology, February. pages 26–28.
1641. Sne, M. (2006). Micro irrigation in arid and semi-arid regions. In: Kulkarni SA (Ed.) *Guidelines for planning and design.* In: International Commission on Irrigation and Drainage. New Delhi, India.
1642. Snellen, Bart, W. (1999). *Irrigation scheme operation and maintenance*. Training Manual, No. 10. Food & Agriculture Organization of the United, 1–48.
1643. Snyder, R. L., Lanini, B. J., Shaw, D. A., Pruitt, W. O. (1989). Using reference evapotranspiration and crop coefficients to estimate crop evapotranspiration for agronomic crops, grasses, and vegetable crops. Cooperative Extension, University of California, Berkeley, CA, Leaflet No. 21427, 12.
1644. Snyder, R. L., Lanini, B. J., Shaw, D. A., Pruitt, W. O. (1989). Using reference evapotranspiration and crop coefficients to estimate crop evapotranspiration for trees and vines. Cooperative Extension, University of California, Berkeley, CA, Leaflet No. 21428, 8.
1645. Snyder, R. L., Pruitt, W. O. (1992). Evapotranspiration data management in California. Proceedings of the Irrigation and Drainage sessions of ASCE Water Forum, ed. Engman, T., New York, NY: ASCE, 128–133.
1646. Soaud, A. A., O. Van Cleemput, Hofman, G. (1992). Uptake and balance of labelled fertilizer nitrogen by potatoes. Fertilizer Research, 31: 351–353.
1647. Software for Planning Corn Production with SDI. Central Plains Irrigation Conference, February 19–20, 2008, Greeley, Colorado.
1648. Soil Conservation Service, (1977). Kansas irrigation guide. USDA- NRCS, Salina, KS, USA.
1649. Solomon, K. H. (1988). Selection of the irrigation system. Publ. Núm. 880702 – Institute of Agricultural Technology, Fresno – of California. 1–11.
1650. Solomon, K. H., Jorgensen, G. (1992). Subsurface drip irrigation. Grounds Maintenance, 27(10): 24, 26.
1651. Solomon, Kenneth H., Allen Dedrick, R. (1995). Standards Development for Micro irrigation. CATI, CA. Publication 950601.
1652. Solomon, Kenneth H., Greg Jorgensen, (1993). Subsurface Drip Irrigation, CATI Publication 930405.
1653. Solomon, Kenneth H., Irrigation System Selection. CATI, CA.
1654. Solomon, Kenneth H., Selection of Irrigation Methods for Agriculture, ASCE.
1655. Solomon, K. H. (1979). Manufacturing variation of emitters in trickle irrigation systems. Trans. of ASAE, 22(5): 1034–1038.
1656. Soltani, A. (1993). Engineering Economics. Shiraz University, 79 pages.
1657. Sommer, R., Fölster, H., Vielhauer, K., Maklouf-Carvalho, E. J., Vlek, P. L. G. (2003). Deep soil water dynamics and depletion by secondary vegetation in the Eastern Amazon.Soil Sci. Soc. of America J., 67: 1672–1686.
1658. Souza, C. F., D. Or and Matsura, E. E. (2004). A variable-volume TDR probe for measuring water content in large soil volumes. Soil Sci. Soc. of America J., 68: 25–31.
1659. Spiegel, B. (2000). Moving wastewater by drip. Irrigation Extra, pages IE1- IE4, Kansas Farmer, Mid-January.
1660. Spiegel, B. (2002). Design SDI for the long haul. Kansas Farmer (April), 140(7): 12–13.
1661. Springer Laosheng, Wu, (2000). Drip irrigation using low-quality water. Irrigation Journal, 50(3): 18–20.

1662. *Sprinklers & Drip Systems*. Sunset Publishing Corporation, Menloe Par – CA – USA, Edition, (2006). 1–98.
1663. Stalcup, L. (2006). Drip irrigation saves water, energy and time without hurting corn yields. Corn and Soybean Digest. December. 2 pages.
1664. Stanghellini, C. (1987). Transpiration of greenhouse crops, an aid to climate management. Ph.dissertation, D., Agric. Univ., Wageningen, The Netherlands.
1665. Stanley, C. D., Clark, G., A. (1991). Water table management using micro irrigation tubing. Soil Crop Sci. Soc. Fla. Proc. 50: 6–8.
1666. Stanley, C. D., Clark, G., A. (1995). Effect of reduced water table and fertilizer levels on sub-irrigated tomato production. Applied Engineering in Agriculture, 11(3): 385–388.
1667. Stanley, C. D., Clark, G. A., Albregts, E. E., and Zazueta, F. S. (1991). Reduction of deep aquifer withdrawals for overhead-irrigated strawberry production using a runoff recovery system. Trans. of ASAE, 7(2): 205–208.
1668. Stanley, C. D., Clark, G. A., Prevatt, J. W., Harbaugh, B. K., and Overman, A. J. (1992). Micro irrigation for flowering ornamental crops in humid regions. Southern Cooperative Series Bulletin 364. Southern Regional Research Project S-143. (also, Univ. of Fla. Coop. Ext. Ser., IFAS Bulletin 887). 9 pages.
1669. Stanley, C. D., Csizinszky, A. A., Clark, G. A., Prevatt, J. W. (1991). Sequential cropping for vegetable production using micro irrigation on sandy soils in southwestern Florida. HortTechnology, 1(1)72–76.
1670. Starck J. C., McCann, I. R., Westermann, D. T., Izadi, B., Tisdall T. A. (1993). Potato response to split N timing with varying amount of excessive irrigation. Am. Potato Journal, 70: 765–777.
1671. Starr, G., Lowery, B., Cooley, E., Hart, G. (1999). Soil water content determination using network analyzer reflectometry methods. Soil Sci. Soc. of America J., 63: 285–289.
1672. Starr, G., Lowery, B., Cooley, E. (1999). Development of a resonant length technique for soil water content measurement. Soil Sci. Soc. of America J., 63: 278–285.
1673. Steduto, P. (1997). Modelling for crop response to water: physiological aspects. Options Méditerranéenes, Sér. A/nr 31: 289–312.
1674. Steele, D. D., Greenland, R. G., Gregor, B., L. (1996). Subsurface drip irrigation systems for speciality crop production in North Dakota. Applied Engr. in Agric., 12(6): 671–679.
1675. Sterret, S. B., Ross, B. B., and Savage, C. P. (1990). Establishment and yield of asparagus as influenced by planting and irrigation method. J. American Soc. Hort. Sci., 115(1): 29–33.
1676. Stevenson, B. A., T. McLendon and Redente, E. F. (2000). Effects of soil fumigation and seeding regimes on secondary succession in a semiarid shrub land. Arid Soil Reseach and Rehabilitation, 14: 87–99.
1677. Stewart, J. B. (1983). A discussion of the relationships between the principal forms of the combination equation for estimating evapotranspiration. Agricultural Meteorology, 30: 111–127.
1678. Stewart, J. B. (1988). Modelling surface conductance of pine forest. Agricultural and Forest Meteorology, 43: 19–35.
1679. Stewart, J. B. (1989). On the use of the Penman-Monteith equation for determining areal evapotranspiration. In Estimation of Areal Evapotranspiration. IAHS, 177: 3–12.
1680. Stewart, J. B., Gay, L. W. (1989). Preliminary modeling of transpiration from the FIFE in Kansas. Agricultural and Forest Meteorology, 48: 305–315.
1681. Stewart, J. I., Hagan, R. M., Pruitt, W. O. (1976). Salinity effects on corn yield, evapotranspiration, leaching fraction, and irrigation efficiency. In: ed. Dregne, H. E., *Managing saline water for irrigation*. Proceedings of the International Conference on Managing Saline Water for Irrigation.
1682. Stewart, J. B., Verma, S. B. (1992). Comparison of surface fluxes and conductances at two contrasting sites within the FIFE area. Journal of Geophysical Research, 97(17): 18623–18628.
1683. Stork, P. R., and Jerie, P. H. (2003). Initial studies of the growth, nitrogen sequestering and dewatering potential of perennial grass selections for use as nitrogen catch crops in orchards. Australian Journal of Agricultural Research, 54: 27–37.

1684. Stork, P. R., Jerie, P. H., and Callinan, A. P.L. (2003). Subsurface drip irrigation in raised bed tomato production. I. Nitrogen and phosphate losses under commercial practice. Australian Journal of Soil Research, 41: 1283–1304.
1685. Stringer, W. C., Wolf, D. D., Baser, R. E. (1981). Summer regrowth of tall fescue: Stubble characteristics and microenvironment. Agronomy Journal, 73: 96–100.
1686. Styles, S. W., Burt, C. M. (1999). *Drip and micro irrigation for trees, vines and row crops*. California: Irrigation Training and Research Center. 1–292.
1687. Suarez-Rey, E., Choi, C. Y., Waller, P. M., Kopec, D. M. (2000). Comparison of subsurface drip irrigation and sprinkler irrigation for Bermuda grass turn in Arizona. Trans. ASAE, 43(3): 631–640.
1688. Subsurface drip and center pivot sprinkler irrigation systems. Appl. Engr. Agric., 14(4): 391–398.
1689. Subsurface drip irrigation (SDI). A. N. G. Ranga Agricultural University, Hyderabad, India, January 2, (2008).
1690. Subsurface drip irrigation (SDI). IARI – India, Delhi, India, January 5, (2008).
1691. Sudan, Falendra Kumar, and Jennifer McKay. Water pricing and participatory irrigation management: experiences and lessons learnt from integrated watershed development project in Jammu and Kashmir, India. International Research Center for Water Policy and Law, University of South Australia, Wa y Lee Building, Room WL3–41, City West Campus, Postal GPO Box 2471, Adelaide SA 5001, South Australia, Australia.
1692. Suleiman, A., Crago, R. (2004). Hourly and daytime evapotranspiration from grassland using radiometric surface temperatures.Soil Sci. Soc. of America J., 68: 384–390.
1693. Sumner, D. M. (2001). Evapotranspiration from a cypress and pine forest subjected to natural fires, Volusia County Florida, 1998–99. Water Resource. Investigations Rep. 01–4245. USGS, Reston, VA.
1694. Sumner, H. R., Dowler, C. C., Garvey, P. M. (2000). Application of agrochemicals by chemigation, pivot attached sprayer systems, and conventional sprayers. Applied Engineering in Agriculture, 16(2): 103–107.
1695. Sunderman, H. D., Lamm, F. R. (1996). Agricultural Research. Report of Progress 764, June, (1996). KAES, Manhattan, KS. 72 pages.
1696. Sunderman, H. D., Lamm, F. R., and Lawless, J. R. (1992). Agricultural Research. Rep. of Progress 660, June, (1992). KAES, Manhattan, KS. 89 pages.
1697. Sung, W. (2002). Corrosion control and chlorination, discolored water and nitrification. Proceedings Water Quality Technology Conference, MWRA, Southborough, MA, USA, 1683–1686.
1698. Suresh, S. J., Naik, V. M. (2002). Theory of dielectric constant of aqueous solutions. The Journal of Chemical Physics, 116: 4212–4220.
1699. Surface Irrigation Systems EM-4828, Washington State University.
1700. Sutton, B. G., Stirzaker, R. J., Doney, C. J., English, S. D. (1985). Solar powered drip irrigation for vegetables. In Proceedings of the 3rd International Drip/Trickle Irrigation Congress, 2: 589–593. St. Joseph, MI: ASAE.
1701. Suwansawat, S., Benson, C. H. (1999). Cell size for water content-dielectric constant calibrations for time domain reflectometry. ASTM Geotechnical Testing Journal, 22: 3–12.
1702. Swaun, G. (1993). Water Quality Control, Rural Water Advisory Services, South East Region Irrigation Association, USA.
1703. Swihart, J. (2002). Canal-lining demonstration project year 10 final report. 1st version. U. S. Dept. of the Interior.
1704. Szeicz, G., Long, I. F. (1969). Surface resistance of crop canopies. Water Resources Research, 5: 622–633.
1705. Tabbagh, A., Camerlynck, C., Cosenza, P. (2000). Numerical modeling for investigating the physical meaning of the relationship between relative dielectric permittivity and water content of soils. Water Resources Research, 36: 2771–2776.

1706. Tajrishy, M. A., Hills, D. J., and Tchobanoglous, G. (1994). Pretreatment of secondary effluent for drip irrigation. J Irrig Drain Eng., 120(4): 716–731.
1707. Takakura, T. (1988). Protected cultivation in Japan. Symposium on High Technology in Protected Cultural. *Acta Horticultural*, 230: 29–37.
1708. Tamaab, T. (2004). Tamaab Database. Water Research Center. Iran.
1709. Tanji, K. K. (1990). *Agricultural salinity assessment and management.* ASCE Manuals and Reports on Engineering Practice No. 71 ASCE, New York, 113–137.
1710. Tanner, C. B., Fuchs, M. (1968). Evaporation from unsaturated surfaces: A generalized combination equation. Journal of Geophysical Research, 73(4): 1299–1304.
1711. Tanner, C. B., Pelton, W. L. (1960). Potential evapotranspiration estimates by the approximate energy balance of Penman. Journal of Geophysical Research, 65(10): 3391–3413.
1712. Tarantino, A., Mongiovì, L. (2001). Experimental methodology and cavitation mechanisms in tensiometer measurements. Geotechnical and Geological Engineering, 19(3): 189–210.
1713. Tarantino, A., Mongiovì, L. (2003). Calibration of tensiometer for direct measurement of matrix suction. Géotechnique Technical Note, 53(1): 137–141.
1714. Tarantino, A. (2003). Direct Measurement of Soil Water Tension. Proceedings 3rd International Conference on Unsaturated Soils (JFT Jucá, TMP de Campos, FAM Marinho, editors.), Recife, Brasil. 44–65.
1715. Tarantino, A. (2004). Discussion on: Tensiometer saturation and the reliable measurement of soil suction, by Take, W. A., M. D. Bolton'. Géotechnique, 54(3): 229–232.
1716. Tarantino, E. (1991). Grass reference measurements in Italy. In: *Lysimeters for evapotranspiration and environmental measurements*, eds. Allen, R. G., Howell, T. A., Pruitt, W. O., Walter, L. A., Jensen, M. E., 200–209 pages. New York, NY: ASCE.
1717. Tarkalson, D. D., Payero, J., O. (2008). Comparison of nitrogen fertilization methods and rates for subsurface drip irrigated corn in the semi-arid Great Plains. Trans. ASABE, 51(5): 1633–1643.
1718. Teixeira, J. L., Fernando, R. M., Pereira, L. S. (1995). Irrigation scheduling alternatives for limited water supply and drought. ICID J., 44: 73–88.
1719. Teixeira, J. L., Farrajota, M. P., and Pereira, L. S. (1995). PROREG simulation software to design demand in irrigation projects. In: *Crop-water simulation models in practice*, eds. Pereira, L. S., B. J. van den Broek, P., Elsevier B. V.
1720. Tejedor, M., Jiménez, C. C., Díaz, F. (2002). Soilmoisture regime changes in tephra-mulched soils: Implications for soil taxonomy. Soil Sci. Soc. of America J., 66(1): 202–206.
1721. Tennakoon, S. B., Milroy, S. P. (2003). Crop water use and water use efficiency on irrigated cotton farms in Australia. Agricultural Water Management, 61: 179–194.
1722. Testezlaf, R., Zazueta, F. S., Yeager, T. H. (1997). A real-time irrigation control system for green houses. Applied Engr. in Agric., 13(3): 329–332.
1723. The Ogallala: Gauging, protecting the aquifer's health. KSU, April, (2008).
1724. Thokal, R., Mahale, D., A. Ganpat Powar, (2004). Drip irrigation systems: Clogging and its prevention. Pointer Publshers. 125–126.
1725. Thokal, R. T., Mahale, D. M., Powar, A. (2004). Drip irrigation systems: Clogging and prevention. Rajasthan (India): Pointer Publisher. 32–40.
1726. Thokal, R. T., Mahale, D. M., Powar, A. G. (2004). Drip Irrigation System: Clogging and its Prevention. Jodhpur-India: Pointer Publisher. 1–107.
1727. Thom, A. S. (1971). Momentum absorption by vegetation. Quarterly Journal of the Royal Meteorological Society, 97: 414–428.
1728. Thom, A. S. (1972). Momentum, mass and heat exchange of vegetation. Quarterly Journal of the Royal Meteorological Society, 98: 124–134.
1729. Thom, A. S., Oliver, H. R. (1977). On Penman's equation for estimating regional evaporation. Quarterly Journal of the Royal Meteorological Society, 103: 345–357.
1730. Thom, A. S., Stewart, J. B., Oliver, H. R., Gash, J. H. C. (1975). Comparison of aerodynamic and energy budget estimates of fluxes over a pine forest. Quarterly Journal of the Royal Meteorological Society, 101: 93–105.

1731. Thom, A. S., Thony, J. L., Vauclin, M. (1981). On the proper employment of evaporation pans and atmometers in estimating potential transpiration. Quarterly Journal of the Royal Meteorological Society, 107: 711–736.
1732. Thomas, A. W., Duke, H. R., Kruse, E. G. (1977). Capillary potential distributions in root zones using subsurface irrigation. Trans. of ASAE, 20(1): 62–67, 75.
1733. Thomas, A. W., Kruse, E. G., Duke, H. R. (1974). Steady infiltration from line sources buried in soil. Trans. of ASAE, 17(1): 125–133.
1734. Thomas, F. (2005). Design and control of sprinkler systems for crop disease research. American Society of Agricultural and Biological Engineers, St. Joseph, Michigan. Vol. 3.
1735. Thomas, J. T., Husein, A. (2003). Application of soil fumigants through micro-irrigation systems. ASAE Annual Meeting. 45–68.
1736. Thomas, L., Doerge, T. A., Godin, R. E. (2002). Subsurface drip irrigation and fertigation of broccoli: II Agronomic, economic, and environmental outcomes. Soil Sci. Soc. of America J., 66: 178–185.
1737. Thompson, H. C., Kelly, W. C. (1957). *Vegetable corps*. 5th ed. New York: McGraw-Hill Book Co, Inc. Chapter 7: 86–106.
1738. Thompson, N., Barrie, L. A., Ayles, M. (1981). The meteorological office rainfall and evaporation calculation system: MORECS. Hydrological Memorandum 45, Hydrometeorological Services, London, Page 66.
1739. Thompson, S. J., Ross, B., B. (1996). Using soil moisture sensors for making irrigation management decisions in Virginia. Extension Agricultural Engineers, Virginia Tech. 67–78.
1740. Thompson, T. L., and Doerge, T. A. (1996). Nitrogen and water interactions in subsurface trickle-irrigated leaf lettuce, I: Plant response. Soil Sci. Soc. of America J., 60: 163–168.
1741. Thompson, T. L., and Doerge, T. A. (1996). Nitrogen and water interactions in subsurface trickle-irrigated leaf lettuce, II: Agronomic, economic and environmental outcomes. Soil Sci. Soc. of America J., 60: 168–173.
1742. Thompson, T. L., Doerge, T. A., and Godin, R. E. (2000). Nitrogen and water interactions in subsurface drip-irrigated cauliflower, I: Plant response. Soil Sci. Soc. of America J., 64: 406–411.
1743. Thompson, T. L., Doerge, T. A., and Godin, R. E. (2000). Nitrogen and water interactions in subsurface drip-irrigated cauliflower, II: Agronomic, economic and environmental outcomes. Soil Sci. Soc. of America J., 64: 412–418.
1744. Thompson, T. L., Doerge, T. A., and Godin, R. E. (2002). Subsurface drip irrigation and fertigation of broccoli, I: Yield, quality and nitrogen uptake. Soil Sci. Soc. of America J., 66: 186–192.
1745. Thompson, T. L., Doerge, T. A., and Godin, R. E. (2002). Subsurface drip irrigation and fertigation of broccoli, II: Agronomic, economic and environmental outcomes. Soil Sci. Society of America Journal 66: 178–185.
1746. Thompson, W. J., Enciso-Medina, J., Multer, W., L. (2002). Why subsurface drip irrigation (SDI)?: An economic analysis. Proc. 2002 Cotton Beltwide Conf., Atlanta, Georgia: National Cotton Council of America. 4 pages.
1747. Thompson, T. L., White, S. A., J. Walworth and Sower, G. J. (2003). Fertigation frequency for subsurface drip-irrigated broccoli. Soil Sci. Soc. of America J., 67: 910–918.
1748. Thornthwaite, C. W. (1948). An approach toward a rational classification of climate. Geography Review, pages 38, 55.
1749. Thorsten, W., Stewart, F., Gupta, H., Bogh, E., Bastidas, L., Nobre, C., Galvao, O. (2005). Regional hydrological impacts of climatic change: Impact assessment and decision making. IAHS Press, 35–55.
1750. Threadgill, E. D., Eisenhauer, D. E., Young, J. R., Bar-Yosef, B. (1999). Chemigation. St. Joseph, MI: American Society of Agricultural Engineers. 749–780.
1751. Tindall, J. A., Petrusak, R. L., and McMahon, P. B. (1995). Nitrate transport and transformation processes in unsaturated porous media. Journal of Hydrology, 169: 51–94.

1752. Tiner, R. W. (1999). Wetland indicators: A guide to wetland identification, delineation, classification, and mapping.CRC Press, Florida – USA. 136 and, 237.

1753. Tinglu, F., Stewart, B. A., Payne, W. A., Wang, Y., Song, S., J. Luo and Robinson, C. A. (2005). Supplemental irrigation and water-yield relationships for plasticure crops in the loess plateau of china. Agronomy Journal, 97(1): 177–188.

1754. Tisdale, S. L., Werner, N. L., Beaton, J. D., and Havlin, J. L. (1993). Soil and fertilizer nitrogen. In: *Soil fertility and fertilizers*. pages 139–144. MacMillan Publishing: New York.

1755. Tiwari, K. N., A. Singh and Mal, P. K. (2002). Effect of drip irrigation on yield of cabbage (Brassica oleracea L. var. capitata) under mulch and non-mulch conditions. Agricultural Water Management, 58(1): 19–28.

1756. Tognetti, R., d'Andria, R., Lavini, A., Morelli, G. (2006). The effect of deficit irrigation on crop yield and vegetative development of Olea europaea L. (cvs. Frantoio and Leccino). European Journal of Agronomy, 25: 356–364.

1757. Tognetti, R., d'Andria, R., Morelli, G., Alvino, A. (2005). The effect of deficit irrigation on seasonal variations of plant water use in Olea europaea L. Plant and Soil, 273: 139–155.

1758. Toker, N. K. (2002). Improvements and reliability of MIT tensiometers and studies on soil moisture characteristic curves. M.Sc. Dissertation, Massachusetts Institute of Technology, Boston, US. 544–566.

1759. Tolk, J. A., Howell, T. A., Evett, S. R. (2006). Nighttime evapotranspiration from alfalfa and cotton in a semiarid climate.Agronomy Journal, 98: 730–736.

1760. Tollefson, S. (1985). The Arizona system: Drip irrigation design for cotton. Proc. Third International Drip/Trickle Irrigation Congress 1: 401–405. St. Joseph, MI: ASAE.

1761. Tollefson, S. (1985). Subsurface drip irrigation of cotton and small grains. In Proc. Third International Drip/Trickle Irrigation Congress, 2: 887–895. St. Joseph, MI: ASAE.

1762. Tollefson, S. (1988). Commercial production of field and vegetable crops with subsurface drip irrigation. Irrigation Association 1988 Technical Conference Proc., pages 144–153.

1763. Tomer, M. D., Clothier, B. E., Vogeler, I., Green, S. (1999). A dielectric-water content relationship for sandy Volcanic Soils in New Zealand. Soil Sci. Soc. of America J., 63: 777–781.

1764. Topp, G. C., Zegelin, S., White, I. (2000). Impacts of the real and imaginary components of relative permittivity on time domain reflectometry measurements in soils. Soil Sci. Soc. of America J., 64: 1244–1252.

1765. Toro Ag Irrigation, (2002). AquaFlow. Software on compact disc from Toro Ag Irrigation, El Cajon, California. Version. 1.0.5.

1766. Trans. R. Soc. London Ser. B. 352, pages 937–947.

1767. Trickle irrigation in the Eastern United States, (1980). Agricultural Cooperative Extension Service, Northeast Regional Agricultural Engineering.

1768. Trooien, T. P. (2002). Management of water and biological effluent for crop production in South Dakota. HATCH Project SD00072-H.

1769. Trooien, T. P., Hills, D. J. (2007). Application of biological effluent. Chapter 9 in *Micro irrigation for Crop Production- Design, Operation, and Management*. Lamm, F. R., Ayars, J. E., and Nakayama, F. S., eds. Elsevier Publications. pages 329–356.

1770. Trooien, T. P., Lamm, F. R., Stone, L. R., Alam, M., Rogers, D. H., Clark, G. A., Schlegel, A., J. (2000). Subsurface drip irrigation using livestock wastewater: Dripline flow rates. Appl. Engr. in Agric. 16(5): 505–508.

1771. Trooien, T. P., Hills D. J., and Lamm, F. R. (2002). Drip Irrigation with biological effluent. Proc. Irrigation Assn. Int. Irrigation Technical Conf., October 24–26, New Orleans, LA. Irrigation Assn, Falls Church VA.

1772. Trooien, T. P., Lamm, F. R., Stone, L. R., Alam, M., Rogers, D. H., Clark, G. A., and Schlegel, A. J. (2000). Using livestock wastewater with SDI. In Proc. Central Plains Irrigation Short Course, Garden City, KS, Feb. 9–10, (2000). CPIA, 970 Fifth, W., Colby, KS. pages 81–87.

1773. Trooien, T. P., Lamm, F. R., Stone, L. R., Alam, M. (1999). Irrigating corn with subsurface drip irrigation and lagoon wastewater. Irrigation Journal, 49(5): 6–7.

1774. Trooien, T. P., Lamm, F. R., Stone, L. R., Alam, M., Rogers, D. H., Clark, G. A., Schlegel, A. J. (2000). Subsurface drip irrigation using livestock wastewater: Dripline flow rates. Apages Engr. in Agr. 16(5): 505–508.
1775. Trooien, T. P., Lamm, F. R., Stone, L. R., Alam, M., Rogers, D. H., Clark, G. A., Schlegel, A. J. (2000). Using subsurface drip irrigation with livestock wastewater. Proceedings of the 4th Decennial National Irrigation Symposium, Phoenix, AZ, Nov. 14–16, (2000). pages 379–384.
1776. Trooien, T. P., Lamm, F. R., Stone, L. R., Alam, M., Rogers, D. H., Clark, G. A., and Schlegel, A. J. (1999). Testing subsurface drip irrigation laterals with lagoon wastewater. In Proceedings of Int'l Irrigation Show and Conference, IA, Fairfax, VA. pages 17–22.
1777. Trooien, T. P., Lamm, F. R., Stone, L. R., Alam, M., Rogers, D. H., Clark, G. A., and Schlegel, A. J. (2000). Two years of subsurface drip irrigation with lagoon wastewater. pages 21–23. In: SWREC Field Day report, Report of Progress 856. K-State Research and Extension, Manhattan, KS.
1778. Trooien, T. P., Lamm, F. R., Stone, L. R., Alam, M., Rogers, D. H., Clark, G. A., and Schlegel, A. J. (1999). Testing subsurface drip irrigation laterals with lagoon wastewater. pages 19–21. In: SWREC Field Day report, Report of Progress 837. K-State Research and Extension, Manhattan, KS.
1779. Trooien, T. P., Lamm, F. R., Stone, L. R., Alam, M., Rogers, D. H., Clark, G. A., and Schlegel, A. J. (2000). Using lagoon wastewater with subsurface drip irrigation. 17th annual Water and the Future of Kansas Conf., Manhattan, KS, March 1, (2000).
1780. Trooien, T. P., Alam, M., and Lamm, F. R. (1998). Filtration and maintenance considerations for SDI systems. In Proc. Central Plains Irrigation Short Course, North Platte, NE, Feb. 17–18, (1998). CPIA, 970 Fifth, W., Colby, KS. pages 22–26.
1781. Trout, T. J., Ajwa, H., A. (2003). Application of soil fumigants through micro-irrigation systems. ASAE Paper No. 032021. ASABE Annual International Mtg., Las Vegas, Nevada, USA, 27–30 July.
1782. Tsai, L. S., Higby, R., Schade, J. (1999). Disinfection of poultry chiller water with chlorine dioxide: consumption and byproduct formation. Journal of Agricultural and Food Chemistry, 43: 2768–2773.
1783. Tupper, G. R., Ebelhar, M., W. (1993). Effect of nutrient balance on cotton yield-soil test K/P ratio. In Proc. Beltwide Cotton Conf., 1301–1303. Memphis, Tenn.: Nat. Cotton Council.
1784. Turbidity and recovering dissolved oxygen from a reclaimed effluent used for micro irrigation. Agric. Water Mgt., 111: 27–33.
1785. Turc, L. (1961). Evaluation des besoins en eau d'irrigation, évapotranspiration potentielle, formule climatique simplifiée et mise a jour. (in French). Annual Agronomy, 12: 13–49.
1786. Turner, J. H., Anderson, C. L. (1980). *Planning an irrigation system*. Athens-GA: American for Association Vocational Instructional Materials, 120.
1787. Tuzet, A., Perrier, A., Masaad, C. (1992). Crop water budget estimation of irrigation requirement. ICID Bulletins, 41(2): 1–17.
1788. Tyagi, N. K., and Minhas, P. S. (Eds.), (1998). *Agricultural Salinity Management in India*. Central Soil Salinity Research Institute, Karnal, India, 526 pages.
1789. Unger, P. W., and Howell, T. A. (1999). Agricultural water conservation – a global perspective. In: Kirkham MB (Ed.), *Water use in crop production*. Food Products Press, New York, pages 1–36.
1790. Unger, P. W., Cassel, D., K. (1991). Tillage implement disturbance effects on soil properties related to soil water conservation: a literature review. Soil Tillage Research, 19: 363–382.
1791. United States Agricultural Research Service, (2001). Irrigation & Drainage: A national research plan to meet competing demands and protect the environment. Washnington – DC: USDA – ARS, 1–27.
1792. United States Bureau of Reclamation. Denver Office, Engineering Division, (2003). Water management workshop session notes.
1793. United States Congress, Senate Committee on Indian Affairs, (2004). The Oglala Sioux Tribe Angostura Irrigation Project Rehabilitation and Development Act [microform]: iii, 41 pages.

1794. United States Environmental Protection Agency, (1999). Integrated Risk Information System (IRIS) on Chlorine.National Center for Environmental Assessment, Office of Research and Development, Washington, DC.
1795. Upchurch, D. R., Wanjura, D. F., Burke, J. J., Mahan, J. R. (1996). Biologically-identified optimal temperature interactive console (BIOTIC) for managing irrigation. U. S. Patent No. 5539637.
1796. USDA-NASS, (2009). *Farm and Ranch Irrigation Survey, (2007).* Census of Agriculture, Volume 3, Special Studies. USDA, Washington DC. 222 pages.
1797. Using livestock wastewater with SDI. University of Missouri Crop Management School, Columbia, Missouri, December 16–17, (2004).
1798. Vaishnava, V. G., Shelke, D. K., Bharambe, P. R. (2002). Drip irrigation and fertigation for sugarcane in deep black soils. Paper number 022203, ASAE Annual Meeting.
1799. Valencia, S. R., Schwankl, L. J., Lanini, W. T. (1988). Weed control by subsurface drip irrigation. Calif. Agric., 42(3): 22–24.
1800. Valiantzas, John D. (2002). Continuous outflow variation along irrigation laterals: Effect of the number of outlets. Journal Irrigation and Drainage Engineering, 128(1): 34–42.
1801. Van Bavel, C. H. M., Ahmed, J., Bhuiyan, S. I., Hiler, E. A., Smajstrla, A. G. (1973). Dynamic simulation of automated subsurface irrigation systems. Trans. of ASAE, 16(6): 1095–1099.
1802. Van Bavel, C. H., Fritschen, L. J., Reeves, W. E. (1963). Transpiration of sudangrass as an externally controlled process. Science, 141: 269–270.
1803. Van der Gulik, T. W. (1999). *B. C. Trickle Irrigation Manual.* B. C. Ministry Agric. and Food Res. Manage. Branch and Irrig. Industry Assoc. of British Columbia, Abbotsford, B. C., Canada. 321 pages.
1804. Van der Meulen, E. S., L. Nol and Cammeraat, L. H. (2006). Effects of irrigation and plastic mulch on soil properties on semiarid abandoned fields. Soil Sci. Soc. of America J., 70(2): 930–939.
1805. Van Keulen, H., Van Laar, H. H. (1986). The relation between water use and crop production. In: Van Keulen, H., Wolf, J. eds., *Modeling of Agricultural Production: weather, soils and crops*. Pudoc, Wageningen, pages 117–129.
1806. Vaz, C. M.P., Hopmans, J. W. (2001). Simultaneous measurement of soil penetration resistance and water content with a combined penetrometer-TDR moisture probe. Soil Sci. Soc. of America J., 65: 4–12.
1807. Vaz, C. M.P., Hopmans, J. W., Macedo, A., Bassoi, L. H., Wildenschild, D. (2002). Soil water retention measurements using a combined tensiometer-coiled time domain reflectometry probe. Soil Sci. Soc. of America J., 66: 1752–1759.
1808. Vaziri, C. M., Gibson, W. (1972). Subsurface and drip irrigation for Hawaiian sugarcane. In: 31st Report Hawaii Sugar Technology Annual Conference, 18–22. Honolulu, Hawaii: Hawaiian Sugar Planters' Association.
1809. Vázquez, Z. R.F., Feyen, J. (2001). Effect of potential evapotranspiration estimates on the performance of the mike she code applied to a medium sized catchment. Annual Meeting of American Society of Agricultural and Biological Engineers, Paper No. 012037.
1810. Veldkamp, E., J. J. O'Brien, (2000). Calibration of a frequency domain reflectometry sensor for humid tropical soils of volcanic origin. Soil Sci. Soc. of America J., 64: 1549–1553.
1811. Velez, J. E., Intrigliolo, D. S., Castel, J. R. (2007). Scheduling deficit irrigation of citrus trees with maximum daily trunk shrinkage. Agricultural Water Management. 90: 197–204.
1812. Verma, S. B. (1989). Aerodynamic resistances to transfers of heat, mass and momentum. In: *Estimation of areal evapotranspiration,* eds. Black, T. A., Spittlehouse, D. L., Novak, M. D., Price, D. T., 13–20 pages. IAHS Pub. No. 177.
1813. Waggoner, P. E., Miller, P. M., Rao, H. C. (1960). Plastic mulching: principles and benefits. Connecticut Agriculture Experiment Station, New Haven, CT. Technical Bulletin 643.
1814. Walker, J. P., Houser, P. R. (2002). Evaluation of the Ohm mapper instrument for soilmoisture measurement.Soil Sci. Soc. of America J., 66: 728–734.

1815. Walker, W. R. *Guidelines for Designing and Evaluating Surface Irrigation Systems*. FAO Pub. 45.
1816. Walker, W. R. (2003). Surface irrigation simulation, evaluation and design – user guide and technical documentation. Utah State University, Logan, Utah.
1817. Wallace, J. S., Roberts, J. M., Sivakuma, M. V. K. (1990). The estimation of transpiration from sparse dryland millet using stomatal conductance and vegetation area indices. Agricultural and Forest Meteorology, 51: 35–49.
1818. Walter, L. A., Siemer, E., Dirks, L. R., Quinian, J. P., Burman, R. D. (1991). Lysimeters vs. buffer areas: evapotranspiration and agronomic comparisons. In: *Lysimeters for evapotranspiration and environmental measurements*, eds. Allen, R. G., Howell, T. A., Pruitt, W. O., Walter, L. A., Jensen, M. E., 10–18. New York, NY: ASCE.
1819. Wang D., Shannon, M. C., Grieve, C. M., Shouse, P. J., Suarez, D. L., 2002, September. Ion partitioning among soil and plant components under drip, furrow, and sprinkler irrigation regimes: field and modeling assessments. Journal of Environmental Quality, 31: 1684–1693.
1820. Wang, C. H. (1999). The capacitive drop tensiometer: A novel multi analysis technique for measuring the properties of liquids. College of Precision Instrument and Optoelectronic Engineering, Tianjin University – China. 345–349.
1821. Wang, D., Shannon, M. C., Grieve, C. M., Shouse, P. J., Suarez, D. L. (2002). Ion partitioning among soil and plant component under drip, furrow, and sprinkler irrigation regimes: Field and modeling assessments. Journal of Environmental Quality, 31(50): 1684–1693.
1822. Wanjura, D. F., Upchurch, D. R., Sassenrath-Cole, G., DeTar, W., R. (1995). Calculating time thresholds for irrigation scheduling. Proc. Beltwide Cotton Conferences, Jan 4–7, San Antonio, Texas. pages 449–452.
1823. WAPDA, (2002). Pakistan Water and Power Development Authority, Water Resources and Hydropower Development Vision- 2025, Planning and Design Division water WAPDA, April 2002, Lahore.
1824. Ward, A.D, Trimble, S. (2004). *Environmental hydrology*.Lewis Publishers, New York. Chapter 4: 83–115.
1825. Warrick, A. W. (2001). *Soil physics companion.* Florida – USA: CRC Press. 1–400.
1826. Warrick, A. W., Shani, U. (1996). Soil-limiting flow from subsurface emitters: 2, Effect on uniformity. J. Irrig. Drain. Eng. Am. Soc. Civ. Eng., 122: 296–300.
1827. Warrick, A. W., Lomen, D. O., Amoozegar-Fard, A. (1980). Linearized moisture flow with root extraction for three dimensional, steady conditions. Soil Sci. Soc. of America J., 44(5): 911–914.
1828. Water management of irrigated-drained fields in the Jordan Valley South of Lake Kinneret. Journal of Irrigation and Drainage Engineering, 131(4): 364–374. July/August, (2005).
1829. Watts, P. J., Hancock, N. H. (1985). Evaporation and potential evaporation: A practical approach for agricultural engineers. Mechanical Engineering Translation, 10(4): 231–240.
1830. Webb, E. K. (1970). Profile relationships: the log-linear range and extension to strong stability. Quarterly Journal of the Royal Meteorological Society, 96: 67–90.
1831. Wehner, D. J., Watschke, T. L. (1981). Heat tolerance of Kentucky bluegrasses, perennial ryegrasses and annual bluegrass. Agronomy Journal, 73: 79–84.
1832. Weise, A. F., Marek, T. H., Harman, W. L. (1998). No tillage increases profit in a limited irrigation: dryland system. Journal of Production Agriculture, 11: 247–252.
1833. Welsh, D. F., Kreuter, U. P., Byles, J. D. (1995). Enhancing subsurface drip irrigation through vector flow. Proceedings of the 5th, International Micro irrigation Congress, ed. Lamm, F. R., 688–693 pages. St. Joseph, MI: ASAE.
1834. Welsh, P. (1999). Pat Welsh's Southern California gardening: A month-by-month guide completely revised and updated. Chronicle Books. 182, 132.
1835. Wendt, C. W., Onken, A. B., Wilke, O. C., Hargrove, R., Bausch, W., Barnes, L. (1977). Effect of irrigation systems on the water requirement of sweet corn. Soil Sci. Soc. of America J., 41(4): 785–788.

1836. Wesseling, J. (1998). Automated farm surface irrigation systems worldwide. International Commission on Irrigation.Agricultural Water Management, 51(1, April): 101.
1837. *Western Fertilizer Hand Book*. California Fertilizer Association. 2222 Watt Avenue, Sacramento, California 95825.
1838. *Western Irrigation*. SDI Management and Maintenance meeting, Garden City, Kansas, February 14, (2006).
1839. Whitaker, R. J., Ritchie, L. G., Bednarzand, W. C., Mills, I. C. (2008). Cotton subsurface drip and overhead irrigation efficiency, maturity, yield, and quality. Am. Soc. of Agronomy 100: 1763–1768.
1840. White, R. E. (2003). *Soils for fine wines*.USA: Oxford University Press. 145–163.
1841. Whitney, L. F. (1970). Review of subsurface irrigation in the Northeast. Proc. National Irrigation Symposium, pages F1-F8. St. Joseph, MI: ASAE.
1842. Whitney, L. F., Lo, K., M. (1969). Plastic orifice inserts for subsurface irrigation. Trans. of the ASAE, 12(5): 602–607.
1843. Wigginton, David W., Steven R. Raine. Measuring Irrigation System Performance in the Queensland Dairy Industry, NCEA, USQ Toowoomba.
1844. Wilde, C., Johnson, J., Bordovsky, J., P. (2009). Economic analysis of subsurface drip irrigation system uniformity. Appl. Engr. Agric. 25(3): 357–361.
1845. Wildman, W. E., Gowans, K. D. (1978). Soil physical environment and how it affects plant production.
1846. Williams, G. S., Hazen, A. (1960). *Hydraulic Tables*. 3rd ed. New York, NY: John Willey and Sons.
1847. Williams, J. R., Llewelyn, R. V., Reed, M. S., Lamm, F. R., and Delano, D. R. (1996). Economic analysis of alternative irrigation systems for continuous corn and grain sorghum in western Kansas. Report of Progress 766, July, (1996). KAES, Manhattan, KS. 46 pages.
1848. Williams, J. R., Llewelyn, R. V., Reed, M. S., Lamm, F. R., and Delano, D. R. (1996). Net returns for grain sorghum and corn under alternative irrigation systems in western Kansas. Dept. of Ag. Economics Staff Paper 96–3, March, (1996). Kansas State University. 31 pages.
1849. Wilson, W. S., Ball, A. S., Hinton, R. H. (1999). Managing risks of nitrates to humans and the environment. Cambridge: The Royal Society of Chemistry.
1850. Winsor, G. W. et al, (1976). The effects of nitrogen, phosphorous, potassium, magnesium and lime in factorial combination on the yield of glasshouse tomatoes. J.Horti.Sci., 42: 277–288.
1851. Withers, B., Vipond, S. (1980). *Irrigation design and practice*. New York: Cornell University.
1852. Witter, S. H. (1993). World-wide use of plastic in horticultural production. Horticulture Technology, 3: 6–27.
1853. Wittwer, S. H., Castilla, N. (1995). Protected cultivation of horticultural crops worldwide. Horticulture Technology, 5: 6–23.
1854. WMO, (1983). *Guide to meteorological instruments and observing practices*, 5th ed., WMO n 8. Geneva: WMO Publication.
1855. Wopereis, M. (2004). Agro-Economic characterization of rice production in a typical irrigation scheme in Burkina Faso. Agronomy Journal, 96: 1314–1322.
1856. Wraith, J. M., Or, D. (2001). Soil water characteristic determination from concurrent water content measurements in reference porous media. Soil Sci. Soc. of America J., 65: 1659–1666.
1857. Wraith, J. M., Or, D. (2001). Soil water characteristic determination from concurrent water content measurements in reference porous media. Soil Sci. Soc. of America J., 65: 1659–1666.
1858. Wright, J. L., Jensen, M. E. (1972). Peak water requirements of crops in southern Idaho. Journal of Irrigation and Drainage Division, ASCE, 96 (IR1): 193–201.
1859. Wright, J. L. (1981). Crop coefficients for estimates of daily crop evapotranspiration. Irrigation Scheduling for Water and Energy Conservation in the 80s, December 1981, ASAE.
1860. Wright, J. L. (1982). New evapotranspiration crop coefficients. Journal of Irrigation and Drainage Division of ASCE, 108 (IR1): 57–74.
1861. Wright, J. L. (1988). Daily and seasonal evapotranspiration and yield of irrigated alfalfa in southern Idaho. Agronomy Journal, 80: 662–669.

1862. Wright, J. L. (1990). Evapotranspiration data for dry, edible beans at Kimberly, Idaho. Unpublished data, USDA-ARS, Kimberly, Idaho.
1863. Wright, J. L. (1991). Using lysimeters to develop evapotranspiration crop coefficients. In: *Lysimeters for evapotranspiration and environmental measurements*, eds. Allen, R. G., Howell, T. A., Pruitt, W. O., L. A.
1864. Wu, I. P., Barragan, J. (2000). Design criteria for micro irrigation systems. Trans. ASAE, 43(5): 1145–1154.
1865. Wu, I. P., Fangmeir, D. C., 1974, December. Hydraulic design of twin-chamber trickle irrigation laterals. Technical Bulletin No. 216, The Agricultural experiment station, Tucson, Ariz.
1866. Wu, I. P., Gitlin, H. M., 1973, June. Hydraulics and uniformity for drip irrigation. Journal of the Irrigation and Drainage Division, ASCE, 99 (IR3): 157–168. Paper, (9786).
1867. Wu, I. P., Gitlin, H. M. (1973). Hydraulics irrigation design based on uniformity. Transactions of America Society American Society Agricultural Engineers, 17(3): 157–168.
1868. Wu, I. P., Gitlin, H. M. (1974). Design of irrigation lines. Technical Bulletin No.96, Hawaii Agricultural Experimental Station, University of Hawaii.
1869. Wu, I. P., Gitlin, H. M. (1975). Energy gradient line for drip irrigation laterals. Journal of the Irrigation and Drainage Division, ASCE, 101 (IR4): 323–326. Paper 11750.
1870. Wu, I. P., Gitlin, H. M. (1977). Design drip irrigation lines varying pipes sizes. Journal of the Irrigation and Drainage Division, ASCE, Vol. 103, No. IR4, Proc. Paper 13384.
1871. Wu, I. P., H. M. Gitlin. Drip irrigation systems design. Bulletin No. 144 and 156 of the Cooperative Extension Service, University of Hawaii.
1872. Wu, I. P. (1975). Design of drip irrigation main lines. Journal of the irrigation and Drainage Division, ASCE. Vol. 101: No. IR4, Proceeding Paper 11803, 265–278.
1873. Wu, I. P.and Gitlin, H. M. (1975). Drip irrigation designs on non uniform slopes. Paper presented at the 1975 Winter Meeting of America SocietyAmerican Society of Agricultural Engineers, Chicago, IL.
1874. Wu, I. P. (1995). Optimal scheduling and minimizing deep seepage in micro irrigation. Trans. ASAE, 38(5): 1385–1392.
1875. Xu, H., Cai, Z., Tsuruta, H. (2003). Soil moisture between rice-growing seasons affects methane emission, production and oxidation. Soil Sci. Soc. of America J., 67: 1147–1157.
1876. Yamauchi, Shinohara, H., Morimoto, T., Shimahara, Y., Yamamoto, M., Yamaoka, Y. (2000). New simple technique for hepatic parenchymal resection using a cavitron ultrasonic surgical Aspirator® and Bipolar Cautery equipped with a channel for water dripping. World Journal of Surgery, 23(10): 1032–1037.
1877. Yang, J., Li, B., Liu, S. (2000). A large weighing lysimeter for evapotranspiration and soil water-groundwater exchange studies. Hydrological Processes, 14: 1887–1897.
1878. Yang, L., Kaleita, A. (2007). Understanding spatio-temporal patterns of soil moisture at the field scales. ASABE Paper No: 072108. Annual International Mtg. of the ASABE, Minneapolis, MN., June 17–20, (2007).
1879. Yang, S., Pei Lu, Okushima, L., Sase, S. (2004). Precision irrigation system based on detection of crop water stress with acoustic emission technique. 444–447. School of Electron. Inf. & Autom., Tianjin University of Science & Technology, China.
1880. Yaron, D., Bresler, E. (1983). Economic analysis of on-farm irrigation using response functions of crops. In: *Advances in Irrigation,* Hillel, D. (ed). Academic Press, New York, 2: 223–255.
1881. Yashima, S, (1987). Water balance dor rice double cropping in the MUDA area, Malaysia. Tropical Agriculture Research Series. No. 20. TARC.
1882. Yasin, M, S.Ahmad, Asghar, M. N., Ahmad, M. M. (2004). Root Zone Salinity Management Using Fractional Skimming Wells with Pressurised Irrigation: Volume- IV Pressurised Irrigation. A report prepared by IWMI, MONA & WRR I for NDP, WAPDA.
1883. Yates, M. V. (1992). Human Health Effects – Pathogens. Proc. of Reclaimed Wastewater: Practical Approaches to Developing an Alt. Water Supply, Davis Ca., May 18, (1992). Davis, U. C., CA 95616.

1884. Yazar, A., Sezen, S. M., Sesveren, S. (2002). LEPA and trickle irrigation of cotton in the Southeast Anatolia. Project (GAP) area in Turkey. Agricultural Water Management, 54: 189–203.
1885. Yoo, S. H., Choi, J. Y., Lee, S. H., Oh, Y. G., Park, N., Y. (2012). The Impacts of Climate Change on Paddy Rice Water Demand and Unit Duty of Water for Paddy Rice Using High-Definition Climate Scenarios. Journal of the Korean Society of Agricultural Engineers.
1886. Yost, R. (2002). Fate of wastewater effluent used for irrigation on turfgrass landscape in the American Pacific. Project No. HAW00849–1012S.
1887. Yu, C., Warrick, A. W., Conklin, M. H. (1999). Derived functions of time domain reflectometry for soil moisture measurement. Water Resources Research, 35: 1789–1796.
1888. Yuan, B. Z., Kang, Y., Nishiyama, S. (2001). Drip irrigation scheduling for tomatoes in unheated greenhouses. Irrigation Science, 20(3): 149–154.
1889. Yuan, Z., Choi, C. Y., Waller, P. M., Colaizzi, P. (2000). Effects of liquid temperature and viscosity on venturi injectors. Trans. of ASAE, 43(6): 1441–1447.
1890. Yuan, Z., Waller, P. M., Choi, C. Y. (1998). Effects of organic acids on salt precipitation in drip emitters and soil. Trans. ASAE, 41(6): 1689–1696.
1891. Yue, R., Phene, C. J., Dale, F., Ayars, J. E., Schoneman, R. A., I.-Wu, P., Kong, L. (1993). Field uniformity of subsurface drip irrigation. In: *Subsurface Drip Irrigation-Theory, Practices and Application*, 181–183 pages. CATI Pub. No. 92–1001. Fresno, CA: California State University.
1892. Yunseop, K., Evans, R. G., Iversen, W. M., Pierce, F. J., Chavez, J. L., 2006, Wireless sensor network, site-specific, control, water management, GPS. ASAE Annual Meeting, St. Joseph – MI.
1893. Yurdem, H., Demir, V., and Degirmencioglu, A. (2008). Development of a mathematical model to predict head losses from disc filters in drip irrigation systems using dimensional analysis. Biosystems Engineering, 100: 14–23.
1894. Yurdem, H., Demir, V., and Degirmencioglu, A. (2010). Development of a mathematical model to predict clean water head losses in hydrocyclone filters in drip irrigation systems using dimensional analysis. Biosystems Engineering, 105: 495–506.
1895. Zachmann, D. W., Thomas, A. W. (1973). A mathematical investigation of steady infiltration from line sources. Soil Sci. Soc. of America J., 37(4): 495–500.
1896. Zaman, W. U., Arshad, M., Saleem, A. (2001). Distribution of nitrate- nitrogen in the soil profile under different irrigation methods. Int. J. Agric. Biol., 2: 208–209.
1897. Zapata, F. (1990). Isotope techniques in soil fertility and plant nutrition studies. In: *Use of Nuclear Techniques in Studies of Soil- Plant Relationships*. Series No 2. Hardarson, G. (eds). IAEA, Vienna.
1898. Zarcinas, B. A., and Cartwright, B. (1983). *Analysis of soil and plant material.* Technical Paper 45, CSIRO Division of Soils, Canberra – AU.
1899. Zarrinkafsh, N. I. (1993). Applied Soil Science. Tehran University Publication No. 1995 Iran, page 147.
1900. Zayani, K., Aloini, A., Lebdi, F. (2001). Design of drip line in irrigation systems using the energy drop ratio approach. Trans. of ASAE, 44(5): 1127–1133.
1901. Zegbe, J. A., Behboudian, M. H., Clothier, B. E. (2004). Emitter and filter tests for waste water reuse by drip irrigation. Agricultural Water Management, 68(2): 135–149.
1902. Zetzsche, J. B., Newman, J. S. (1966). Subirrigation with plastic pipe. Agric. Eng. 47(1): 74–75.
1903. Zhang H, and Oweis, T. (1999). Water-yield relations and optimal irrigation scheduling of wheat in the Mediterranean region. Agric. Water Manage., 38: 195–211.
1904. Zhang, H., Oweis, T. Y., Garabet, S., and Pala, M. (1998). Water-use efficiency and transpiration efficiency of wheat under rain-fed conditions and supplemental irrigation in a Mediterranean type environment. Plant Soil, 201: 295–305.
1905. Zhang, N., Fan, G., Lee, K. H., Kluitenberg, G. J., Loughin, T. M. (2004). Simultaneous measurement of soil water content and salinity using a frequency-response method. Soil Sci. Soc. of America J., 68: 1515–1525.

1906. Zhang, Y., Yu, Q., Liu, C., Jiang, J., Zhang, X. (2004). Estimation of winter wheat evapotranspiration under water stress with two semiempirical approaches.Soil Sci. Soc. of America J., 68: 159–168.
1907. Zhu, H., Butt, C. L., Lamb, M. C., Blankenship, P. D. (2004). An implement to install and retrieve surface drip irrigation laterals. Applied Engineering in Agriculture, 20(1): 17–23.
1908. Zhu, Y., Fox, R. H., Toth, J. D. (2002). Leachate collection efficiency of zero-tension pan and passive capillary fiberglass wick lysimeters.Soil Sci. Soc. of America J., 66: 37–43.
1909. Zimmer, A. L., McFarland, M. J., Moore, J. (1988). Upward free water movement from buried trickle emitters. ASAE Paper No. 88–2063. St. Joseph, MI: ASAE.
1910. Zoldoske, D. F. (1993). The future of irrigation is buried. In Proceedings of the International Irrigation Expo and Technical Conference, 86–88. Arlington, VA.: Irrigation Association.
1911. Zoldoske, D. F., Norum, E. M. (1997). Final Report. Progress report on the Lehman Farms. Project: A case study in the conversion of an old vineyard from flood to surface drip (SDI) and subsurface drip (SSDI) irrigation. Center for Irrig. Tech., Pub. No. 970702. Fresno, CA: California State University.
1912. Zoldoske, D. F., S. Genito and Jorgensen, G. S. (1995). Subsurface drip irrigation (SDI) on turfgrass: A university experience. Proceedings of the 5th International Micro irrigation Congress, ed. Lamm, F. R., 300–302. St. Joseph, MI: ASAE.
1913. Zur, B. (1976). The pulsed irrigation principle for controlled soil wetting. Soil Sci. 122(5): 282–291.

WEBLINKS

1. http://aenews.wsu.edu
2. http://ag.arizona.edu/azmet/et1.htm
3. http://agbiopubs.sdstate.edu/articles/FS876.pdf#search=%22methods%20%20measure%20soil%20moisture%22
4. http://aggie-horticulture.tamu.edu/extension/newsletters/vpmnews/feb03/art5feb.html
5. http://aggie-horticulture.tamu.edu/greenhouse/hydroponics/drip.html
6. http://agnews.tamu.edu/dailynews/stories/SOIL/Oct2505a.htm
7. http://agsyst.wsu.edu/MulchReport03.pdf
8. http://agsyst.wsu.edu/MulchReport04.pdf
9. http://asae.frymulti.com/abstract.asp
10. http://ascelibrary.aip.org/vsearch/servlet
11. http://attra.ncat.org/attra-pub/soil_moisture.html
12. http://biomet.ucdavis.edu/evapotranspiration.html
13. http://bse.unl.edu/Research/evapo.htm
14. http://bse.unl.edu/Research/subsurface.htm
15. http://cahe.nmsu.edu/pubs/_h/h-637.html
16. http://cals.arizona.edu/crops/irrigation/azdrip/azdripindex.html
17. http://cals.arizona.edu/crops/irrigation/azdrip/SDI.htm
18. http://cati.csufresno.edu/cit/rese/90/900606/index.html
19. http://cigr-ejournal.tamu.edu/submissions/volume.htm
20. http://cr4.globalspec.com/thread/2318/Measuring-HUmidity-Using-Capacitive
21. http://doultonusa.com/commercial_industrial_filters/commercial_and_industrial_water_filters.htm
22. http://driptips.toro.com/
23. http://dripwatering.com/os-load.asp?load=os-catalog&catalogID=162&trail=7,162
24. http://eartheasy.com/grow_drip-irrigation.htm
25. http://edis.ifas.ufl.edu
26. http://edis.ifas.ufl.edu/AE077
27. http://edis.ifas.ufl.edu/AE080
28. http://edis.ifas.ufl.edu/AE093

29. http://edis.ifas.ufl.edu/AE093
30. http://edis.ifas.ufl.edu/AE094
31. http://edis.ifas.ufl.edu/AE143
32. http://edis.ifas.ufl.edu/AE146
33. http://edis.ifas.ufl.edu/AE171
34. http://edis.ifas.ufl.edu/AE256
35. http://edis.ifas.ufl.edu/CH156
36. http://edis.ifas.ufl.edu/WI007
37. http://edis.ifas.ufl.edu/WI009
38. http://en.wikipedia.org/wiki/Drip_irrigation
39. http://en.wikipedia.org/wiki/Plastic_mulch
40. http://en.wikipedia.org/wiki/Talk: Drip_irrigation
41. http://environmentalrisk.cornell.edu/C&ER/PlasticsDisposal/AgPlasticsRecycling/References/Garthe2002b.pdf
42. http://esce.ucr.edu/soilwater/summer2000/soilwatersummer_2000.htm
43. http://extension.oregonstate.edu/catalog/pdf/ec/ec1368.pdf
44. http://extension.oregonstate.edu/catalog/pdf/ec/ec1565.pdf
45. http://extension.oregonstate.edu/catalog/pdf/em/em8846-e.pdf
46. http://extension.oregonstate.edu/catalog/pdf/em/em8880-e.pdf
47. http://extension.oregonstate.edu/catalog/pdf/em/em8900.pdf
48. http://extension.oregonstate.edu/catalog/pdf/em/em8901.pdf
49. http://extension.oregonstate.edu/catalog/pdf/em/em8902.pdf
50. http://findarticles.com/p/articles/mi_qa4038/is_200210/ai_n9090867
51. http://fl.water.usgs.gov/PDF_files/wri96_4244_sumner.pdf
52. http://floriculture.osu.edu/archive/apr05/Sensors.html
53. http://fruitsandnuts.ucdavis.edu/crops/papers/Chapter_06.pdf#search='drip%20irrigation%20filtering'
54. http://ga.water.usgs.gov/edu/irdrip.html
55. http://ga.water.usgs.gov/edu/irmethods.html
56. http://geochange.er.usgs.gov/sw/changes/natural/et/
57. http://www.aces.edu/precisionag/staticpages/index.php?page=20081217203945319
58. http://grounds-mag.com/irrigation/grounds_maintenance_innovations_irrigation/
59. http://grounds-mag.com/irrigation/grounds_maintenance_irrigating_difficult_spaces/
60. http://grounds-mag.com/irrigation/grounds_maintenance_irrigating_steepsloped_landscapes/
61. http://grounds-mag.com/mag/grounds_maintenance_install_drip_irrigation/ http://www.dripirrigation.ca/HowTo_Maintain.asp
62. http://grounds-mag.com/mag/grounds_maintenance_irrigationfiltration_systems/index.html
63. http://hgic.clemson.edu/factsheets/HGIC1705.htm
64. http://isw03.cityofmesa.org/utilities/conservation/pdf/dripguide-v701–0503.pdf#search=%22monthly%20average%20drip%20irrigation%22
65. http://landscaping.about.com/cs/cheaplandscaping1/f/drip_irrigation.htm
66. http://landscaping.suite101.com/article.cfm/landscape_fabric_pros_co
67. http://lieth.ucdavis.edu/Extension/CANTensiomSum98/index.htm
68. http://lubbock.tamu.edu/cottondvd/content/cottondvd/Fertility/nutmgmtdripirrcot.pdf
69. http://microlab.berkeley.edu/labmanual/chap8/8.53.html
70. http://newindianexpress.com/cities/bangalore/Micro-irrigation-to-be promoted/2013/08/17/article1738597.ece
71. http://news.thomasnet.com/fullstory/464484/954
72. http://newswatch.nationalgeographic.com/2012/06/25/drip-irrigation-expanding-worldwide/
73. http://nfrec-sv.ifas.ufl.edu/mulch.htm
74. http://nimss.umd.edu/homepages/pub.cfm?trackID=11236#4
75. http://orgprints.org/6770/
76. http://osuextra.okstate.edu/pdfs/F-1511web.pdf

77. http://osuextra.okstate.edu/pdfs/F-1511web.pdf#search=%22trickle%20irrigation%20maintenance%2
78. http://pods.dasnr.okstate.edu/docushare/dsweb/Get/Document-1443/f-1511%20web.pdf
79. http://pubs.caes.uga.edu/caebspubs/pubcd/B1130.htm
80. http://soil.scijournals.org/cgi/content/full/65/4/1081
81. http://soils.usda.gov/technical/ classification/osd/>
82. http://southcenters.osu.edu/hort/plastic.htm
83. http://texaserc.tamu.edu
84. http://timesofindia.indiatimes.com/topic/drip-irrigation/news/
85. http://topirrigation.co.uk/copyright.html
86. http://twri.tamu.edu/water_resources_research/2004/leskovar_proposal.pdf#search='Automation%20in%20Drip%20Irrigation
87. http://ucanr.org/alf_symp/2000/00–119.pdf
88. http://ucce.ucdavis.edu.files.filelibrary/40/975.pdf
89. http://web.mit.edu/2.25/www/pdf/DA_unified.pdf
90. http://winegrapes.tamu.edu/grow/scheduling.shtml
91. http://wizard.arsusda.gov/acsl/acslhome.html
92. http://www.1-hydroponics.co.uk/top-tips/drip-irrigation.htm
93. http://www.actahort.org/books/2/2_12.htm
94. http://hinduwebsite.com/utilities/webDir/webDir.asp?/Science/Agriculture/Practices_and_Systems/Irrigation
95. http://www.aem.cornell.edu/research/researchpdf/rb9901.pdf http://www.dripdepot.com/planning_installation_guide_1.html?id=wfjajmWE
96. http://www.ag.ohio-state.edu/
97. http://www.agf.gov.bc.ca/resmgmt/publist/500series/5771001.pdf#search=%22methods%20%20measure%20soil%20moisture%22
98. http://www.agnet.org/library/article/eb333.html
99. http://www.agweb.com/article/irrigation_efficiency_below_the_surface/
100. http://www.almegcontrols.com/continuous.htm
101. http://www.alsaustin.org/tech-evapo.htm
102. http://www.americanplasticscouncil.org/benefits/about_plastics/history.html
103. http://www.ams.usda.gov/science/pdp
104. http://www.asabe.org/membership/students/Roberts.doc
105. http://www.attra.org/downloads/water_quality/irrigation.pdf
106. http://www.austagnetwork.com.au/articlesitem.aspx?parentid=7&pageid=7&articlesitemid=78
107. http://www.bae.ncsu.edu/programs/extension/evans/ebae-91–153.html
108. http://www.bae.umn.edu/extens/ennotes/enmay02/mwps30.htm
109. http://www.bioline.org.br/abstract?id=at03019&lang=en
110. http://www.bioline.org.br/request?cg01007
111. http://www.bioline.org.br/request?cg02008
112. http://www.bioline.org.br/request?cg05012
113. http://www.c3.org/chlorine_issues/new_perspective.html
114. http://www.c3.org/chlorine_knowledge
115. http://www.ces.ncsu.edu/depts/hort/hil/hil-33.html
116. http://www.ces.ncsu.edu/depts/hort/hil/hil-33-a.html
117. http://www.ces.ncsu.edu/depts/hort/hil/hil-33-b.html
118. http://www.ces.ncsu.edu/depts/hort/hil/hil-8033.html
119. http://www.cimis.water.ca.gov/
120. http://www.clw.csiro.au/aclep/asc_re_on_line/soilhome.htm
121. http://www.clw.csiro.au/publications/science/2006/sr18–06.pdf
122. http://www.clw.csiro.au/publications/technical2002/tr12–02.pdf

123. http://www.control.com/1026151936/index http://www.aces.edu/department/extcomm/publications/anr/anr-1169/anr-1169.html
124. http://www.cprl.ars.usda.gov/programs
125. http://www.cropinfo.net/AnnualReports/1996/ondrip96.htm
126. http://www.cropinfo.net/AnnualReports/1997/instrumentation.wq.htm
127. http://www.cropinfo.net/AnnualReports/2002/Hansen2002.htm
128. http://www.cropinfo.net/drip.htm
129. http://www.cropinfo.net/drip.htm
130. http://www.cropinfo.net/drip.htm#ComponentsDo-it-yourself drip irrigation
131. http://www.vintagerosery.com/dripirrigation.htmDrip-irrigation
132. http://www.irrigationtutorials.com/dripguide.htmEartheasy.com (c) 2000–2006
133. http://www.eartheasy.com/grow_drip-irrigation.htm#d
134. http://www.cseindia.org/dtesupplement/water20031115/drip_irrigation.pdf
135. http://www.ctahr.hawaii.edu/oc/freepubs/pdf/F_N-12.pdf
136. http://www.daff.qld.gov.au/26_4364.htm
137. http://www.distagenomics.unibo.it/wuemed/M_bittelli_innovative_methods.pdf
138. http://www.dpi.vic.gov.au/dpi/nrenfa.nsf/LinkView
139. http://www.dpi.vic.gov.au/dpi/nreninf.nsf/childdocs/-2BAF4D73531CD1544A-2568B3000505AF-FFB44D93F7BB6C37CA256C4A0083CDB7–0123B04778E33B-D84A256DEA0027B8C3–2FE8144CEA3D02E0CA256C1A0020FAA8?
140. http://www.dripirr.com/services/system_guide/specifications/trees.htm
141. http://www.dripirrigation.ca/HowTo.asp
142. http://www.dripirrigation.ca/HowTo_Plan.asp
143. http://www.dripirrigation.com/drip_tutorial.php
144. http://www.dripirrigation.com/drip_tutorial.php?page_view=head
145. http://www.dripworksusa.com/tech.html
146. http://www.ds.worldbank.org/servlet/WDSContentServer/WDSP/IB/2003/08/02/000094946_03071804010122/Rendered/PDF/multi0page.pdf
147. http://www.eartheasy.com/grow_drip-irrigation.htm
148. http://www.edie.net/news/news_story.asp?id=8149
149. http://www.eere.energy.gov/inventions/pdfs/nwpreag.pdf
150. http://www.egr.msu.edu?age/aenewsletter/ae_7_99/sprinkler_irrigation_systems_han.htm
151. http://www.ehponline.org/members/2006/8526/8526.html
152. http://www.eidis.ifas.ufl.edu/scrips/htmlgen.exe?DOCUMENT_AE080
153. http://www.ent.iastate.edu/Ipm/Icm/2000/5–29–2000/wateruse.html
154. http://www.ent.iastate.edu/Ipm/Icm/2000/5–29–2000/wateruse.html
155. http://www.ext.colostate.edu/drought/soilmoist.html
156. http://www.ext.colostate.edu/PUBS/crops/04704.html
157. http://www.ext.colostate.edu/pubs/crops/04716.html
158. http://www.ext.colostate.edu/Pubs/garden/04702.html
159. http://www.ext.nodak.edu/extpubs/ageng/irrigate/ae1243w.htm
160. http://www.ext.nodak.edu/extpubs/ageng/irrigate/ae91w.htm
161. http://www.ext.vt.edu/pubs/bse/442–757/442–757.html
162. http://www.ext.vt.edu/pubs/bse/442–757/figure1.html
163. http://www.ext.vt.edu/pubs/envirohort/426–322/426–322.html
164. http://www.ext.vt.edu/pubs/rowcrop/442–024/442–024.html
165. http://www.extension.usu.edu/files/engrpubs/biewm27.pdf
166. http://www.extension.usu.edu/files/factsheets/irrigation.pdf
167. http://www.extento.hawaii.edu/kbase/reports/dripirrigation.htm
168. http://www.fao.org/docrep/s8684e/s8684e07.htm
169. http://www.fao.org/docrep/S8684E/s8684e07.htm#6.1%20when%20to%20use%20drip%20irrigation
170. http://www.fao.org/docrep/X0490E/X0490E00.htm

171. http://www.fao.org/documents/show_cdr.asp?url_file=/docrep/005/y8163e/y8163e00.htm
172. http://www.fao.org/nr/water/infores_databases_cropwat.html
173. http://www.farmradio.org/english/radio-scripts/71–10script_en.asp
174. http://www.fngla.org/reports/25/finalreport.pdf
175. http://www.frc.ri.cmu.edu/~mcm/seminar.html
176. http://www.fs.fed.us/psw/rsl/projects/water/Ziemer67.PDF#search=%22methods%20%20measure%20soil%20moisture%22
177. http://www.gardeners.com/The-Right-Mulch-Makes-a-Difference/default/5013.page?SC=
178. http://www.geoflow.com/agriculture.html
179. http://www.geoflow.com/landscape%20(inclu%20golf)/design_general.htm
180. http://www.greenindustry.com/ij/1999/0299/299
181. http://www.greenmediaonline.com/ij/2001/0110/
182. http://www.growermagazine.com/home/02–03filters.html
183. http://www.growermagazine.com/home/02–03filters.html
184. http://www.growingmagazine.com/article-8341.aspx
185. http://www.helvetasnepal.org.np/lisp.htm>
186. http://www.hfrr.kstate.edu/DesktopModules/ViewDocument.aspx?DocumentID=1093#search=%22plastic%20mulch%22
187. http://www.robertmarvel.com/Plastic_Mulch.html
188. http://www.highbeam.com/library/docFree.asp?DOCID=1G1: 105480610
189. http://www.highbeam.com/library/docFree.asp?DOCID=1G1: 108550679
190. http://www.homeautomationforum.com/ubb/Forum1/HTML/000067.html
191. http://www.hydroponics.com/gardens/grownintechniques.html
192. http://www.ianr.unl.edu
193. http://www.ictinternational.com.au/appnotes/ICT227.htm
194. http://www.ictinternational.com.au/faqjetfill.htm
195. http://www.idrc.ca/en/ev-42826–201–1-DO_TOPIC.html
196. http://www.interscience.com
197. http://www.iop.org/EJ/abstract/0508–3443/2/4/301
198. http://www.ipm.ucdavis.edu/TOOLS/TURF/SITEPREP/irrdes.html
199. http://www.irrig8right.com.au/banners/bibliography.htm
200. http://www.irrig8right.com.au/Irrigation_Methods/Micro_Irrigation/Drip/Drip_Trickle/Details_DR.htm
201. http://www.irrigation.learnabout.info/
202. http://www.irrigation.org/ibt/_9911/p57.html
203. http://www.irrigationtutorials.com/dripguide.htm
204. http://www.irrigationtutorials.com/install.html
205. http://www.irrigationtutorials.com/sprinkler00.htm
206. http://www.isprs.org/publications/related/ISRSE/html/papers/594.pdf
207. http://www.itrc.org/reports/salinity/treecropsalinity.pdf
208. http://www.iuss.org/bull108files/WCSS.htm
209. http://www.iwmi.cgiar.org/Assessment/files_new/publications/Workshop
210. http://www.jains.com/irrigation/drip%20irrigation%20system.htm
211. http://www.ksre.ksu.edu/ sdi/Reports/2004/LS100104.pdf
212. http://www.ksre.ksu.edu/library/agec2/mf585.pdf
213. http://www.ksre.ksu.edu/library/agec2/mf836.pdf
214. http://www.ksre.ksu.edu/library/ageng2/mf2575.pdf
215. http://www.ksre.ksu.edu/sdi/Abstracts/asae9500.htm
216. http://www.ksre.ksu.edu/sdi/Reports/1995/WaterReq.pdf
217. http://www.ksre.ksu.edu/sdi/Reports/1998/EconSDICP.pdf
218. http://www.ksre.ksu.edu/sdi/Reports/2000/SDILWaste.pdf
219. http://www.ksre.ksu.edu/sdi/Reports/2002/MWIAPaper.pdf
220. http://www.ksre.ksu.edu/sdi/Reports/2003/mf2576.pdf

221. http://www.ksre.ksu.edu/sdi/Reports/2004/LS100104.pdf
222. http://www.ksre.ksu.edu/sdi/Reports/2005/DepthSDI.pdf
223. http://www.ksre.ksu.edu/sdi/Reports/2005/IA05–1209.pdf
224. http://www.ksre.ksu.edu/sdi/Reports/2005/SDIFreqIA.pdf
225. http://www.ksre.ksu.edu/sdi/Reports/2007/AppUtlSDIWW.pdf.
226. http://www.ksre.ksu.edu/sdi/Reports/2009/FRLUnique09.pdf
227. http://www.ksre.ksu.edu/sdi/Reports/2009/IA/Enciso101709.pdf
228. http://www.ksre.ksu.edu/sdi/Reports/2009/IA/FRL20Yr09IA.pdf
229. http://www.ksre.ksu.edu/sdi/Reports/2009/Roger09SDI.pdf
230. http://www.ksre.ksu.edu/sdi/Reports/2010/DDepth10.pdf
231. http://www.ksre.ksu.edu/sdi/Reports/2010/ESpace10.pdf
232. http://www.ksvltd.com/content/index/sigma703D
233. http://www.laspilitas.com/drip.htm
234. http://www.microirrigationforum.com
235. http://www.microirrigationforum.com/new/archives
236. http://www.mofga.org/mofga/other/mofgj05es.html
237. http://www.ncea.org.au
238. http://www.ncea.org.au/Irrigation/downloads/DripIrrigation.pdf
239. http://www.ncwcd.org/ims/ims_info/usings1d.pdf
240. http://www.nespal.org/SIRP/Research/2005.03.Fact_Sheet_02SW.pdf
241. http://www.netafim.com/Business_Divisions/Engineering_and_Technical Support/
242. http://www.netafimusa.net/downloads/LND/LSUBGD_Drip_Irrigation_Guide.pdf
243. http://www.nrcc.cornell.edu/reports/RR_94–1.html
244. http://www.nres.uiuc.edu/research/r-masiunas.html
245. http://www.nrri.umn.edu/gla/pet.htm
246. http://www.nzwine.com/assets//evaluating_irrigation_systems.pdf
247. http://www.oclandscape.com/articles/irrigationprocess_article.htm
248. http://asae.frymulti.com/abstract.asp?aid=5803&t=1Pepsi drip/Easy drip irrigation
249. http://www.practicafoundation.nl/smartwater/EN/pepsidrip.htm
250. http://www.omafra.gov.on.ca/english/crops/hort/news/vegnews/2005/vg0805a3.htm
251. http://www.orival.com
252. http://www.oznet.k-state.edu/irrigate/MDI.htm
253. http://www.oznet.ksu.edu/sdi/Reports/2002/DIBioEff.pdf
254. http://www.oznet.kstate.edu/sdi/Abstracts/Drip Irrigation of Row Crops.htm
255. http://www.oznet.k-state.edu/sdi/Abstracts/Drip%20Irrigation%20of%20Row%20Crops.htm
256. http://ohioline.osu.edu/sc173/sc173_13.html
257. http://www.oznet.ksu.edu/library/ageng2/l796.pdf
258. http://www.oznet.ksu.edu/library/ageng2/mf2361.pdf
259. http://www.oznet.ksu.edu/sdi/REPORTS/2002
260. http://www.oznet.ksu.edu/sdi/Reports/2002/ADofSDI.pdf
261. http://www.oznet.ksu.edu/sdi/Reports/2002/MWIAPaper.pdf
262. http://www.pesticide.org/BasicRegistration.pdf
263. http://www.physicalgeography.net/fundamentals/8j.html
264. http://www.plasticulture.com
265. http://www.plasticulture.org/history_global_community.htm
266. http://www.plt.org/ irrigation
267. http://www.plumbingsupply.com/ed-micro.html
268. http://www.progardenbiz.com/issues/v1issue3/Irrigation-Installation.html
269. http://www.public.iastate.edu/~taber/Extension/tensiometer%20tips/tens_tips.html
270. http://www.pubs.asce.org/
271. http://www.quantumlynx.com/water/back/vol2no1/v21st4.htm
272. http://www.rainbird.com/ag/du.htm
273. http://www.rainbird.com/documents/drip/LDLDesignGuide.pdf

274. http://www.rainbird.com/documents/drip/XFSeriesDesignGuide.pdf
275. http://www.rainbird.com/dripdetails
276. http://www.regional. org.au/au/asa/2001/p/3/barber.htm
277. http://www.sabi.co.za/html/designnorms.html
278. http://www.sandia.gov/energy-water/EastPresentations/IrrigWaterPower.pdf
279. http://www.scielo.br/scielo.php
280. http://www.smallfarms.wsu.edu/crops/dripIrrigation.html
281. http://www.snwa.com/html/land_irrig.html
282. http://www.snwa.com/html/land_irrig_drip.html
283. http://www.soilsci.com/pt/re/soilsci/abstract
284. http://www.sonorapacific.com/files/Chlorine_Injection.pdf
285. http://www.sowacs.com
286. http://www.srs.fs.usda.gov/pubs
287. http://www.srs.fs.usda.gov/pubs/rn/rn_srs011.pdf
288. http://www.swfwmd.state.fl.us/waterres/drought/articles/drip.htmDrip irrigation
289. http://www.swrcb.ca.gov/ab885/docs/techonsite/chapter10.pdf
290. http://www.sydneywater.com.au/SavingWater/InYourGarden/WateringSystems/
291. http://www.taunton.com/finegardening/pages/g00005.asp
292. http://www.toolbase.org/Technology-Inventory/Sitework/drip-irrigation-leach-field
293. http://www.twri.tamu.edu/./reports/1975/65.html
294. http://www.uae.gov.ae/uaeagricent/wateranddam/moderirrigation_e.stmAnsari
295. http://www.un.org/waterforlifedecade/water_cooperation_2013/madagascar.shtml
296. http://www.unce.unr.edu/publications/FS97/FS9713.htm
297. http://www.urbanfarmerstore.com/drip/dripintro.html#applications
298. http://www.uregina.ca/~sauchyn/geog327/Farnsworth, R. K
299. http://www.engineering.usu.edu/uwrl/atlas/ch3/index.html
300. http://www.agu.org/revgeophys/engman00/node6.html
301. http://www.uri.edu/ce/factsheets/sheets/mulch.html
302. http://www.usu.edu/cpl/PDF/TensiometerOperatingInstructions2.pdf
303. http://www.virtualsciencefair.org/2004/kwol4s0/public_html/adv.dis.htm
304. http://www.waterboysprinkler.com/ieq-ctrl.html
305. http://www.wateright.org/site2/publications/880105.asp
306. http://www.wca-infonet.org
307. http://www.wca-infonet.org/cds_upload/1058151636725_SUBSURFACE_IRRIGATION.pdf
308. http://www.wcainfonet.org/servlet/CDSServlet?status=ND0xMjY1LjEyNjE1JjY9ZW4mMzM9ZG9jdW1lbnRzJjM3PWluZm8~
309. http://www.westlandswater.org/wtrcon/handbook/eval6.htm
310. http://www.westlandswater.org/wtrcon/handbook/eval6.htm
311. http://www.cvwd.org/lush&eff/lsh&ef7.htm
312. http://www.wisegeek.com/what-is-drip-irrigation.htm
313. http://www.woodycrops.org/paducah/bar.html
314. http://www.wsl.ch/staff/niklaus.zimmermann/programs/aml3_3.html
315. http://www1.agric.gov.ab.ca/$department/deptdocs.nsf/all/faq8266?opendocument
316. http://www2.dpi.qld.gov.au/fieldcrops/17650.html
317. http://www.cortezjournal.com/archives/1news1661.htm
318. http://www2.dpi.qld.gov.au/fieldcrops/17653.html
319. http://www2.hawaii.edu/~hector/vegcropupdate/1996/August96.html
320. http://wwwww.umsmuc.de/datenmanagment/news_en/singlenews_english.html?L=1tx_ttnews%5Btt_news%5D=42&tx_ttnews%5BbackPid%5D=46&Hash=ofe8ecef62
321. http:/www.gardenstew.com/blog/e3–15-mulching-benefits—organic-and-inorganic-mulch-types.html
322. https: //www.soils.org/publications/sssaj/articles/74/5/1518

APPENDICES

(Modified and reprinted with permission from: Goyal, Megh R., 2012. Appendices. Pages 317–332. In: *Management of Drip/Trickle or Micro Irrigation* edited by Megh R. Goyal. New Jersey, USA: Apple Academic Press Inc.)

APPENDIX A

CONVERSION SI AND NON-SI UNITS

To convert the	Column 1	Column 2	To convert the Column
Column 1 in the	Unit	Unit	2 in the Column 1
Column 2,	SI	Non-SI	Multiply by
Multiply by			

LINEAR

0.621	kilometer, km (10^3 m)	miles, mi	1.609
1.094	meter, m	yard, yd	0.914
3.28	meter, m	feet, ft	0.304
3.94×10^{-2}	millimeter, mm (10^{-3})	inch, in	25.4

SQUARES

2.47	hectare, he	acre	0.405
2.47	square kilometer, km^2	acre	4.05×10^{-3}
0.386	square kilometer, km^2	square mile, mi^2	2.590
2.47×10^{-4}	square meter, m^2	acre	4.05×10^{-3}
10.76	square meter, m^2	square feet, ft^2	9.29×10^{-2}
1.55×10^{-3}	mm^2	square inch, in^2	645

CUBICS

9.73×10^{-3}	cubic meter, m^3	inch-acre	102.8
35.3	cubic meter, m^3	cubic-feet, ft^3	
2.83×10^{-2}			
6.10×10^4	cubic meter, m^3	cubic inch, in^3	1.64×10^{-5}
2.84×10^{-2}	liter, L (10^{-3} m^3)	bushel, bu	35.24
1.057	liter, L	liquid quarts, qt	0.946
3.53×10^{-2}	liter, L	cubic feet, ft^3	28.3
0.265	liter, L	gallon	3.78
33.78	liter, L	fluid ounce, oz	2.96×10^{-2}
2.11	liter, L	fluid dot, dt	0.473

WEIGHT

2.20×10^{-3}	gram, g (10^{-3} kg)	pound,	454
3.52×10^{-2}	gram, g (10^{-3} kg)	ounce, oz	28.4
2.205	kilogram, kg	pound, lb	0.454
10^{-2}	kilogram, kg	quintal (metric), q	100
1.10×10^{-3}	kilogram, kg	ton (2000 lbs), ton	907
1.102	mega gram, mg	ton (US), ton	0.907
1.102	metric ton, t	ton (US), ton	0.907

YIELD AND RATE

0.893	kilogram per hectare	pound per acre	1.12
7.77×10^{-2}	kilogram per cubic meter	pound per fanega	12.87
1.49×10^{-2}	kilogram per hectare	pound per acre, 60 lb	67.19
1.59×10^{-2}	kilogram per hectare	pound per acre, 56 lb	62.71
1.86×10^{-2}	kilogram per hectare	pound per acre, 48 lb	53.75
0.107	liter per hectare	galloon per acre	9.35
893	ton per hectare	pound per acre	1.12×10^{-3}
893	mega gram per hectare	pound per acre	1.12×10^{-3}
0.446	ton per hectare	ton (2000 lb) per acre	2.24
2.24	meter per second	mile per hour	0.447

SPECIFIC SURFACE

10	square meter per kilogram	square centimeter per gram	0.1
10^{3}	square meter per kilogram	square millimeter per gram	10^{-3}

PRESSURE

9.90	megapascal, MPa	atmosphere	0.101
10	megapascal	bar	0.1
1.0	megagram per cubic meter	gram per cubic centimeter	1.00
2.09×10^{-2}	pascal, Pa	pound per square feet	47.9
1.45×10^{-4}	pascal, Pa	pound per square inch	6.90×10^{3}

To convert the	Column 1	Column 2	To convert the column
column 1 in the	Unit	Unit	2 in the column 1
Column 2,	SI	Non-SI	Multiply by
Multiply by			

TEMPERATURE

1.00 (K-273)	Kelvin, K	centigrade, °C	1.00 (C+273)
(1.8 C + 32)	centigrade, °C	Fahrenheit, °F	(F--32)/1.8

ENERGY

9.52×10^{-4}	Joule J	BTU	1.05×10^{3}
0.239	Joule, J	calories, cal	4.19
0.735	Joule, J	feet-pound	1.36
2.387×10^{5}	Joule per square meter	calories per square centimeter	4.19×10^{4}
10^{5}	Newton, N	dynes	10^{-5}

WATER REQUIREMENTS

9.73×10^{-3}	cubic meter	inch acre	102.8
9.81×10^{-3}	cubic meter per hour	cubic feet per second	101.9
4.40	cubic meter per hour	galloon (US) per minute	0.227
8.11	hectare-meter	acre-feet	0.123
97.28	hectare-meter	acre-inch	1.03×10^{-2}
8.1×10^{-2}	hectare centimeter	acre-feet	12.33

CONCENTRATION

1	centimol per kilogram	milliequivalents per 100 grams	1
0.1	gram per kilogram	percents	10
1	milligram per kilogram	parts per million	1

NUTRIENTS FOR PLANTS

2.29	P	P_2O_5	0.437
1.20	K	K_2O	0.830
1.39	Ca	CaO	0.715
1.66	Mg	MgO	0.602

NUTRIENT EQUIVALENTS

Column A	Column B	Conversion A to B	Equivalent B to A
N	NH_3	1.216	0.822
	NO_3	4.429	0.226
	KNO_3	7.221	0.1385
	$Ca(NO_3)_2$	5.861	0.171

Column A	Column B	Conversion A to B	Equivalent B to A
	$(NH_4)_2SO_4$	4.721	0.212
	NH_4NO_3	5.718	0.175
	$(NH_4)_2 HPO_4$	4.718	0.212
P	P_2O_5	2.292	0.436
	PO_4	3.066	0.326
	KH_2PO_4	4.394	0.228
	$(NH_4)_2 HPO_4$	4.255	0.235
	H_3PO_4	3.164	0.316
K	K_2O	1.205	0.83
	KNO_3	2.586	0.387
	KH_2PO_4	3.481	0.287
	Kcl	1.907	0.524
	K_2SO_4	2.229	0.449
Ca	CaO	1.399	0.715
	$Ca(NO_3)_2$	4.094	0.244
	$CaCl_2 \times 6H_2O$	5.467	0.183
	$CaSO_4 \times 2H_2O$	4.296	0.233
Mg	MgO	1.658	0.603
	$MgSO_4 \times 7H_2O$	1.014	0.0986
S	H_2SO_4	3.059	0.327
	$(NH_4)_2 SO_4$	4.124	0.2425

Column A	Column B	Conversion A to B	Equivalent B to A
	K_2SO_4	5.437	0.184
	$MgSO_4 \times 7H_2O$	7.689	0.13
	$CaSO_4 \times 2H_2O$	5.371	0.186

APPENDIX B

PIPE AND CONDUIT FLOW

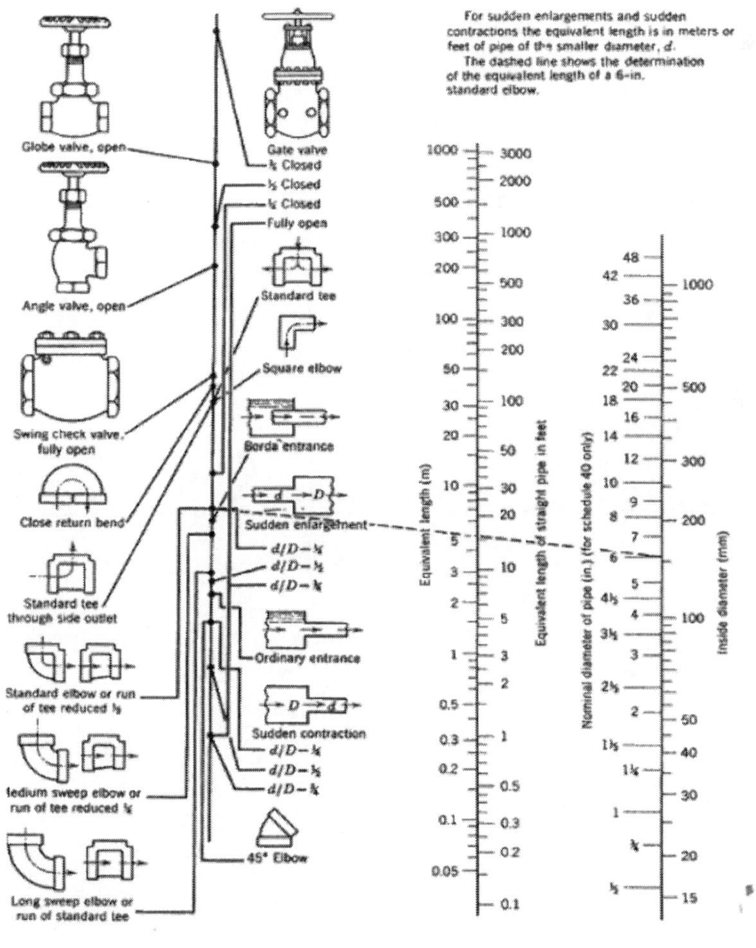

APPENDIX C

PERCENTAGE OF DAILY SUNSHINE HOURS: FOR NORTH AND SOUTH HEMISPHERES

Latitude	Jan	Feb	Mar	Apr	May	Jun	Jul	Aug	Sep	Oct	Nov	Dec
						NORTH						
0	8.50	7.66	8.49	8.21	8.50	8.22	8.50	8.49	8.21	8.50	8.22	8.50
5	8.32	7.57	8.47	3.29	8.65	8.41	8.67	8.60	8.23	8.42	8.07	8.30
10	8.13	7.47	8.45	8.37	8.81	8.60	8.86	8.71	8.25	8.34	7.91	8.10
15	7.94	7.36	8.43	8.44	8.98	8.80	9.05	8.83	8.28	8.20	7.75	7.88
20	7.74	7.25	8.41	8.52	9.15	9.00	9.25	8.96	8.30	8.18	7.58	7.66
25	7.53	7.14	8.39	8.61	9.33	9.23	9.45	9.09	8.32	8.09	7.40	7.52
30	7.30	7.03	8.38	8.71	9.53	9.49	9.67	9.22	8.33	7.99	7.19	7.15
32	7.20	6.97	8.37	8.76	9.62	9.59	9.77	9.27	8.34	7.95	7.11	7.05
34	7.10	6.91	8.36	8.80	9.72	9.70	9.88	9.33	8.36	7.90	7.02	6.92
36	6.99	6.85	8.35	8.85	9.82	9.82	9.99	9.40	8.37	7.85	6.92	6.79
38	6.87	6.79	8.34	8.90	9.92	9.95	10.1	9.47	3.38	7.80	6.82	6.66
40	6.76	6.72	8.33	8.95	10.0	10.1	10.2	9.54	8.39	7.75	6.72	7.52
42	6.63	6.65	8.31	9.00	10.1	10.2	10.4	9.62	8.40	7.69	6.62	6.37
44	6.49	6.58	8.30	9.06	10.3	10.4	10.5	9.70	8.41	7.63	6.49	6.21
46	6.34	6.50	8.29	9.12	10.4	10.5	10.6	9.79	8.42	7.57	6.36	6.04
48	6.17	6.41	8.27	9.18	10.5	10.7	10.8	9.89	8.44	7.51	6.23	5.86
50	5.98	6.30	8.24	9.24	10.7	10.9	11.0	10.0	8.35	7.45	6.10	5.64
52	5.77	6.19	8.21	9.29	10.9	11.1	11.2	10.1	8.49	7.39	5.93	5.43
54	5.55	6.08	8.18	9.36	11.0	11.4	11.4	10.3	8.51	7.20	5.74	5.18
56	5.30	5.95	8.15	9.45	11.2	11.7	11.6	10.4	8.53	7.21	5.54	4.89
58	5.01	5.81	8.12	9.55	11.5	12.0	12.0	10.6	8.55	7.10	4.31	4.56
60	4.67	5.65	8.08	9.65	11.7	12.4	12.3	10.7	8.57	6.98	5.04	4.22
						SOUTH						
0	8.50	7.66	8.49	8.21	8.50	8.22	8.50	8.49	8.21	8.50	8.22	8.50
5	8.68	7.76	8.51	8.15	8.34	8.05	8.33	8.38	8.19	8.56	8.37	8.68
10	8.86	7.87	8.53	8.09	8.18	7.86	8.14	8.27	8.17	8.62	8.53	8.88
15	9.05	7.98	8.55	8.02	8.02	7.65	7.95	8.15	8.15	8.68	8.70	9.10
20	9.24	8.09	8.57	7.94	7.85	7.43	7.76	8.03	8.13	8.76	8.87	9.33
25	9.46	8.21	8.60	7.74	7.66	7.20	7.54	7.90	8.11	8.86	9.04	9.58
30	9.70	8.33	8.62	7.73	7.45	6.96	7.31	7.76	8.07	8.97	9.24	9.85
32	9.81	8.39	8.63	7.69	7.36	6.85	7.21	7.70	8.06	9.01	9.33	9.96
34	9.92	8.45	8.64	7.64	7.27	6.74	7.10	7.63	8.05	9.06	9.42	10.1
36	10.0	8.51	8.65	7.59	7.18	6.62	6.99	7.56	8.04	9.11	9.35	10.2
38	10.2	8.57	8.66	7.54	7.08	6.50	6.87	7.49	8.03	9.16	9.61	10.3

40	10.3	8.63	8.67	7.49	6.97	6.37	6.76	7.41	8.02	9.21	9.71	10.5
42	10.4	8.70	8.68	7.44	6.85	6.23	6.64	7.33	8.01	9.26	9.8	10.6
44	10.5	8.78	8.69	7.38	6.73	6.08	6.51	7.25	7.99	9.31	9.94	10.8
46	10.7	8.86	8.90	7.32	6.61	5.92	6.37	7.16	7.96	9.37	10.1	11.0

APPENDIX D

PSYCHOMETRIC CONSTANT (Γ) FOR DIFFERENT ALTITUDES (Z)

$$\gamma = 10^{-3}\,[(C_p.P) \div (\varepsilon.\lambda)] = (0.00163) \times [P \div \lambda]$$

γ, psychrometric constant [kPa C^{-1}] c_p, specific heat of moist air = 1.013

[kJ kg^{-1}°C^{-1}] P, atmospheric pressure [kPa].

ε, ratio molecular weight of water

vapor/dry air = 0.622 λ, latent heat of vaporization [MJ kg^{-1}]

= 2.45 MJ kg^{-1} at 20°C.

Z (m)	γ kPa/°C	z (m)	γ kPa/°C	z (m)	γ kPa/°C	z (m)	γ kPa/°C
0	0.067	1000	0.060	2000	0.053	3000	0.047
100	0.067	1100	0.059	2100	0.052	3100	0.046
200	0.066	1200	0.058	2200	0.052	3200	0.046
300	0.065	1300	0.058	2300	0.051	3300	0.045
400	0.064	1400	0.057	2400	0.051	3400	0.045
500	0.064	1500	0.056	2500	0.050	3500	0.044
600	0.063	1600	0.056	2600	0.049	3600	0.043
700	0.062	1700	0.055	2700	0.049	3700	0.043
800	0.061	1800	0.054	2800	0.048	3800	0.042
900	0.061	1900	0.054	2900	0.047	3900	0.042
1000	0.060	2000	0.053	3000	0.047	4000	0.041

APPENDIX E

SATURATION VAPOR PRESSURE [e_s] FOR DIFFERENT TEMPERATURES (T)

Vapor pressure function = e_s = [0.6108]*exp{[17.27*T]/[T + 237.3]}

T °C	e_s kPa	T °C	e_s kPa	T °C	e_s kPa	T °C	e_s kPa
1.0	0.657	**13.0**	1.498	**25.0**	3.168	**37.0**	6.275
1.5	0.681	**13.5**	1.547	**25.5**	3.263	**37.5**	6.448
2.0	0.706	**14.0**	1.599	**26.0**	3.361	**38.0**	6.625
2.5	0.731	**14.5**	1.651	**26.5**	3.462	**38.5**	6.806

3.0	0.758	**15.0**	1.705	**27.0**	3.565	**39.0**	6.991
3.5	0.785	**15.5**	1.761	**27.5**	3.671	**39.5**	7.181
4.0	0.813	**16.0**	1.818	**28.0**	3.780	**40.0**	7.376
4.5	0.842	**16.5**	1.877	**28.5**	3.891	**40.5**	7.574
5.0	0.872	**17.0**	1.938	**29.0**	4.006	**41.0**	7.778
5.5	0.903	**17.5**	2.000	**29.5**	4.123	**41.5**	7.986
6.0	0.935	**18.0**	2.064	**30.0**	4.243	**42.0**	8.199
6.5	0.968	**18.5**	2.130	**30.5**	4.366	**42.5**	8.417
7.0	1.002	**19.0**	2.197	**31.0**	4.493	**43.0**	8.640
7.5	1.037	**19.5**	2.267	**31.5**	4.622	**43.5**	8.867
8.0	1.073	**20.0**	2.338	**32.0**	4.755	**44.0**	9.101
8.5	1.110	**20.5**	2.412	**32.5**	4.891	**44.5**	9.339
9.0	1.148	**21.0**	2.487	**33.0**	5.030	**45.0**	9.582
9.5	1.187	**21.5**	2.564	**33.5**	5.173	**45.5**	9.832
10.0	1.228	**22.0**	2.644	**34.0**	5.319	**46.0**	10.086
10.5	1.270	**22.5**	2.726	**34.5**	5.469	**46.5**	10.347
11.0	1.313	**23.0**	2.809	**35.0**	5.623	**47.0**	10.613
11.5	1.357	**23.5**	2.896	**35.5**	5.780	**47.5**	10.885
12.0	1.403	**24.0**	2.984	**36.0**	5.941	**48.0**	11.163
12.5	1.449	**24.5**	3.075	**36.5**	6.106	**48.5**	11.447

APPENDIX F

SLOPE OF VAPOR PRESSURE CURVE (Δ) FOR DIFFERENT TEMPERATURES (T)

$$\Delta = [4098.\ e^0(T)] \div [T + 237.3]^2$$
$$= 2504\{\exp[(17.27T) \div (T + 237.2)]\} \div [T + 237.3]^2$$

T °C	**Δ kPa/°C**	**T °C**	**Δ kPa/°C**	**T °C**	**Δ kPa/°C**	**T °C**	**Δ kPa/°C**
1.0	0.047	13.0	0.098	25.0	0.189	37.0	0.342
1.5	0.049	13.5	0.101	25.5	0.194	37.5	0.350
2.0	0.050	14.0	0.104	26.0	0.199	38.0	0.358
2.5	0.052	14.5	0.107	26.5	0.204	38.5	0.367
3.0	0.054	15.0	0.110	27.0	0.209	39.0	0.375
3.5	0.055	15.5	0.113	27.5	0.215	39.5	0.384
4.0	0.057	16.0	0.116	28.0	0.220	40.0	0.393
4.5	0.059	16.5	0.119	28.5	0.226	40.5	0.402
5.0	0.061	17.0	0.123	29.0	0.231	41.0	0.412
5.5	0.063	17.5	0.126	29.5	0.237	41.5	0.421
6.0	0.065	18.0	0.130	30.0	0.243	42.0	0.431

6.5	0.067	18.5	0.133	30.5	0.249	42.5	0.441
7.0	0.069	19.0	0.137	31.0	0.256	43.0	0.451
7.5	0.071	19.5	0.141	31.5	0.262	43.5	0.461
8.0	0.073	20.0	0.145	32.0	0.269	44.0	0.471
8.5	0.075	20.5	0.149	32.5	0.275	44.5	0.482
9.0	0.078	21.0	0.153	33.0	0.282	45.0	0.493
9.5	0.080	21.5	0.157	33.5	0.289	45.5	0.504
10.0	0.082	22.0	0.161	34.0	0.296	46.0	0.515
10.5	0.085	22.5	0.165	34.5	0.303	46.5	0.526
11.0	0.087	23.0	0.170	35.0	0.311	47.0	0.538
11.5	0.090	23.5	0.174	35.5	0.318	47.5	0.550
12.0	0.092	24.0	0.179	36.0	0.326	48.0	0.562
12.5	0.095	24.5	0.184	36.5	0.334	48.5	0.574

APPENDIX G

NUMBER OF THE DAY IN THE YEAR (JULIAN DAY)

Day	Jan	Feb	Mar	Apr	May	Jun	Jul	Aug	Sep	Oct	Nov	Dec
1	1	32	60	91	121	152	182	213	244	274	305	335
2	2	33	61	92	122	153	183	214	245	275	306	336
3	3	34	62	93	123	154	184	215	246	276	307	337
4	4	35	63	94	124	155	185	216	247	277	308	338
5	5	36	64	95	125	156	186	217	248	278	309	339
6	6	37	65	96	126	157	187	218	249	279	310	340
7	7	38	66	97	127	158	188	219	250	280	311	341
8	8	39	67	98	128	159	189	220	251	281	312	342
9	9	40	68	99	129	160	190	221	252	282	313	343
10	10	41	69	100	130	161	191	222	253	283	314	344
11	11	42	70	101	131	162	192	223	254	284	315	345
12	12	43	71	102	132	163	193	224	255	285	316	346
13	13	44	72	103	133	164	194	225	256	286	317	347
14	14	45	73	104	134	165	195	226	257	287	318	348
15	15	46	74	105	135	166	196	227	258	288	319	349
16	16	47	75	106	136	167	197	228	259	289	320	350
17	17	48	76	107	137	168	198	229	260	290	321	351
18	18	49	77	108	138	169	199	230	261	291	322	352
19	19	50	78	109	139	170	200	231	262	292	323	353
20	20	51	79	110	140	171	201	232	263	293	324	354
21	21	52	80	111	141	172	202	233	264	294	325	355

Day	Jan	Feb	Mar	Apr	May	Jun	Jul	Aug	Sep	Oct	Nov	Dec
22	22	53	81	112	142	173	203	234	265	295	326	356
23	23	54	82	113	143	174	204	235	266	296	327	357
24	24	55	83	114	144	175	205	236	267	297	328	358
25	25	56	84	115	145	176	206	237	268	298	329	359
26	26	57	85	116	146	177	207	238	269	299	330	360
27	27	58	86	117	147	178	208	239	270	300	331	361
28	28	59	87	118	148	179	209	240	271	301	332	362
29	29	(60)	88	119	149	180	210	241	272	302	333	363
30	30	—	89	120	150	181	211	242	273	303	334	364
31	31	—	90	—	151	—	212	243	—	304	—	365

APPENDIX H

STEFAN-BOLTZMANN LAW AT DIFFERENT TEMPERATURES (T):

$[\sigma^*(T_K)^4] = [4.903 \times 10^{-9}]$, MJ K^{-4} m^{-2} day^{-1}
Where: $T_K = \{T[°C] + 273.16\}$

T	$\sigma^*(T_K)^4$	T	$\sigma^*(T_K)^4$	T	$\sigma^*(T_K)^4$
			Units		
°C	MJ m^{-2} d^{-1}	°C	MJ m^{-2} d^{-1}	°C	MJ m^{-2} d^{-1}
1.0	27.70	**17.0**	34.75	**33.0**	43.08
1.5	27.90	**17.5**	34.99	**33.5**	43.36
2.0	28.11	**18.0**	35.24	**34.0**	43.64
2.5	28.31	**18.5**	35.48	**34.5**	43.93
3.0	28.52	**19.0**	35.72	**35.0**	44.21
3.5	28.72	**19.5**	35.97	**35.5**	44.50
4.0	28.93	**20.0**	36.21	**36.0**	44.79
4.5	29.14	**20.5**	36.46	**36.5**	45.08
5.0	29.35	**21.0**	36.71	**37.0**	45.37
5.5	29.56	**21.5**	36.96	**37.5**	45.67
6.0	29.78	**22.0**	37.21	**38.0**	45.96
6.5	29.99	**22.5**	37.47	**38.5**	46.26
7.0	30.21	**23.0**	37.72	**39.0**	46.56
7.5	30.42	**23.5**	37.98	**39.5**	46.85
8.0	30.64	**24.0**	38.23	**40.0**	47.15
8.5	30.86	**24.5**	38.49	**40.5**	47.46
9.0	31.08	**25.0**	38.75	**41.0**	47.76
9.5	31.30	**25.5**	39.01	**41.5**	48.06
10.0	31.52	**26.0**	39.27	**42.0**	48.37

T	$\sigma^*(T_K)^4$	T	$\sigma^*(T_K)^4$	T	$\sigma^*(T_K)^4$
		Units			
10.5	31.74	**26.5**	39.53	**42.5**	48.68
11.0	31.97	**27.0**	39.80	**43.0**	48.99
11.5	32.19	**27.5**	40.06	**43.5**	49.30
12.0	32.42	**28.0**	40.33	**44.0**	49.61
12.5	32.65	**28.5**	40.60	**44.5**	49.92
13.0	32.88	**29.0**	40.87	**45.0**	50.24
13.5	33.11	**29.5**	41.14	**45.5**	50.56
14.0	33.34	**30.0**	41.41	**46.0**	50.87
14.5	33.57	**30.5**	41.69	**46.5**	51.19
15.0	33.81	**31.0**	41.96	**47.0**	51.51
15.5	34.04	**31.5**	42.24	**47.5**	51.84
16.0	34.28	**32.0**	42.52	**48.0**	52.16
16.5	34,52	**32.5**	42.80	**48.5**	52.49

APPENDIX I

THERMODYNAMIC PROPERTIES OF AIR AND WATER

1. Latent Heat of Vaporization (λ)

$$\lambda = [2.501 - (2.361 \times 10^{-3})\, T]$$

Where: λ = latent heat of vaporization [MJ kg^{-1}]; and T = air temperature [°C].

The value of the latent heat varies only slightly over normal temperature ranges. A single value may be taken (for ambient temperature = 20°C): λ = 2.45 MJ kg^{-1}.

2. Atmospheric Pressure (P)

$$P = P_o\ [\{T_{Ko} - \alpha(Z - Z_o)\} \div \{T_{Ko}\}]^{(g/(\alpha.R))}$$

Where: P, atmospheric pressure at elevation *z* [kPa]

P_o, atmospheric pressure at sea level = 101.3 [kPa]

z, elevation [m]

z_o, elevation at reference level [m]

g, gravitational acceleration = 9.807 [m s^{-2}]

R, specific gas constant == 287 [J kg^{-1} K^{-1}]

α, constant lapse rate for moist air = 0.0065 [K m^{-1}]

T_{Ko}, reference temperature [K] at elevation z_o = 273.16 + T

T, means air temperature for the time period of calculation [°C]

When assuming P_o = 101.3 [kPa] at z_o = 0, and T_{Ko} = 293 [K] for T = 20 [°C], above equation reduces to:

$$\mathbf{P = 101.3[(293–0.0065Z)\ (293)]^{5.26}}$$

3. Atmospheric Density (ρ)

$$\rho = [1000P] \div [T_{Kv}\ R] = [3.486P] \div [T_{Kv}], \text{ and } T_{Kv} = T_K[1–0.378(e_a)/P]^{-1}$$

Where: ρ, atmospheric density [kg m^{-3}]

R, specific gas constant = 287 [J $kg^{-1\,K-1}$]

$T_{Kv,}$ virtual temperature [K]

$T_{K,}$ absolute temperature [K]: $T_K = 273.16 + T$ [°C]

$e_{a,}$ actual vapor pressure [kPa]

T, mean daily temperature for 24-hour calculation time steps.

For average conditions (e_a in the range 1–5 kPa and P between 80–100 kPa), T_{Kv} can be substituted by: $T_{Kv} \approx 1.01\ (T + 273)$

4. Saturation Vapor Pressure function (es)

$$e_s = [0.6108]*\exp\{[17.27*T]/[T + 237.3]\}$$

Where: e_s, saturation vapor pressure function [kPa]

T, air temperature [°C]

5. Slope Vapor Pressure Curve (Δ)

$$\Delta = [4098.\ e°(T)] \div [T + 237.3]^2$$
$$= 2504\{\exp[(17.27T) \div (T + 237.2)]\} \div [T + 237.3]^2$$

Where: **Δ**, slope vapor pressure curve [kPa C^{-1}]

T, air temperature [°C]

$e^0(T)$, saturation vapor pressure at temperature T [kPa]

In 24-hour calculations, **Δ** is calculated using mean daily air temperature. In hourly calculations T refers to the hourly mean, T_{hr}.

6. Psychrometric Constant (γ)

$$\gamma = 10^{-3}\ \mathbf{[(C_p.P) \div (\varepsilon.\lambda)] = (0.00163) \times [P \div \lambda]}$$

Where: γ, psychrometric constant [kPa C^{-1}]

c_p, specific heat of moist air = 1.013 [kJ $kg^{-1\,0}C^{-1}$]

P, atmospheric pressure [kPa]: equations 2 or 4

ε, ratio molecular weight of water vapor/dry air = 0.622

λ, latent heat of vaporization [MJ kg^{-1}]

7. Dew Point Temperature (Tdew)

When data is not available, T_{dew} can be computed from e_a by:

$$T_{dew} = [\{116.91 + 237.3\mathrm{Log}_e(e_a)\} \div \{16.78–\mathrm{Log}_e(e_a)\}]$$

Where: T_{dew}, dew point temperature [°C]

e_a, actual vapor pressure [kPa]

For the case of measurements with the Assmann psychrometer, T_{dew} can be calculated from:

$$T_{dew} = (112 + 0.9T_{wet})[e_a \div (e^0\ T_{wet})]^{0.125}–[112–0.1T_{wet}]$$

8. Short Wave Radiation on a Clear-Sky Day (Rso)

The calculation of R_{so} is required for computing net long wave radiation and for checking calibration of pyranometers and integrity of R_{so} data. A good approximation for R_{so} for daily and hourly periods is:

$$R_{so} = (0.75 + 2 \times 10^{-5} z)R_a$$

Where: z, station elevation [m]

$R_{a,}$ extraterrestrial radiation [MJ m^{-2} d^{-1}]

Equation is valid for station elevations less than 6000 m having low air turbidity. The equation was developed by linearizing Beer's radiation extinction law as a function of station elevation and assuming that the average angle of the sun above the horizon is about 50°.

For areas of high turbidity caused by pollution or airborne dust or for regions where the sun angle is significantly less than 50° so that the path length of radiation through the atmosphere is increased, an adoption of Beer's law can be employed where P is used to represent atmospheric mass:

$$\mathbf{R_{so} = (R_a) \exp[(-0.0018P) \div (K_t \sin(\Phi))]}$$

Where: K_t, turbidity coefficient, $0 < K_t \leq 1.0$ where $K_t = 1.0$ for clean air and $K_t = 1.0$ for extremely turbid, dusty or polluted air.

P, atmospheric pressure [kPa]

Φ, angle of the sun above the horizon [rad]

$R_{a,}$ extraterrestrial radiation [MJ m^{-2} d^{-1}]

For hourly or shorter periods, Φ is calculated as:

$$\sin \Phi = \sin \varphi \sin \delta + \cos \varphi \cos \delta \cos \omega$$

Where: φ, latitude [rad]

δ, solar declination [rad] (Eq. (24) in Chapter 3)

ω, solar time angle at midpoint of hourly or shorter period [rad]

For 24-hour periods, the mean daily sun angle, weighted according to R_a, can be approximated as:

$$\sin(\Phi_{24}) = \sin[0.85 + 0.3\ \varphi \sin\{(2\pi J/365) - 1.39\} - 0.42\ \varphi^2]$$

Where: $\Phi_{24,}$ average Φ during the daylight period, weighted according to R_a [rad]

φ, latitude [rad]

J, day in the year

The Φ_{24} variable is used to represent the average sun angle during daylight hours and has been weighted to represent integrated 24-hour transmission effects on 24-hour R_{so} by the atmosphere. Φ_{24} should be limited to ≥0. In some situations, the estimation for R_{so} can be improved by modifying to consider the effects of water vapor on short wave absorption, so that: $R_{so} = (K_B + K_D) R_a$ where:

$$K_B = 0.98\exp[\{(-0.00146P) \div (K_t \sin \Phi)\} - 0.091\{w/\sin \Phi\}^{0.25}]$$

Where: K_B, the clearness index for direct beam radiation

K_D, the corresponding index for diffuse beam radiation

$K_D = 0.35 - 0.33\ K_B$ for $K_B \geq 0.15$

$K_D = 0.18 + 0.82\ K_B$ for $K_B < 0.15$

R_a, extraterrestrial radiation [MJ m^{-2} d^{-1}]

K_t, turbidity coefficient, $0 < K_t \leq 1.0$ where $K_t = 1.0$ for clean air and $K_t = 1.0$ for extremely turbid, dusty or polluted air.

P, atmospheric pressure [kPa]

Φ, angle of the sun above the horizon [rad]

W, perceptible water in the atmosphere [mm] $= 0.14\ e_a\ P + 2.1$

e_a, actual vapor pressure [kPa]

P, atmospheric pressure [kPa]

APPENDIX J

PSYCHROMETRIC CHART AT SEA LEVEL

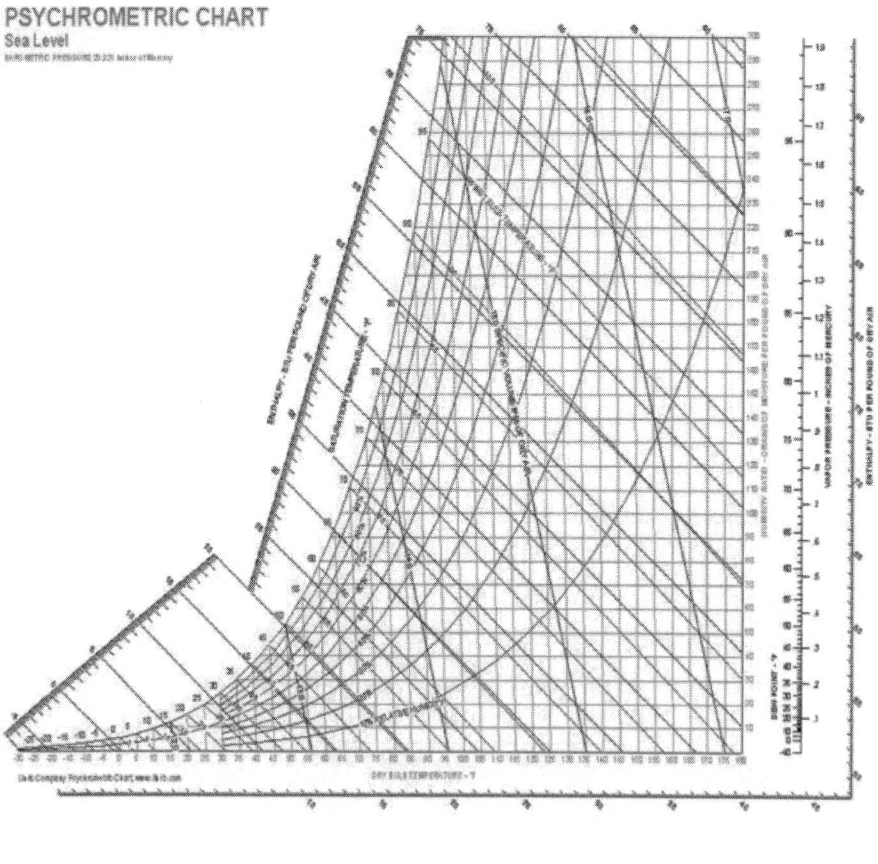

INDEX

J

K

L

O

P